An Introduction to Real Analysis

The Sally Series

Pure and Applied UNDERGRADUATE TEXTS · 65

An Introduction to Real Analysis

Yitzhak Katznelson
Yonatan Katznelson

AMS AMERICAN MATHEMATICAL SOCIETY
Providence, Rhode Island USA

The cover photo of Karl Weierstrass (top photo) is in the Public Domain: https://commons.wikimedia.org/w/index.php?curid=324146.

The cover photo of Augustin Louis Cauchy (second from the top) is in the Public Domain: Library of Congress Prints and Photographs Division. From an illustration in: Das neunzehnte Jahrhundert in Bildnissen / Karl Werckmeister, ed. Berlin Kunstverlag der photographische gesellschaft, 1901, vol. V, no. 581., Public Domain, https://commons.wikimedia.org/w/index.php?curid=7059486.

The cover photo of Bernhard Riemann (third from the top) is in the Public Domain: http://www.sil.si.edu/digitalcollections/hst/scientific-identity/explore.htm according to the German Wikipedia, Public Domain, https://commons.wikimedia.org/w/ index.php?curid=27383.

The cover photo of Henri Lebesgue (bottom photo) is in the Public Domain: https://commons.wikimedia.org /w/index.php?curid=336482.

2020 *Mathematics Subject Classification.* Primary 26-XX.

For additional information and updates on this book, visit
www.ams.org/bookpages/amstext-65

Library of Congress Cataloging-in-Publication Data

Cataloging-in-Publication Data has been applied for by the AMS.
See http://www.loc.gov/publish/cip/.
DOI: https://doi.org/10.1090/amstext/65

∞ The paper used in this book is acid-free and falls within the guidelines
established to ensure permanence and durability.
Visit the AMS home page at `https://www.ams.org/`

10 9 8 7 6 5 4 3 2 1 29 28 27 26 25 24

To our students

Contents

Preface

This book provides the essential elements of real analysis with which mathematicians, and those who use mathematics at more than a superficial level, may want to be familiar. It covers the foundations of differential and integral calculus in one variable, beginning with the structure of the real numbers as a complete ordered field and concluding with a chapter on Lebesgue measure and integration. Along the way, we discuss sequences and series of real numbers, the metric topology of the real numbers, continuous functions, sequences and series of functions, differentiation, and the Riemann integral.

While the material is developed "from scratch", we have written the book primarily for students of mathematics (and related fields) that have several semesters of calculus under their belts. In particular we expect that the reader has attained a basic level of "mathematical maturity" and is prepared to handle some abstraction. The book is appropriate, for example, as a text for a one or two term advanced undergraduate course in analysis, or as a supplementary text for a first year graduate course.

Needless to say, none of the material presented here is new or original, and all of it can be found in any number of well-known texts (to say nothing of an increasingly rich trove of online sources),[1] and our goal in writing this book is not to provide another compendium of this material. Rather our main purpose here is to provide an introduction to the methods of real analysis and a way of thinking about it. As such, there are many common topics that we do not cover at all. For example, while we do introduce the metric topology of $\mathbb{R}^n$ and some of its features, we don't discuss calculus in several variables.

We strongly believe that students learn more by filling in some of the details themselves, and as we progress through the book, we provide them with more opportunities to do so. Sometimes we do this explicitly, by leaving proofs or parts of proofs as exercises, and at other times, the proofs we give are complete, but "terse". Our aim is to

[1] As a consequence, we do not include references with two exceptions, as footnotes.

provide a text that sits a level between the more elementary introductions to analysis, and those that are meant for graduate-level courses in mathematics.

The text focuses primarily on the core material, with few numerical examples and no applications to other fields. If used in a course where such examples are useful or necessary, we leave it to the instructor to provide them. On the other hand, we do include sections that explore topics in real analysis that are either not commonly discussed in undergraduate texts on real analysis, or are slightly more advanced. Such sections are marked with asterisks and may be omitted without affecting the flow of the material. We include them because the topics they cover are interesting (to us), and because they provide an opportunity to practice using the analytic tools that have been developed. If this text were used as a course textbook, these sections could be used as topics for students to present in class, "seminar style", for example.

The material is divided into eight chapters, with an appendix that covers the basic elements of set theory and a construction of the real numbers. Each chapter is divided into sections, and each section is divided into subsections. Definitions, lemmas, theorems, etc., are not individually numbered and are referred to according to the subsection in which they appear, where no subsection contains more than one of each type of statement. For example, Theorem* 4.3.3 is the theorem that appears in Chapter 4, Section 4.3, Subsection* 4.3.3. Exercises are collected at the end of each section, and are numbered to refer to the chapter and section in which they appear. E.g., **ex5.4.2** is the second exercise for Section 5.4 of Chapter 5.

Chapter 1

The Real Numbers

Real analysis is largely concerned with the study of functions whose domain is $\mathbb{R}$, the set of real numbers, subsets of $\mathbb{R}$, or more generally subsets of Euclidean space ($\mathbb{R}^n$).[1] We begin this book therefore by studying the underlying "theater of operations": the set $\mathbb{R}$ of real numbers itself and its characteristic properties. We assume basic familiarity with these properties, including the properties of the integers $\mathbb{Z}$ and the rational numbers $\mathbb{Q}$, such as you would learn in an introductory calculus course, or indeed in middle and high school, though we do summarize (many of) these properties in the first few pages.

The set of real numbers can be described formally as *the unique ordered field that is complete with respect to the order.* In this chapter we explain what this means by clarifying what an ordered field is, what completeness is, and in what sense $\mathbb{R}$ is unique.

On the other hand, we do not prove in this chapter that $\mathbb{R}$ has all of these properties. Such a proof entails constructing $\mathbb{R}$ from the set $\mathbb{Q}$ of rational numbers.[2] We do sketch one such construction in the Appendix, but for the purpose of studying real analysis, we accept that the set of real numbers (exists and) has the properties that we describe.

We begin with a quick review of the natural numbers, the integers, and the rational numbers.

1.1. The natural numbers

The natural numbers are the positive integers: 1, 2, 3, ... We denote the set of natural numbers by $\mathbb{N}$,[3] and summarize below some of the principle features of $\mathbb{N}$.

[1] See Section 1.7.

[2] The rationals are constructed in turn from the set $\mathbb{Z}$ of integers, which themselves are constructed from the natural numbers, $\mathbb{N}$. As Leopold Kronecker reportedly said: *"God made the integers; all else is the work of man."*

[3] Some texts include 0 in $\mathbb{N}$, others do not. Ultimately it is not terribly important, but we do not.

1.1.1. Arithmetic in $\mathbb{N}$

The natural numbers come equipped with the binary operations of addition and multiplication, which we write as $n+m$ and $n \cdot m$ (or nm), respectively, and which satisfy the familiar axioms below.

Axioms for addition:

add 1. For all $n, m \in \mathbb{N}$, $n+m \in \mathbb{N}$, ($\mathbb{N}$ is *closed* under addition).

add 2. For all $n, m, l \in \mathbb{N}$, $n+(m+l)=(n+m)+l$, (addition is *associative*).

add 3. For all $n, m \in \mathbb{N}$, $n+m=m+n$, (addition is *commutative*).

Axioms for multiplication:

mult 1. For all $n, m \in \mathbb{N}$, $nm \in \mathbb{N}$, ($\mathbb{N}$ is *closed* under multiplication).

mult 2. For all $n, m, l \in \mathbb{N}$, $n(ml)=(nm)l$, (multiplication is *associative*).

mult 3. For all $n, m \in \mathbb{N}$, $nm=mn$, (multiplication is *commutative*).

mult 4. There is a unit element $1 \in \mathbb{N}$ such that $1 \cdot n = n$ for all $n \in \mathbb{N}$.

Finally, linking addition and multiplication is the *distributive law:*

dist For all $n, m, l \in \mathbb{N}$, $n(m+l)=nm+nl$.

1.1.2. The natural order on $\mathbb{N}$

Besides its arithmetic properties, the set $\mathbb{N}$ is also equipped with a *linear order*.

Definition. A *linear order* on a set E is a binary relation, denoted by "<" satisfying the axioms

ord 1. For all $x \in E$, $x \not< x$, (the order is *irreflexive*).

ord 2. For all $x, y, z \in E$, if $x<y$ and $y<z$, then $x<z$, (the order is *transitive*).

ord 3. For all $x, y \in E$, $x<y$, $x=y$ or $y<x$, (the order is *total* or *linear*).

A set E equipped with a linear order is called an *ordered set.*[4]

We define a linear order on $\mathbb{N}$, by declaring that for $m, n \in \mathbb{N}$, $m<n$ if and only if there exists $k \in \mathbb{N}$ such that $n=m+k$. The facts that this order is transitive and irreflexive are not hard to show. The fact that the order is *total* requires more work, and relies on the formal, set theoretic construction of the natural numbers, which we omit.

The next proposition shows that the order "plays nicely" with addition and multiplication.

Proposition. *For all $n, m, l \in \mathbb{N}$, if $m<n$, then $m+l<n+l$, and $ml<nl$.*

Proof. Since $m<n$, then there is a $k \in \mathbb{N}$ such that $n=m+k$ and it follows that $n+l=(m+k)+l=(m+l)+k$, so $n+l>m+l$. Now, by the distributive property, $nl=(m+k)l=ml+kl$, so $nl>ml$. □

[4]As defined here, E is a *strictly* ordered set. See Appendix 1.3 for more details.

This interaction between the order and the arithmetic in $\mathbb{N}$ is a precursor for the concept of an *ordered field.*

1.1.3. Induction

The natural numbers are *well ordered* by the natural order: every nonempty subset $\mathbb{N}$ contains a least element with respect to this order. This is also called the *least natural number principle.* As with the totality of the order, this follows from the set theoretic construction of the natural numbers.

The well-ordering of the natural numbers is equivalent to the *inductive property.* If $S \subset \mathbb{N}$, $1 \in S$ and for every $n \in S$, $n+1 \in S$, then $S = \mathbb{N}$. This is often phrased in terms of statements about natural numbers: *Suppose that $P(n)$ is a statement about the natural number n. If $P(1)$ is true, and if $P(k)$ implies $P(k+1)$, then $P(n)$ is true for all $n \in \mathbb{N}$.*

As an example, we use induction to establish the identity below.

Proposition. *For any numbers x and y, and any $n \in \mathbb{N}$,*

$$x^n - y^n = (x-y)\sum_{k=1}^{n} x^{n-k}y^{k-1}. \tag{1.1.1}$$

We write this less formally as

$$x^n - y^n = (x-y)(x^{n-1} + x^{n-2}y + x^{n-3}y^2 + \cdots + xy^{n-2} + y^{n-1}).$$

Proof. When $n = 1$, the identity reduces to $x - y = (x-y)\cdot 1$, so the identity is true in this case. Now suppose that $n \geq 1$ and equation (1.1.1) is true for n. To show that the identity holds for $n+1$, we observe that

$$x^{n+1} - y^{n+1} = x^{n+1} - x^n y + x^n y - y^{n+1} = x^n(x-y) + y(x^n - y^n).$$

From the induction hypothesis it follows that

$$y(x^n - y^n) = y(x-y)\sum_{k=1}^{n} x^{n-k}y^{k-1} = (x-y)\sum_{k=1}^{n} x^{n-k}y^{k}$$

and therefore

$$x^{n+1} - y^{n+1} = (x-y)\left(x^n + \sum_{k=1}^{n} x^{n-k}y^k\right) = (x-y)\sum_{k=1}^{n+1} x^{n+1-k}y^{k-1}.$$

I.e., if the identity is true for n, then it is true for $n+1$, and since it is true for $n=1$, the identity holds for all $n \in \mathbb{N}$. □

Induction can also be used to prove the *binomial theorem.* Recall that the binomial coefficients $\binom{n}{k}$ are defined for $1 \leq k \leq n-1$ by

$$\binom{n}{k} = \frac{n!}{k!\,(n-k)!},$$

where $n!$ is defined recursively for $n \in \mathbb{N}$: $1! = 1$, and $n! = n \cdot (n-1)!$ for $n > 1$. We also define $0! = 1$, from which it follows that $\binom{n}{0} = \binom{n}{n} = 1$.

Theorem. *For any numbers x and y, and any $n \in \mathbb{N}$,*

$$(x+y)^n = \sum_{k=0}^{n} \binom{n}{k} x^{n-k} y^k, \tag{1.1.2}$$

We leave the proof of the binomial theorem as an exercise (with hints) for the motivated reader.

Another formulation of the inductive property is called *strong induction*. In this formulation, we use the stronger assumption that $P(m)$ is true for *all* $m \leq n$ to prove that $P(n+1)$ is true. It is an exercise in logic to show that strong induction and "ordinary" induction are in fact equivalent.

1.1.4. Prime numbers and unique factorization

For natural numbers n and m, we say that n *divides* m if there is a natural number k such that $nk = m$. If n divides m, we say that n is a *factor* or a *divisor* of m. Note that 1 divides every natural number, but the only natural number that divides 1 is 1 itself.

If d divides both m and n, then we say that d is a *common divisor* of m and n. If m and n have no common divisors greater than 1, then we say that they are *relatively prime*.

A natural number $p > 1$ is *prime* if $nm = p$ implies that $n = 1$ or $m = 1$ (i.e., p is only divisible by itself and by 1).

Lemma. *If $n \in \mathbb{N}$, then either $n = 1$ or there is a prime number p such that p divides n.*

Proof. The statement is true for $n = 1$ and also for $n = 2$ (since 2 is prime). Using strong induction, we assume that $n \geq 2$ and that for all $1 < m \leq n$, m is divisible by a prime.

If $n+1$ is prime, then the statement is true for $n+1$. If not, then there exist $m, k \in \mathbb{N}$ such that $1 < m, k$ and $n+1 = mk$. Since $1 < k$, it follows from Proposition 1.1.2 that $m < n+1$, so $m = pm'$ for some prime p and $m' \in \mathbb{N}$. Hence, $n+1 = (pm')k = p(m'k)$, so $n+1$ is divisible by a prime. □

Proposition. *There are infinitely many prime numbers.*

Proof. If $p_1 < p_2 < \ldots < p_n$ are primes, then $N = p_1 p_2 \cdots p_n + 1$ is a natural number and therefore has a prime divisor q, by the fundamental theorem of arithmetic. It follows that $N = qk$ for some $k \in \mathbb{N}$.

If $q = p_j$ for some $1 \leq j \leq n$, then $p_1 p_2 \cdots p_n = qm$, where m is the product of the $n-1$ primes in $\{p_1, \ldots, p_n\}$ different from p_j. It follows that

$$1 = N - p_1 p_2 \cdots p_n = qk - qm = q(k-m),$$

which implies that q divides 1, which is impossible. Hence, no finite set of primes contains all primes, so the set of primes is infinite. □

The importance of the prime numbers lies in the fact that they are the "multiplicative building blocks" of $\mathbb{N}$ (and $\mathbb{Z}$).

Theorem. *If n is a natural number greater than one, then n can be written as a product of prime factors, $n = p_1 p_2 \cdots p_m$ in one and only one way (ignoring the order of the factors).*

This is called the *fundamental theorem of arithmetic*, and can also be proved by induction.

Corollary. *Every natural number n greater than 1 has a unique representation as a product of the form*

$$n = p_1^{l_1} p_2^{l_2} \cdots p_k^{l_k},$$

where $p_1 < p_2 < \cdots < p_k$ are primes and $l_j \in \mathbb{N}$ for $1 \leq j \leq k$.

This product is called the *prime power decomposition* of n.

Exercises

ex1.1.1. Prove by induction that for all $n \in \mathbb{N}$

$$\sum_{k=1}^{n} k = \frac{n(n+1)}{2}.$$

ex1.1.2. Prove by induction that $2^n > n$ for all $n \in \mathbb{N}$.

ex1.1.3. Prove that for any number $q \neq 1$,

$$\sum_{k=0}^{n} q^k = \frac{q^{n+1} - 1}{q - 1}.$$

You can do this by induction, or by using identity (1.1.1).

ex1.1.4.

(a) Show that

$$\binom{n}{k-1} + \binom{n}{k} = \binom{n+1}{k}$$

for $n \geq k \geq 1$.

(b) Use induction to prove the binomial theorem.

Hints:

(i) Express $(x+y)\sum_{k=0}^{n}\binom{n}{k}x^{n-k}y^k$ as a sum of two sums.

(ii) Show that $\sum_{k=0}^{n}\binom{n}{k}x^{n-k}y^{k+1} = \sum_{k=1}^{n+1}\binom{n}{k-1}x^{n+1-k}y^k$.

(iii) Use (a).

1.2. The integers

The set of integers $\mathbb{Z}$ extend the natural numbers by including 0 and the negative integers, $-1, -2, -3, \ldots$. We may write

$$\mathbb{Z} = \mathbb{N} \cup \{0\} \cup \{-n : n \in \mathbb{N}\}.$$

The arithmetic and order on $\mathbb{N}$ extend naturally to all of $\mathbb{Z}$.

1.2.1. Arithmetic in $\mathbb{Z}$

Addition and multiplication on $\mathbb{Z}$ extend these operations on $\mathbb{N}$, and satisfy the axioms **add** 1. - 3, **mult** 1. - 4. and **dist**. In $\mathbb{Z}$, addition also satisfies two more axioms:

add 4. There exists $0 \in \mathbb{Z}$ such that $m + 0 = m$ for all $m \in \mathbb{Z}$.

add 5. For all $m \in \mathbb{Z}$, there is a $-m \in \mathbb{Z}$ such that $m + (-m) = 0$.

We call $-m$ the *additive inverse* of m (or negative m) and observe that $n-m$ is shorthand for $n + (-m)$.

The following property of 0 is often mistaken as axiomatic.

Proposition. $m \cdot 0 = 0$ *for all* $m \in \mathbb{Z}$.

Proof. By the distributive law and the defining property of 0,

$$m \cdot 0 = m \cdot (0 + 0) = m \cdot 0 + m \cdot 0.$$

adding $-(m \cdot 0)$ to both sides, shows that $0 = m \cdot 0$, as claimed. □

The notion of "divisibility" also extends to $\mathbb{Z}$. That is m divides n in $\mathbb{Z}$, if $m \neq 0$ and there is a $k \in \mathbb{Z}$ such that $mk = n$. We say (as before) that m is a factor of n. The proposition above implies that 0 is divisible by every nonzero integer.

Both -1 and 1 divide every integer, but have only themselves as divisors. If m and n have no common factors other than -1 and 1, we say (as before) that they are relatively prime.

1.2.2. The natural order on $\mathbb{Z}$

The order on $\mathbb{N}$ is extended to all of $\mathbb{Z}$ by declaring that $m < n$ if and only if $n-m \in \mathbb{N}$. This definition makes $\mathbb{Z}$ an ordered set, moreover the order continues to play nicely with the arithmetic, i.e., Proposition 1.1.2 extends to $\mathbb{Z}$ (with a qualification).

Proposition. *For all* $n, m \in \mathbb{Z}$, *(a) if* $m < n$, *then for all* $l \in \mathbb{Z}$, $m + l < n + l$, *and (b) if* $m < n$ *and* $0 < k$, *then* $mk < nk$.

Proof. We leave the proof of (a) as an exercise. If $m < n$, then $n - m \in \mathbb{N}$, by the definition of the order on $\mathbb{Z}$, so $kn - km = k(n-m) \in \mathbb{N}$, because $k \in \mathbb{N}$ and $\mathbb{N}$ is closed under multiplication. □

Exercises

ex1.2.1. Prove Proposition 1.2.2 (a).

In the following exercises, n and m are integers.

ex1.2.2. Show that if $n + m = m$, then $n = 0$. This proves that the 0 element in $\mathbb{Z}$ is unique.

ex1.2.3. Show that if $n + m = 0$, then $n = -m$. This shows that for every $m \in \mathbb{Z}$, $-m$ is unique.

ex1.2.4. Show that $(-1) \cdot m = -m$.

ex1.2.5. Show that if $m > 0$, then $-m < 0$ and vice versa.

ex1.2.6. Show that if $m < 0$ and $n < 0$, then $mn > 0$.

ex1.2.7. Show that if $m \neq 0$, then $m^2 > 0$. Conclude that $1 > 0$.

ex1.2.8. Show that if $m < 0$ and $n > 0$, then $mn < 0$.

ex1.2.9. Show that if $m \neq 0$ and $n \neq 0$, then $mn \neq 0$. Conclude that if $mn = 0$, then $m = 0$ or $n = 0$.

Hint: Proposition 1.2.2 (b) and the conclusions of exercises **ex1.2.6** and **ex1.2.8** will be useful here.

1.3. The rational numbers

We think of the rational numbers as quotients of integers. I.e., if $m, n \in \mathbb{Z}$ with $n \neq 0$, then the *ratio* m/n is a rational number with *numerator* m and *denominator* n. We denote the set of rational numbers by $\mathbb{Q}$.

Every rational number can be expressed as a quotient of integers in infinitely many ways, e.g., the quotients $2/3 = (-4)/(-6) = 6/9$ all represent the same rational number. In general, the quotients m/n and l/k represent the same rational number if and only if $mk = nl$. In this case we say that m/n and k/l are *representatives* of the rational number $r = m/n = k/l$.

If $r = m/n$ and $n < 0$, then we also have $r = (-m)/(-n)$, i.e., every rational number has a representative with a positive denominator, and without loss of generality, we will always assume that $n > 0$ when we write $r = m/n$. Moreover, for every $r \in \mathbb{Q}$ there is a unique pair of relatively prime integers m, n, with $n > 0$, such that $r = m/n$. We say that the quotient m/n is *reduced* in this case.

Identifying the integer n with the rational number $n/1$ allows us to think of $\mathbb{Z}$ as a subset of $\mathbb{Q}$.[5]

[5]Technically, the set $Z = \{n/1 : n \in \mathbb{Z}\} \subset \mathbb{Q}$ is not the same set as $\mathbb{Z}$, but it is structurally identical. We say that Z is isomorphic to $\mathbb{Z}$.

1.3.1. $\mathbb{Q}$ is a field

Addition and multiplication in $\mathbb{Z}$ are extended to $\mathbb{Q}$ by setting

$$\frac{m}{n}+\frac{k}{l}=\frac{ml+kn}{nl} \quad \text{and} \quad \frac{m}{n}\cdot\frac{k}{l}=\frac{mk}{nl}.$$

The same rational number can be represented by different quotients of integers, so we need to verify that the results of addition and multiplication of rational numbers do not depend on the particular quotients of integers that we choose to represent them. We say in this case that the operations are *well-defined.*

Proposition. *Suppose that k_1, k_2, l_1, l_2, m_1, m_2, n_1 and n_2 are integers, with l_1, l_2, n_1, $n_2 \neq 0$. If $l_1k_2 = l_2k_1$ and $m_1n_2 = m_2n_1$ (so that $k_1/l_1 = k_2/l_2$ and $m_1/n_1 = m_2/n_2$), then*

$$\frac{k_1}{l_1}+\frac{m_1}{n_1}=\frac{k_2}{l_2}+\frac{m_2}{n_2} \quad \textit{and} \quad \frac{k_1}{l_1}\cdot\frac{m_1}{n_1}=\frac{k_2}{l_2}\cdot\frac{m_2}{n_2}.$$

Proof. We check multiplication first. By assumption $k_1l_2 = k_2l_1$ and $m_1n_2 = m_2n_1$, so $k_1l_2m_1n_2 = k_2l_1m_2n_1$ which means that

$$\frac{k_1m_1}{l_1n_1}=\frac{k_2m_2}{l_2n_2}.$$

Checking that addition is well-defined is similar, if a tiny bit longer. We want to show that

$$\frac{k_1n_1+m_1l_1}{l_1n_1}=\frac{k_2n_2+m_2l_2}{l_2n_2}$$

so we need to show that

$$n_2l_2(m_1l_1+k_1n_1)=n_1l_1(m_2l_2+k_2n_2).$$

Starting with the left-hand side, expanding, rearranging the terms and using the assumptions, we have

$$\begin{aligned} n_2l_2(m_1l_1+k_1n_1) &= n_2m_1l_1l_2+l_2k_1n_1n_2 \\ &= n_1m_2l_1l_2+l_1k_2n_1n_2 \\ &= n_1l_1(m_2l_2+k_2n_2), \end{aligned}$$

which is the right-hand side. □

We also note that addition and multiplication of integers is consistent with addition and multiplication of rational numbers. I.e.,

$$\frac{n}{1}+\frac{m}{1}=\frac{n\cdot 1+m\cdot 1}{1\cdot 1}=\frac{n+m}{1}=n+m$$

and

$$\frac{n}{1}\cdot\frac{m}{1}=\frac{nm}{1}=nm.$$

If we set $0 = 0/1$ ($= 0/n$ for all $n \neq 0$) and $1 = 1/1$ ($= n/n$ for all $n \neq 0$), then the rational numbers satisfy all the axioms for addition and multiplication that are satisfied by the integers. And one more: every nonzero rational number has a multiplicative inverse.

Definition. A *field* is a nonempty set $\mathbb{F}$ together with two *binary operations* — maps from $\mathbb{F} \times \mathbb{F} \to \mathbb{F}$ — *addition* and *multiplication*, written $(x, y) \mapsto x + y$ for addition and $(x, y) \mapsto xy$ or $x \cdot y$, for multiplication, satisfying the following axioms.

Axioms for addition:

add 1. For all $x, y, z \in \mathbb{F}$, $x + (y + z) = (x + y) + z$ (addition is *associative*)

add 2. For all $x, y \in \mathbb{F}$, $x + y = y + x$ (addition is *commutative*)

add 3. There is an element $0 \in \mathbb{F}$ such that $0 + x = x$ for all $x \in \mathbb{F}$.

add 4. For every $x \in \mathbb{F}$ there exists $y \in \mathbb{F}$ such that $x + y = 0$. The corresponding number y is denoted $-x$ and referred to as "negative x" or "minus x".

Axioms for multiplication:

mult 1. For all $x, y, z \in \mathbb{F}$, $x \cdot (y \cdot z) = (x \cdot y) \cdot z$, (multiplication is *associative*)

mult 2. For all $x, y \in \mathbb{F}$, $x \cdot y = y \cdot x$ (multiplication is *commutative*)

mult 3. There is a unit element $1 \in \mathbb{F}$ such that $1 \neq 0$ and $1 \cdot x = x$ for all $x \in \mathbb{F}$,

mult 4. For every $x \in \mathbb{F}$, if $x \neq 0$, then there exists $y \in \mathbb{F}$ such that $x \cdot y = 1$. The corresponding number y is denoted x^{-1} or $1/x$, and referred to as the *reciprocal* of x or the *inverse* of x.

And once again, linking addition and multiplication we have the *distributive law:*

dist For all $x, y, z \in \mathbb{F}$, $x(y + z) = xy + xz$.

Theorem. *The set $\mathbb{Q}$ of rational numbers with addition and multiplication defined as above and with $0 = 0/1$ and $1 = 1/1$ form a field.*

Proof. The proof amounts to verifying that all of the axioms are satisfied. We do this for our favorites and leave the rest (associativity and commutativity of addition and multiplication and the distributive law) as an exercise.

($*$) If $r \in \mathbb{Q}$, then $r = m/n$ for integers m, n with $n \neq 0$. We have

$$0 + r = \frac{0}{1} + \frac{m}{n} = \frac{0 \cdot n + 1 \cdot m}{n \cdot 1} = \frac{m}{n} = r,$$

so axiom **add** 3. is satisfied.

($*$) If $r = m/n \in \mathbb{Q}$, then setting $-r = (-m)/n$, we have

$$r + (-r) = \frac{m}{n} + \frac{-m}{n} = \frac{mn - mn}{mn} = \frac{0}{mn} = 0,$$

so **add** 4 is satisfied.

($*$) For $r = m/n \in \mathbb{Q}$, we have

$$r \cdot 1 = \frac{m}{n} \cdot \frac{1}{1} = \frac{m \cdot 1}{n \cdot 1} = r,$$

so **mult** 3 is satisfied.

(∗) If $r \neq 0$, then $r = m/n$ with $n \neq 0$ *and* $m \neq 0$. It follows that n/m is rational and

$$\frac{m}{n} \cdot \frac{n}{m} = \frac{mn}{mn} = 1,$$

i.e., $m/n = (m/n)^{-1}$ and **mult** 4 is satisfied. □

Fields come with two operations (addition and multiplication). "Subtraction" and "division" are just shorthand for adding the negative and multiplying by the inverse. Nonetheless they are very convenient shorthands and we will continue to use the notation $x - y = x + (-y)$ and $x/y = x \cdot (1/y)$ without further comment.

Comment. The sum of $n = 3$ or more elements, $a_1, a_2, \ldots, a_n$ in the field $\mathbb{F}$ is denoted using the usual shorthand (summation notation):

$$a_1 + a_2 + \cdots + a_n = \sum_{j=1}^{n} a_j.$$

The associative and commutative laws guarantee that such a sum does not depend on the order of the terms nor on how they are grouped.

We also use the usual exponential notation for repeated multiplication: if $n \in \mathbb{N}$ and $a \in \mathbb{F}$, then we write

$$a^n = \overbrace{a \cdot a \cdots a}^{n}.$$

Many of the properties of the rational (and real) numbers that we may take for granted (or consider axiomatic) are consequences of the axioms. In particular, the uniqueness of 0, 1 and inverses are not asserted in the field axioms, nor is the fact that $0 \cdot x = 0$ for all x.

Proposition. *Suppose that* $\mathbb{F}$ *is a field and* $x, y, z \in \mathbb{F}$.

- **(a)** $0 \cdot x = 0$.
- **(b)** *If* $x + y = x$, *then* $y = 0$. *(This implies the uniqueness of* 0.*)*
- **(c)** *If* $x + y = 0$, *then* $y = -x$. *(This implies the uniqueness of* $-x$.*)*
- **(d)** *If* $x \neq 0$ *and* $xy = x$, *then* $y = 1$. *(This implies the uniqueness of* 1.*)*
- **(e)** *If* $xy = 1$, *then* $y = 1/x$. *(This implies the uniqueness of* $1/x$.*)*

Proof.

(a) See the proof of Proposition 1.2.1.

(b) If $x + y = x$, then using the definitions of 0 and $-x$ and the associative property, we see that

$$y = 0 + y = (-x + x) + y = -x + (x + y) = -x + x = 0.$$

(c) If $x + y = 0$, then as above

$$y = 0 + y = (-x + x) + y = -x + (x + y) = -x + 0 = -x.$$

(d) If $x \neq 0$, then x has a multiplicative inverse, and if $xy = x$, then

$$y = 1 \cdot y = (1/x \cdot x)y = 1/x(xy) = 1/x \cdot x = 1.$$

(e) If $xy = 1$, then $x \neq 0$ by (a) so $1/x$ exists, and analogously to (b),

$$y = 1 \cdot y = (1/x \cdot x)y = 1/x(xy) = 1/x \cdot 1 = 1/x. \qquad \square$$

Note that as a corollary to the uniqueness of additive and multiplicative inverses, it follows that $-(-x) = x$ and $1/(1/x) = x$.

The fact that $\mathbb{Q}$ is a field is just a formal affirmation of the fact that the set of rational numbers has the familiar arithmetic structure that most of us rely on without much reflection.[6]

1.3.2. Algebra in a field

Thanks to the existence of multiplicative inverses, any equation of the form $ax + b = c$, with $a \neq 0$, can be solved in $\mathbb{Q}$ or any other field, with the solution given by $x = a^{-1}(c - b)$. More generally, the field axioms imply a variety of properties that can be used to simplify and solve more complicated equations.

Theorem. *Let $\mathbb{F}$ be a field, then for all $r, s, t \in \mathbb{F}$*

(i) *If $r + s = r + t$, then $s = t$ (cancellation of like terms).*

(ii) *If $r \neq 0$ and $rs = rt$, then $s = t$ (cancellation of like factors).*

(iii) *If $rs = 0$, then $r = 0$ or $s = 0$ (a field has no zero dividers).*

(iv) $(-1) \cdot r = -r$.

(v) $(-r)(-s) = rs$.

Proof.

(i) Add $-r$ to both sides of the equation $r + s = r + t$.

(ii) Multiply both sides of the equation $rs = rt$ by $1/r$, which exists because $r \neq 0$.

(iii) If $r = 0$, then there is nothing more to prove. If $r \neq 0$, then $1/r$ exists and

$$s = 1 \cdot s = (1/r \cdot r) \cdot s = (1/r)(rs) = (1/r) \cdot 0 = 0.$$

(iv) We have $(-1) \cdot r + r = (-1) \cdot r + 1 \cdot r = (-1 + 1) \cdot r = 0 \cdot r = 0$. It follows from the uniqueness of $-r$ that $(-1) \cdot r = -r$.

(v) Since $(-1)(-1) = -(-1) = 1$, by (iv), we have

$$(-r)(-s) = ((-1) \cdot r)((-1) \cdot s) = ((-1)(-1)) \cdot (rs) = rs. \qquad \square$$

1.3.3. Ordered fields

Definition. A *linear order* on a nonempty set S is a relation $<$ such that for all $x, y, z \in S$

- $x \not< x$,
- either $x < y$ or $y < x$, and
- if $x < y$ and $y < z$, then $x < z$.

[6] Monsieur Jourdain: *"il y a plus de quarante ans que je dis de la prose sans que j'en susse rien,... "*—Molière, *Le Bourgeois Gentilhomme*

A field $\mathbb{F}$ that is equipped with a linear order $<$ is an *ordered field* if the following axioms relating the order to the arithmetic in the field are satisfied.

ordf 1. For all $x, y, z \in \mathbb{F}$, if $x < y$, then $x + z < y + z$.

ordf 2. For all $x, y \in \mathbb{F}$, if $0 < x$ and $0 < y$, then $0 < xy$.

You may have noticed that these are the same properties satisfied by $\mathbb{Z}$, as shown in Proposition 1.2.2. This makes $\mathbb{Z}$ an example of an *ordered ring*.

The axioms of an ordered field imply a variety of other familiar properties.[7]

Theorem. *Let $\mathbb{F}$ be an ordered field and let $r, s, t \in \mathbb{F}$*

(i) *If $0 < r$, then $-r < 0$.*

(ii) *If $0 < r$ and $s < t$, then $rs < rt$.*

(iii) *If $r \neq 0$, then $0 < r^2$.*

(iv) *If $0 < r$, then $0 < 1/r$.*

(v) *If $0 < r < s$, then $1/s < 1/r$.*

(vi) *If $r < 0 < s$, then $rs < 0$.*

(vii) *For any $n \in \mathbb{N}$, if $0 < r, s$ then $r^n < s^n$ if and only if $r < s$.*

Proof.

(i) If $0 < r$, then $-r = 0 + (-r) < r + (-r) = 0$, by **ordf** 1.

(ii) If $s < t$, then $0 < (t - s)$ and $0 < r(t - s)$, by **ordf** 2, so $rs < rt$.

(iii) If $0 < r$, then $0 < r^2$ by **ordf** 2, and if $r < 0$, then $0 < -r$, by (i), so $0 < r^2 = (-r)^2$, by Theorem 1.3.2 (v).

It follows that $0 < 1^2 = 1$ in any ordered field.

(iv) Suppose that $0 < r$ and $1/r < 0$. Then $0 < -1/r$, so

$$0 < r(-1/r) = r \cdot (-1) \cdot (1/r) = -1.$$

But this contradicts (i), since $0 < 1$.

(v) If $0 < r < s$, then $0 < 1/r$ and $0 < 1/s$, so $0 < (1/r)(1/s)$. It follows then from (ii) that

$$1/s = (1/s)(1/r) \cdot r < s \cdot (1/s)(1/r) = 1/r.$$

(vi) If $r < 0$, then $0 < -r$ so $0 < (-r)s = -(rs)$ and it follows that $rs = -(-rs) < 0$.

(vii) If $n = 1$, then there is nothing to prove. If $n > 1$, then by the identity (1.1.1),

$$s^n - r^n = (s - r)(s^{n-1} + s^{n-2}r + \cdots + sr^{n-2} + r^{n-1}).$$

We have $0 < (s^{n-1} + s^{n-2}r + \cdots + sr^{n-2} + r^{n-1})$, since all the terms in the sum are positive. This means that the differences $s^n - r^n$ and $s - r$ are either both positive or both negative, by (vi), which completes the proof. □

[7]The point of the following theorem and other similar theorems in this chapter is not to teach the reader something new, but rather to establish a solid foundation for things the reader knows, but perhaps never questioned.

The set $\mathbb{F}^+ = \{r \in \mathbb{F} : 0 < r\}$ is the set of *positive* elements in the ordered field $\mathbb{F}$ and $\mathbb{F}^- = \{-r : r \in \mathbb{F}^+\}$ is called the set of *negative* elements. It follows from **ordf** 1 and 2 that $\mathbb{F}^+$ is closed under addition and multiplication, and it follows from part (i) of the preceding theorem that $\mathbb{F} = \mathbb{F}^+ \cup \{0\} \cup \mathbb{F}^-$.

1.3.4. The natural order on $\mathbb{Q}$

We extend the order on $\mathbb{Z}$ to all of $\mathbb{Q}$ by first defining the set $\mathbb{Q}^+$ of *positive* rationals to be the set of rational numbers that can be expressed as quotients of positive integers:

$$\mathbb{Q}^+ = \{t \in \mathbb{Q} : t = m/n \text{ for some } n, m \in \mathbb{N}\}.$$

Then, for all $r, s \in \mathbb{Q}$, $r < s$ if and only if $s - r \in \mathbb{Q}^+$.

Equivalently, if $r = m/n$ and $s = k/l$, with $n, l > 0$, then $r < s$ if and only if $lm < nk$.

Proposition. *With the order defined above, $\mathbb{Q}$ is an* ordered field.

Proof. Exercise. □

In fact, $\mathbb{Q}$ is the smallest ordered field. To make sense of this, we need the following definitions.

Definition.

(i) A *subfield* of a field $\mathbb{F}$ is a nonempty subset $\mathbb{F}_1$ of $\mathbb{F}$ that satisfies the field axioms with respect to the operations of addition and multiplication in $\mathbb{F}$. In particular, a subfield contains 0 and 1, and is "closed under" addition, multiplication, negatives, and inverses. In other words, $\mathbb{F}_1$ is a subfield of $\mathbb{F}$ if and only if $0, 1 \in \mathbb{F}_1$ and for every $x, y \in \mathbb{F}_1$
  - $x + y \in \mathbb{F}_1$ and $xy \in \mathbb{F}_1$,
  - $-x \in \mathbb{F}_1$ and
  - if $x \neq 0$, then $x^{-1} \in \mathbb{F}_1$.

(ii) If $\mathbb{F}$ and $\mathbb{K}$ are ordered fields, then the map $\varphi : \mathbb{F} \to \mathbb{K}$ is an *ordered field isomorphism* if for all $x, y \in \mathbb{F}$, $\varphi(x + y) = \varphi(x) + \varphi(y)$, $\varphi(xy) = \varphi(x)\varphi(y)$ and if $x < y$ then $\varphi(x) < \varphi(y)$. If such a map exists, we say that $\mathbb{F}$ and $\mathbb{K}$ are *isomorphic*.

Theorem. *If $\mathbb{F}$ is an ordered field, then $\mathbb{F}$ contains a subfield $\mathbb{F}_1$ that is isomorphic to $\mathbb{Q}$ (with its natural order).*

We leave the proof as a series of exercises for the motivated reader.

Exercises

ex1.3.1. Suppose that $\mathbb{F}$ is a field and that $\mathbb{F}_+$ is a subset of $\mathbb{F}$ with the following properties.

(i) $0 \notin \mathbb{F}^+$

(ii) For all $x \in \mathbb{F}$, if $x \neq 0$, then $x \in \mathbb{F}^+$ or $-x \in \mathbb{F}^+$, but not both.

(iii) For all $x, y \in \mathbb{F}^+$, $x + y \in \mathbb{F}^+$ and $xy \in \mathbb{F}^+$.

Define a relation $<$ on $\mathbb{F}$ by setting $x < y$ if and only if $y - x \in \mathbb{F}^+$. Show that $<$ is a linear order on $\mathbb{F}$ and that equipped with this order, $\mathbb{F}$ is an ordered field.

ex1.3.2. Verify that the two formulations defining the order on $\mathbb{Q}$ are in fact equivalent.

ex1.3.3. Prove Proposition 1.3.4.

ex1.3.4. Show that the equation $x^2 = 2$ has no solutions in $\mathbb{Q}$.

Hints: Suppose to the contrary that $r^2 = 2$ for some $r \in \mathbb{Q}$. (i) Show that you may assume that $r > 0$, so $r = m/n$, with $m, n \in \mathbb{N}$. (ii) You may assume that m/n is *reduced* (what does this mean?). (iii) Use the unique factorization theorem to arrive at a contradiction.

ex1.3.5. Suppose that $\mathbb{F}$ is an ordered field, $x, y \in \mathbb{F}$ and $x, y > 0$. Show that $x < y$ if and only if $x^2 < y^2$.

ex1.3.6. Suppose that $\mathbb{F}$ is an ordered field, $a, b, c \in \mathbb{F}$ and $a > 0$. Show that $ax^2 + bx + c \geq 0$ for all $x \in \mathbb{F}$ if and only if $4ac \geq b^2$.

Hint: Complete the square.

ex1.3.7. Show that for any two elements a and b in the ordered field $\mathbb{F}$ $a \leq b$ if and only if $a < b + \varepsilon$ for every $\varepsilon > 0$ in $\mathbb{F}$.

ex1.3.8. Suppose that $\mathbb{F}$ is an ordered field, with $0_{\mathbb{F}}$ and $1_{\mathbb{F}}$ as its additive and multiplicative units.

(i) Show that $1 > 0$.

(ii) For $n \in \mathbb{N}$, let $n_{\mathbb{F}} = \overbrace{1_{\mathbb{F}} + \cdots + 1_{\mathbb{F}}}^{n}$. Show that for all $n, m \in \mathbb{N}$, $n_{\mathbb{F}} > m_{\mathbb{F}}$ if and only if $n > m$.

(iii) Let $Q_{\mathbb{F}}^+ = \{n_{\mathbb{F}} \cdot m_{\mathbb{F}}^{-1} : n, m \in \mathbb{N}\}$ and $Q_{\mathbb{F}}^- = \{-x : x \in Q_{\mathbb{F}}^+\}$. Show that

$$Q_{\mathbb{F}} = \{0_{\mathbb{F}}\} \cup Q_{\mathbb{F}}^+ \cup Q_{\mathbb{F}}^-$$

is an ordered subfield of $\mathbb{F}$ that is isomorphic to $\mathbb{Q}$.

1.4. Least upper bounds and greatest lower bounds

In the previous sections we discussed properties of the rational numbers (and the integers) that should be familiar. Only the language that we used to describe these properties was possibly new. In this section, we introduce concepts that themselves may be unfamiliar. These concepts—least upper bounds and greatest lower bounds—can be defined in any linearly ordered set, but the ordered field structure of $\mathbb{Q}$ (and later $\mathbb{R}$) will also play an important role in the discussion.

This section is also where we begin to practice in earnest one of the most basic skills in real analysis: working with inequalities.

1.4.1. Upper and lower bounds

Let A be a linearly ordered set, and $B \subset A$.

Definition. The set $B \subset A$ is *bounded above* if there exists an element $t \in A$ such that $b \leq t$ for every element b of B. Such an element t is called an *upper bound* for B.

Likewise, $B \subset A$ is *bounded below* if there exists an element $s \in A$ such that $b \geq s$ for every element b of B. This element s is called a *lower bound* for B.

If the subset B is bounded both above and below, we say that B is *bounded* in A.

If an upper bound for B belongs to B it is called the *maximal element* in B. It is commonly denoted $\max_{x \in B} x$, or $\max\{x : x \in B\}$, or simply $\max B$. Similarly, if a lower bound for B belongs to B it is the *minimal element* in B, denoted $\min_{x \in B} x$, $\min\{x : x \in B\}$, or $\min B$ for short. We observe that every nonempty *finite* subset $B \subset A$ has both a maximal element and a minimal element.

1.4.2. Supremum and infimum

If A is an ordered set and $B \subset A$ is bounded above, then we denote by $\mathscr{UB}(B, A)$ the nonempty set of *upper bounds* for B in A. When the set A is assumed known, the reference to the set A is often omitted in this notation. I.e., we write $\mathscr{UB}(B, A) = \mathscr{UB}(B)$. Likewise, if B is bounded below we denote by $\mathscr{LB}(B, A)$ (or by $\mathscr{LB}(B)$) the set of *lower bounds* for B in A.

Definition. If $\mathscr{UB}(B, A)$ has a smallest element we refer to it as the *least upper bound of B* or as the supremum of B, denoted $\sup B$.

If $\mathscr{LB}(B, A)$ has a biggest element we refer to it as the *greatest lower bound of B* or as the *infimum* of B, denoted $\inf B$.

If $\inf B \in B$, then it is the smallest element in B so $\inf B = \min B$. Likewise, if $\sup B \in B$, then $\sup B = \max B$.

Finally, an ordered set A is *order-complete* or *Dedekind-complete* if every nonempty subset of A that has an upper bound in A, has a *least* upper bound in A.

The following simple characterization of the supremum is useful in practice. There is an analogous characterization of the infimum whose proof we leave as an exercise.

Proposition. *$\beta = \sup B$ if and only if β is an upper bound of B and for all $x < \beta$, there is $b \in B$ such that $x < b$.*

Proof. Suppose first that $\beta = \sup B$. By definition $\beta \in \mathscr{UB}(B)$, so β is an upper bound of B. If $x < \beta$, then x is not an upper bound of B, because β is the smallest element in $\mathscr{UB}(B, A)$, so there is a $b \in B$ such that $x < b$.

If $\beta \in \mathscr{UB}(B, A)$ and $x < \beta$ implies $x < b$ for some $b \in B$, then β must be the least element in $\mathscr{UB}(B, A)$, because for all $x < \beta$, $x \notin \mathscr{UB}(B, A)$. □

As an example, consider the set $E = \{x \in \mathbb{Q} : 0 < x \text{ and } x^2 < 1\}$. If $1 \leq x$, then $1 \leq x^2$, so $x \notin E$. This means that E is bounded and that $1 \in \mathscr{UB}(E, \mathbb{Q})$. It might seem clear that $1 = \sup E$, but it is worth demonstrating this fact, if only to practice using the proposition above.

We already know that 1 is an upper bound of E, so to prove that $1 = \sup E$, it remains to show that if $s < 1$, then $s \notin \mathcal{UB}(E, \mathbb{Q})$. We first observe that if $s < 0$, then $s < 1/2$, so s is not an upper bound of E because $1/2 \in E$. Suppose then that $0 < s < 1$ and let $t = (s + 1)/2$. It follows that $s < t < 1$ because $1 - t = t - s = (1 - s)/2 > 0$. Hence $t^2 < 1^2 = 1$, so $t \in E$, and since $s < t$, it follows that $s \notin \mathcal{UB}(E, \mathbb{Q})$.

It is worth recapping this result: 1 is the *least* upper bound of E, because (i) 1 is a (rational) upper bound of E and (ii) every (rational) number smaller than 1 is *not* an upper bound of E. In this example, the least upper bound of E *exists* as an element of $\mathbb{Q}$. This is not always the case. In other words, $\mathbb{Q}$ is *not* order-complete.

1.4.3. A bounded set in $\mathbb{Q}$ with no supremum

Consider the set $F = \{x \in \mathbb{Q} : 1 < x \text{ and } x^2 < 2\}$. We first note that F is nonempty because $1.1 \in F$, for example. The set F is also bounded above because if $2 \leq x$, then $2 < 4 \leq x^2$, so $x \notin F$. I.e., F is bounded above by 2.

It might be tempting to proceed as with the previous example and try to prove that $\sup F = \sqrt{2}$. The problem is that the equation $x^2 = 2$ has no solution in $\mathbb{Q}$ (see **ex1.3.4**), so that $\sqrt{2}$ does not exist in $\mathbb{Q}$.

Proposition. *The set F has no supremum in $\mathbb{Q}$.*

Proof. We will show that given any *rational* upper bound of F, there is a smaller rational upper bound, so that no upper bound is the least one.

We begin by showing that if $t \in F$, then t is not an upper bound for F. If $t \in F$, then $u = 2 - t^2$ is rational, and $0 < u < 1$ (because $1 < t$, so $2 - t^2 < 2 - 1^2 = 1$). Consider the number $t_1 = t + (u/4t)$: it is rational and $t_1 > t > 1$. To show that t is not an upper bound of F, it is enough to show that $t_1 \in F$.

Since $1 < t$ and $0 < u < 1$, we have $u^2 < u$ and $t < t^2$ so $16 < 16t^2$, and therefore $u^2/16t^2 < u/16$. It follows that

$$t_1^2 = \left(t + \frac{u}{4t}\right)^2 = t^2 + \frac{u}{2} + \frac{u^2}{16t^2} < t^2 + \frac{u}{2} + \frac{u}{16} < t^2 + u = 2.$$

In other words, $t_1 \in F$, and t is not an upper bound of F.

Now suppose that $s \in \mathcal{UB}(F, \mathbb{Q})$. Then $s^2 - 2 > 0$, because $s^2 \not< 2$ and $s^2 \neq 2$ (since $s \in \mathbb{Q}$). It follows that $v = s^2 - 2$ is a positive rational number. If $s_1 = s - (v/4s)$, then s_1 is rational and $s_1 < s$. To show that s is not the *least* upper bound of F, we will show that s_1 is also an upper bound of F. To that end, we first note that

$$s_1 = \frac{4s^2 - v}{4s} > \frac{s^2 - v}{4s} = \frac{2}{4s} > 0$$

and that $s_1^2 > 2$ because

$$s_1^2 - 2 = s^2 - \frac{v}{2} + \frac{v^2}{16s^2} - 2 > (s^2 - 2) - \frac{v}{2} = \frac{v}{2} > 0.$$

Finally, suppose that $x \in F$ so that $x^2 < 2$. It follows that

$$(s_1 - x)(s_1 + x) = s_1^2 - x^2 > s_1^2 - 2 > 0.$$

Now, $s_1 + x > 0$ because both terms are positive and it follows that

$$s_1 - x = \frac{s_1^2 - x^2}{s_1 + x} > 0,$$

so $s_1 > x$. Since x was an arbitrary member of F, it follows that s_1 is an upper bound of F, as claimed. □

Exercises

ex1.4.1. Show that if A is a linearly ordered set and $B \subset A$, then $\alpha = \inf B$ if and only if $\alpha \in \mathscr{LB}(B, A)$ and $\alpha < x$ implies that $b < x$ for some $b \in B$.

ex1.4.2. Let $G = \{s \in \mathbb{Q} : 0 < s \text{ and } 2 < s^2\}$, and F be the set defined in Section 1.4.3.

(i) Use arguments similar to the ones used in the proof of Proposition 1.4.3 to show that the set G has no greatest lower bound in $\mathbb{Q}$.

(ii) Show that every $s \in G$ is an upper bound of F and every $t \in F$ is a lower bound of G.

(iii) Show that if u is an upper bound of F, then u is *not* a lower bound of G.

(iv) Show that there is no $u \in \mathbb{Q}$ that is a lower bound of G and an upper bound of F. I.e., there is no $u \in \mathbb{Q}$ between F and G: there is a "gap" in $\mathbb{Q}$ between F and G.

1.5. The real numbers

Exercise **ex1.4.2**, above, shows that the sets $F = \{t \in \mathbb{Q} : 1 < t \text{ and } t^2 < 2\}$ and $G = \{x \in \mathbb{Q} : 0 < x \text{ and } 2 < x^2\}$ are separated by a "gap" in $\mathbb{Q}$. There are many such "gaps" in $\mathbb{Q}$, due to the fact that there are many bounded sets in $\mathbb{Q}$ that have no rational supremum or infimum.

For the purposes of analysis, these holes are problematic. What we want is an ordered field that "fills" these gaps — a field that includes $\mathbb{Q}$ as a subfield and contains enough additional elements to make it *order-complete*.

Definition. A linearly ordered set S is *order-complete* if every bounded set in S has a supremum in S. An ordered field that is *order-complete* is also called *Dedekind complete*.

Theorem. *There exists a unique complete ordered field.*

This unique, complete ordered field is the field $\mathbb{R}$ of real numbers.

The proof of the theorem entails constructing the set $\mathbb{R}$ from subsets of $\mathbb{Q}$, defining addition, multiplication, and an order on this set, and then showing that $\mathbb{R}$ has the properties claimed above. There are different ways of doing this and we will sketch one such construction in the appendix. Uniqueness means that if $\mathbb{F}$ is a complete ordered field then it is isomorphic to $\mathbb{R}$.

1.5.1. Intervals

In the context of analysis, the subsets of $\mathbb{R}$ we encounter most often are *intervals*. We assume that the reader is familiar with intervals and the associated notation, but for the sake of thoroughness, we define them formally below.

Definition. Intervals may be *bounded* or *unbounded*.

If $a, b \in \mathbb{R}$ with $a < b$, then the set $(a, b) = \{x \in \mathbb{R} : a < x < b\}$ is a bounded *open* interval; the set $[a, b] = \{x \in \mathbb{R} : a \leq x \leq b\}$ is a bounded *closed* interval; and the sets $(a, b] = \{x \in \mathbb{R} : a < x \leq b\}$ and $[a, b) = \{x \in \mathbb{R} : a \leq x < b\}$ are bounded *half-open* (or *half-closed*) intervals.

For $a \in \mathbb{R}$, the sets $(a, \infty) = \{x \in \mathbb{R} : a < x\}$, $(-\infty, a) = \{x \in \mathbb{R} : x < a\}$, $[a, \infty) = \{x \in \mathbb{R} : a \leq x\}$ and $(-\infty, b] = \{x \in \mathbb{R} : x \leq b\}$ are *unbounded* intervals, as is the set $(-\infty, \infty) = \mathbb{R}$.

The intervals (a, ∞), $(-\infty, a)$ and $(-\infty, \infty)$ are also open intervals.

Comment. $[a, \infty)$ and $(-\infty, b]$ are sometimes called half-open intervals as well. In Chapter 3, when we study the metric topology on $\mathbb{R}$, we will see that it is perhaps more accurate to call them closed.

Proposition. *A nonempty set $E \subset \mathbb{R}$ is an interval if and only if E contains at least two points and for all $x, y \in E$, if $x < z < y$, then $z \in E$.*[8]

Proof. We leave the proof that intervals have the stated property as an exercise. Suppose that E has the stated property and that E is bounded. Let $a = \inf E$ and $b = \sup E$. Then for all $z \in E$, we have $a \leq z \leq b$, so $E \subset [a, b]$. If $a < z < b$, then from the definitions of the supremum and the infimum, there must exist $x, y \in E$ such that $a \leq x < z$ and $z < y \leq b$. It follows that $x < z < y$, so $z \in E$, by the assumption on E, which implies that $(a, b) \subset E$. It follows that $E = (a, b)$ if $a, b \notin E$, that $E = [a, b)$ if $a \in E$ and $b \notin E$, that $E = (a, b]$ if $a \notin E$ and $b \in E$, and that $E = [a, b]$ if $a, b \in E$.

Suppose now that E is bounded below and not bounded above, and let $a = \inf E$. If $a < z$, then there must exist an $x \in E$, such that $a \leq x < z$, and since E is not bounded above, there must be a $y \in E$ such that $z < y$. By the assumed property of E, it follows that $z \in E$, and hence $z \in E$ for all $a < z$. If $a \in E$, then $E = [a, \infty)$, otherwise $E = (a, \infty)$.

The proofs that E is an interval if it is bounded above and not below, or if it is not bounded above or below are analogous and we leave them as exercises. □

1.5.2. The Archimedean property

Informally, the ordered field $\mathbb{F}$ has the *Archimedean property* if given a positive number x in the field, no matter how small, you can travel any distance in the field by taking a sufficiently large number of steps of size x.

Definition. An ordered field $\mathbb{F}$ has the *Archimedean property* if the set $\{nx : n \in \mathbb{N}\}$ is *not* bounded above for any $x > 0$ in $\mathbb{F}$.[9] Equivalently, if $x > 0$ and $y \in \mathbb{F}$, then there

[8]We see in Chapter 3 that this property means that intervals are (the only) *connected* subsets of $\mathbb{R}$.

[9]If $\mathbb{F}$ is a field, $n \in \mathbb{N}$ and $x \in \mathbb{F}$ then nx is shorthand for the sum $x + x + \cdots + x$, with n terms.

exists a natural number n such that $nx > y$. An ordered field with the Archimedean property is called an *Archimedean field.*

Both $\mathbb{Q}$ and $\mathbb{R}$ are Archimedean. The proof for $\mathbb{Q}$ relies on the explicit representation of $\mathbb{Q}$ as quotients of integers. The proof for $\mathbb{R}$ relies on order-completeness.

Proposition. *$\mathbb{Q}$ is an Archimedean field.*

Proof. Let $x, y \in \mathbb{Q}$ with $x > 0$. We want to show that there exists a natural number n, such that $nx > y$. If $y \leq 0$, then $1 \cdot x > y$ and there is nothing more to prove. If $y > 0$, then we proceed as follows.

We can write $x = a/b$ and $y = c/d$, with $a, b, c, d \in \mathbb{N}$. From $d \geq 1$ it follows that $y = c/d \leq c$. If we set $n = (c+1)b$, then

$$nx = (c+1)b \cdot \frac{a}{b} = a(c+1) \geq c+1 > c \geq y,$$

since $a \geq 1$. □

Theorem. *$\mathbb{R}$ is an Archimedean field.*

Proof. Suppose that $\mathbb{R}$ does not have the Archimedean property. Then there exists an $x \in \mathbb{R}$, $x > 0$, such that the set $A = \{nx : n \in \mathbb{N}\}$ is bounded. This implies that A has a supremum, $s_A \in \mathbb{R}$, by the order completeness of $\mathbb{R}$. In this case, it follows that $s_A - x$ is not an upper bound of A, because $s_A - x < s_A$. Hence there exists $n \in \mathbb{N}$ such that $nx > s_A - x$. But then $(n+1)x = nx + x > s_A$, contradicting the fact that s_A is an upper bound of A. Hence, the assumption that $\mathbb{R}$ does not have the Archimedean property must be false. □

1.5.3. The density of $\mathbb{Q}$ in $\mathbb{R}$

The real numbers can be thought of as the *order-completion* of $\mathbb{Q}$. By this we mean that $\mathbb{R}$ fills in all the "gaps" in $\mathbb{Q}$. On the other hand, these "gaps" don't have any "width" in the sense that between every two real numbers, there is a rational number. We say that $\mathbb{Q}$ is *order-dense* in $\mathbb{R}$.

To begin, we need a lemma.

Lemma. *If $y > 0$, then there is an $n \in \mathbb{N}$ such that $0 < 1/n < y$.*

Proof. If $y > 0$, then $1/y > 0$ as well, and by the Archimedean property (applied to $1 > 0$) there exists $n \in \mathbb{N}$ such that $n = n \cdot 1 > 1/y$. It follows that $0 < 1/n < 1/(1/y) = y$. □

Theorem. *For any $x, y \in \mathbb{R}$, with $x < y$, there is an $r \in \mathbb{Q}$ such that $x < r < y$.*

In other words, if $x < y$, then the interval (x, y) contains a rational point.

Proof. If $y > 0$ and $x < 0$, then $x < 0 < y$.

If $0 < x < y$, then $0 < y - x$, so there exists an $n \in \mathbb{N}$ such that $0 < 1/n < y - x$, by the preceding lemma. By the Archimedean property, there exists a natural number m such that $x < m \cdot (1/n) = m/n$, because $0 < 1/n$. Hence the set $A = \{m \in \mathbb{N} : x < m/n\}$ is not empty and so contains a least member, $m_0 > 1$.

Since $m_0 - 1 < m_0$, it follows that $m_0 - 1 \notin A$, so $(m_0 - 1)/n \leq x$ and therefore

$$x < \frac{m_0}{n} \leq x + \frac{1}{n} < x + (y - x) = y.$$

Finally, if $x < y < 0$, then $0 < -y < -x$ and there exists an $r \in \mathbb{Q}$ such that $-x < r < -y$, and It follows that $x < -r < y$. □

1.5.4. The existence of $\sqrt{2}$ in $\mathbb{R}$

The equation $x^2 = 2$ has no rational solutions and as a consequence, the bounded set $F = \{x \in \mathbb{Q} : 1 < x \text{ and } x^2 < 2\}$ has no *rational* least upper bound, as we saw in Proposition 1.3.2. On the other hand, F does have a supremum in $\mathbb{R}$, and this supremum is a natural candidate for $\sqrt{2}$.

Proposition. *Let $F_R = \{x \in \mathbb{R} : 1 < x \text{ and } x^2 < 2\}$, then F_R is bounded and $(\sup F_R)^2 = 2$.*

Proof. The proof relies on Proposition 1.4.3 (and its proof), and we note that if $1 < x$ and $x^2 < 2$, then $x < 2$, so F_R is indeed bounded. Write $\sup F_R = \alpha$, for convenience, and to prove that $\alpha^2 = 2$, we show that $\alpha^2 \geq 2$ and $\alpha^2 \leq 2$.

In the proof of Proposition 1.4.3., we saw that if $t^2 < 2$ then t is *not* an upper bound of F, and if $s^2 > 2$, then s is *not* the *least* upper bound of F. The arguments used there can be repeated here, word-for-word, replacing $t, s \in \mathbb{Q}$ by $t, s \in \mathbb{R}$ as necessary. From these arguments it follows that if $t^2 < 2$, then $t \in \mathbb{R}$ is not an upper bound of F_R and if $2 < s^2$, then $s \in \mathbb{R}$ is not the *least* upper bound of F_R.

Therefore $\alpha^2 \geq 2$, because α is an upper bound of F_R, so $\alpha^2 \not< 2$. Likewise, $\alpha^2 \leq 2$ because α is the least upper bound of F_R, so $\alpha^2 \not> 2$. This means that $\alpha^2 = 2$, as claimed.

We add that α is the only positive solution to $x^2 = 2$. Suppose that β is another positive solution, the $\alpha^2 = 2 = \beta^2$, so

$$\alpha^2 - \beta^2 = (\alpha - \beta)(\alpha + \beta) = 0,$$

and since $\alpha + \beta > 0$, it follows that $\alpha - \beta = 0$. □

1.5.5. The absolute value in $\mathbb{R}$

Definition. The *absolute value* of a real number x, denoted $|x|$, is defined by $|x| = x$ if $0 \leq x$ and $|x| = -x$ if $x < 0$.

The absolute value of a real number is a measure of its *size* and the absolute value of the difference of two real numbers is a measure of the distance between them. We will return to distance functions in Chapter 2.

Lemma. *For all $x, y \in \mathbb{R}$*

(i) $|x| = |-x|$.

(ii) $|x| \geq 0$ *and* $|x| = 0$ *if and only if* $x = 0$.

(iii) $x \leq |x|$.

(iv) $|x|^2 = x^2$.

(v) $|xy| = |x||y|$

Proof. Exercise. □

Theorem. *For all* $u, v \in \mathbb{R}$, $|u + v| \le |u| + |v|$.

This is called the *triangle inequality*.

Proof. From (ii) and (iv) above, we have $uv \le |u||v|$ for all $u, v \in \mathbb{R}$. It follows then from (iii) that

$$\begin{aligned} |u + v|^2 &= (u + v)^2 \\ &= u^2 + 2uv + v^2 \\ &= |u|^2 + 2uv + |v|^2 \\ &\le |u|^2 + 2|u||v| + |v|^2 = (|u| + |v|)^2 . \end{aligned}$$

The claim now follows from the conclusion of **cx1.4.4**. □

Proposition. *For all* $u, v, \delta \in \mathbb{R}$ *with* $\delta > 0$, $|u - v| < \delta$ *if and only if* $u - \delta < v < u + \delta$.

Proof. If $u - \delta < v < u + \delta$, then $-\delta < v - u < \delta$. If $v - u \ge 0$, then $|u - v| = v - u < \delta$ and if $v - u < 0$, then $0 < u - v < \delta$ and once again $|u - v| < \delta$.

In the other direction, if $|u - v| < \delta$, then $|v - u| < \delta$ as well (by Theorem 1.6.4(i)), so $u - v < \delta$ and $v - u < \delta$ (by Theorem 1.6.4(iii)). It follows that $u - \delta < v$ and $v < u + \delta$. □

1.5.6. The extended real numbers

Definition. The set of *extended* real numbers is the set $\mathbb{R} \cup \{-\infty, \infty\}$, with the arithmetic and order on $\mathbb{R}$ extended to $\mathbb{R} \cup \{-\infty, \infty\}$ by the following rules.

For all $a \in \mathbb{R}$,

(i) $-\infty < a < \infty$.

(ii) $a + \infty = \infty$.

(iii) $a - \infty = -\infty$.

(iv) $a \cdot \infty = \infty$, for $a > 0$ and $a \cdot \infty = -\infty$ for $a < 0$.

(v) $\infty + \infty = \infty \cdot \infty = \infty$.

(vi) $a/\infty = 0$.

Note that $0 \cdot \infty$, $\infty - \infty$ and ∞/∞ are not defined.

The main purpose of extending the real numbers in this way is for notational coherence. For example, the statement $\sup A = \infty$ can now be interpreted as a statement about the extended real numbers, as opposed to shorthand for "the set A is not bounded above". The extended real numbers also play an important role in the definition of the Lebesgue integral, in Chapter 8.

Exercises

ex1.5.1. Show that $\mathbb{R}$ has the *greatest lower bound property*: if A is a nonempty subset of $\mathbb{R}$ that is bounded below, then A has an infimum in $\mathbb{R}$.[10]

ex1.5.2. Complete the missing cases in the proof of Proposition 1.5.1.

ex1.5.3. Without referring to the completeness of $\mathbb{R}$, show that between any two rational numbers we can find another rational number.

ex1.5.4. Following the pattern of the proof of Propositions 1.5.4 and 1.4.3, show that the equation $x^2 = 3$ has a unique positive solution in $\mathbb{R}$.

ex1.5.5. Suppose that $B \subset \mathbb{R}$ is bounded above and $a > 0$. Show that $\sup\{ab : b \in B\} = a \cdot \sup B$.

ex1.5.6. Suppose that $A, B \subset \mathbb{R}$ are bounded, with $a > 0$ for all $a \in A$ and $b > 0$ for all $b \in B$. Show that

$$S_{AB} = \sup\{ab : a \in A \text{ and } b \in B\} = (\sup A)(\sup B) = S_A S_B.$$

Hint: Use the previous exercise (twice) to show that $S_{AB} \leq S_A S_B$. In the other direction, show that for any $\varepsilon > 0$, you can find $a \in A$ and $b \in B$ such that $ab > S_A S_B - \varepsilon$. Now what?

ex1.5.7. Prove Proposition 1.5.5.

*1.6. Rational powers of real numbers

In standard calculus classes, the algebraic properties of the functions $f(x) = x^u$ and $g(x) = a^x$, e.g., $x^{u+v} = x^u x^v$, are often introduced without much justification beyond the intuition coming from the case of (positive) integer exponents. Their analytic properties — the continuity of $g(x)$, for example — are given even shorter shrift. In this section and the next, we provide some of this missing background both for the sake of completeness and to practice using some of the analytic concepts that we have introduced.[11]

1.6.1. Integer powers

For $a \in \mathbb{R}$ and $n \in \mathbb{N}$, a^n is defined as repeated multiplication,

$$a^n = \overbrace{a \cdot a \cdots a}^{n \text{ factors}}.$$

For $a \neq 0$, We extend exponentiation to all integer powers as usual: we set $a^0 = 1$ and for $m < 0$, we set $a^m = (1/a)^{-m}$.

With these definitions, integer exponentiation has the following familiar algebraic properties.

Theorem. *If $a, b \neq 0$ and $l, k \in \mathbb{Z}$, then $(ab)^k = a^k b^k$, $a^{k+l} = a^k a^l$ and $\left(a^k\right)^l = a^{kl}$.*

[10] Assuming, as we are, that $\mathbb{R}$ has the least upper bound property.

[11] These sections can be safely omitted without loss of continuity, though we do refer to some of the results collected here in later chapters.

Proof. Exercise. □

Corollary. *If $a \neq 0$, then $a^{-k} = 1/a^k$ for all $k \in \mathbb{Z}$.*

Proof. Exercise. □

Integer exponentiation also has important analytic, i.e., order-related properties.

Lemma. *Let $a \in \mathbb{R}$ and $k, l \in \mathbb{Z}$.*

(i) *If $a > 0$, then $a^k > 0$ for all $k \in \mathbb{Z}$.*

(ii) *If $a > 1$ and $k < l$, then $a^k < a^l$.*

(iii) *If $k > 0$, then for any $\varepsilon > 0$, there is a $\delta > 0$ such that for any $b \in \mathbb{R}$, if $|a - b| < \delta$, then $|a^k - b^k| < \varepsilon$.*

Proof. If $a > 0$, then $1/a > 0$ and (i) follows from the fact that the product of positive numbers is positive.

For (ii), we have $l - k > 0$, so $a^l a^{-k} = a^{l-k} > 1^{l-k} = 1$, by Theorem 1.3.3 (vii). Therefore,

$$a^l = a^l(a^{-k}a^k) = (a^l a^{-k})a^k = a^{l-k}a^k > 1 \cdot a^k = a^k,$$

by **ordf** 2 since $a^k > 0$.

For (iii), if $k = 1$, then we may take $\delta = \varepsilon$. If $k \geq 2$, then

$$|a^k - b^k| = |a - b|\,|a^{k-1} + a^{k-2}b + \cdots + b^{k-1}|,$$

by (1.1.1). If $|b - a| < 1$, then $|b| < |a| + 1$ and

$$\begin{aligned}|a^{k-1} + a^{k-2}b + \cdots + b^{k-1}| &\leq |a|^{k-1} + |a|^{k-2}|b| + \cdots + |b|^{k-1} \\ &< (|a| + 1)^{k-1} + \cdots + (|a| + 1)^{k-1} \\ &= k(|a| + 1)^{k-1}.\end{aligned}$$

Setting $\delta = \min\left(1, \frac{\varepsilon}{k(|a|+1)^{k-1}}\right)$, if $|a - b| < \delta$, then $|a - b| < 1$ and therefore

$$|a^k - b^k| < \delta \cdot k(|a| + 1)^{k-1} \leq \varepsilon.$$ □

1.6.2. The existence of nth roots in $\mathbb{R}$

If you read the proofs of Propositions 1.5.4 and 1.4.3 carefully, you will see that the only property of the number 2 that we used in showing that $\sqrt{2}$ exists in $\mathbb{R}$ is the fact that $2 > 1$. In much the same way, we can prove that for any $a > 0$, the equation $x^2 = a$ has a unique positive solution in $\mathbb{R}$. More is true: we can do this for any natural exponent.

Theorem. *For any $a > 0$ and any $n \in \mathbb{N}$, the equation $x^n = a$ has a unique positive solution in $\mathbb{R}$.*

The unique positive solution of $x^n = a$ is called the nth *root of a*, and denoted by $\sqrt[n]{a}$ or $a^{1/n}$.

Proof. First, if $x, y > 0$ and $x \neq y$ then either $x < y$ or $x > y$. In the first case, $x^n < y^n$ and in the second one, $x^n > y^n$. In either case, $x^n \neq y^n$, so if there is a positive solution to $x^n = a$, then it must be unique.

Next, if $n = 1$, then the unique positive solution to $x^1 = a$ is a itself, so we may assume that $n > 1$.

One last observation before getting to the meat of the proof. If $0 < a < 1$, then $1 < 1/a$. If $x^n = 1/a$, then $(1/x)^n = 1/x^n = a$, so if the theorem is true for $a > 1$, then it is also true for $0 < a < 1$. Since the theorem is trivially true for $a = 1$, we may assume without loss of generality that $a > 1$. The proof now follows that same pattern as the proof that $x^2 = 2$ has a solution in $\mathbb{R}$ (though some of the technical details are a little *more* technical).

Consider the set $A = \{x \in \mathbb{R} : 0 < x \text{ and } x^n < a\}$. Since $a > 1$, the set A is not empty because $1 \in A$. Also, if $a < x$, then $a < a^n < x^n$, which means that $x \notin A$. In other words, a is an upper bound of A. Thus A is bounded and nonempty, so A has a least upper bound, $\sup A = \eta$. We will prove that $\eta^n = a$ by showing that $\eta^n \geq a$ and $\eta^n \leq a$. We do this as we did in the proof that $\sqrt{2}$ exists, by showing that if $t^n < a$, then t is not an upper bound of A and if $a < s^n$, then s is not the *least* upper bound of A.

Suppose first that $t^n < a$. To show that t is not an upper bound of A, we will find an $\varepsilon > 0$ such that $(t + \varepsilon)^n < a$. To do so, we first note that

$$(1.6.1) \qquad (t+\varepsilon)^n = \sum_{k=0}^{n} \binom{n}{k} t^{n-k}\varepsilon^k = t^n + \sum_{k=1}^{n} \binom{n}{k} t^{n-k}\varepsilon^k,$$

by the binomial theorem.

In order for $(t + \varepsilon)^n < a$, we need to find an $\varepsilon > 0$ small enough so that the sum on the right of (1.6.1) is smaller that $a - t^n$. To this end, we observe that if $0 < \varepsilon < 1$, then $\varepsilon^k \leq \varepsilon$ for $1 \leq k$, and therefore

$$(1.6.2) \qquad \sum_{k=1}^{n} \binom{n}{k} t^{n-k}\varepsilon^k \leq \varepsilon \cdot \sum_{k=1}^{n} \binom{n}{k} t^{n-k} < \varepsilon(t+1)^n,$$

because

$$\sum_{k=1}^{n} \binom{n}{k} t^{n-k} < \sum_{k=0}^{n} \binom{n}{k} t^{n-k} \cdot 1^k = (t+1)^n.$$

If we set $\varepsilon = \dfrac{a - t^n}{(t+1)^n}$ or $1/2$, whichever is smaller (guaranteeing that $\varepsilon < 1$), it follows from (1.6.1) and (1.6.2) that

$$(t+\varepsilon)^n < t^n + \varepsilon \cdot (t+1)^n \leq t^n + (a - t^n) = a.$$

I.e., $t + \varepsilon \in A$, so t is not an upper bound of A.

Now suppose that $s > 0$ and $s^n > a$. Similar to the argument above, for any $\varepsilon > 0$ we have

$$(1.6.3) \qquad (s-\varepsilon)^n = \sum_{k=0}^{n} \binom{n}{k} s^{n-k}(-\varepsilon)^k = s^n - \sum_{k=1}^{n} (-1)^{k+1} \binom{n}{k} s^{n-k}\varepsilon^k.$$

The terms in the sum on the right-hand side of (1.6.3) are *alternate in sign*, because of the $(-1)^{k+1}$ factors, so if $1 > \varepsilon > 0$, then

$$\sum_{k=1}^{n}(-1)^{k+1}\binom{n}{k}s^{n-k}\varepsilon^{k} < \sum_{k=1}^{n}\binom{n}{k}s^{n-k}\varepsilon^{k} < \varepsilon(s+1)^{n},$$

where the right-hand inequality follows from the same argument that justified (1.6.2). This together with (1.6.3) imply that

$$(s-\varepsilon)^{n} > s^{n} - \sum_{k=1}^{n}\binom{n}{k}s^{n-k}\varepsilon^{k} > s^{n} - \varepsilon(s+1)^{n},$$

and letting ε be the smaller of $\dfrac{s^{n}-a}{(s+1)^{n}}$ or $1/2$, it follows that

$$(s-\varepsilon)^{n} > s^{n} - \varepsilon(s+1)^{n} \geq s^{n} - (s^{n}-a) = a.$$

Now, if $x \in A$, then $x^{n} < a < (s-\varepsilon)^{n}$, so $s-\varepsilon > x$ (by Theorem 1.4.3(vii)). Therefore $s-\varepsilon$ is an upper bound of A, so s is not the least upper bound of A.

It follows that $\eta^{n} = a$ because $\eta^{n} \not< a$ (since η is an upper bound of A) and $\eta^{n} \not> a$ (since η is the least upper bound of A). □

We note that it follows from the uniqueness of nth roots that

$$\left(a^{n}\right)^{1/n} = a = \left(a^{1/n}\right)^{n}$$

for all $n \in \mathbb{N}$ and $a > 0$.

The algebraic properties of integer exponentiation extend to n^{th} roots, however we will leave the proof of this to the case of general rational powers in the next section.

The claims of Lemma 1.7.1 also have analogues for n^{th} roots, which appear as parts (i) and (iii) of the following Lemma. Part (ii) below is an analogue of Theorem 1.4.3(vii). Part (iv) does not have an integer power analogue.

Lemma. *Let $a \in \mathbb{R}$.*

(i) *If $a > 1$ and $n, m \in \mathbb{N}$ with $m < n$, then $a^{1/n} < a^{1/m}$.*

(ii) *If $0 < a < b$ and $n \in \mathbb{N}$, then $a^{1/n} < b^{1/n}$.*

(iii) *If $a > 0$ and $n \in \mathbb{N}$, then for any $\varepsilon > 0$ there is a $\delta > 0$ such that if $|a-b| < \delta$, then $|a^{1/n} - b^{1/n}| < \varepsilon$.*

(iv) *If $a > 1$ and $\varepsilon > 0$, then there is an $n_{\varepsilon} \in \mathbb{N}$ such that $a^{1/n} < 1+\varepsilon$ for all $n > n_{\varepsilon}$.*

Proof. We first note that if $0 < x \leq 1$, then $x^{n} \leq 1^{n} = 1$ for all $n \in \mathbb{N}$ by Theorem 1.4.3(vii). It follows that if $x > 0$ and $x^{n} = a > 1$, then $a^{1/n} = x > 1$.

To prove (i), suppose to the contrary that $a^{1/m} \leq a^{1/n}$. Since $1 < a^{1/m}$, it would follow then from Lemma 1.7.1 and Theorem 1.4.3(vii) that

$$a = (a^{1/m})^{m} < (a^{1/m})^{n} \leq (a^{1/n})^{n} = a.$$

The contradiction $a < a$ shows that the assumption $a^{1/m} \leq a^{1/n}$ must be false.

To prove (ii), we set $a^{1/n} = x$ and $b^{1/n} = y$. Then $0 < x, y$, so $0 < y^k x^{n-1-k}$ for all $0 \le k \le n-1$, and $x^n < y^n$. It then follows from the identity (1.1.1) that

$$y - x = \frac{y^n - x^n}{y^{n-1} + y^{n-2}x + \cdots + yx^{n-2} + x^{n-1}} > 0,$$

so $a^{1/n} = x < y = b^{1/n}$.

To prove (iii), we write $x = a^{1/n}$ and $y = b^{1/n}$ once again, and note that it follows from (ii) that $(a/2)^{1/n} < x$. Furthermore, if $|a - b| < a/2$, then $a/2 < b$ in which case we also have $(a/2)^{1/n} < y$.

Invoking (1.1.1) again, we note that if $a > 0$ and $b > a/2 > 0$, then

$$|x - y| = \frac{|x^n - y^n|}{|x^{n-1} + x^{n-2}y + \cdots + xy^{n-2} + y^{n-1}|} < \frac{|a-b|}{n\left((a/2)^{1/n}\right)^{n-1}},$$

because $((a/2)^{1/n})^{n-1} < x^k y^{n-1-k}$ for $0 \le k \le n-1$ and there are n such terms in the denominator of the expression in the middle.

Setting $\delta = \min\left(a/2, n\left((a/2)^{1/n}\right)^{n-1} \cdot \varepsilon\right)$, if $|a - b| < \delta$, then $a/2 < b$ and

$$|a^{1/n} - b^{1/n}| = |x - y| < \varepsilon.$$

Finally, if $\varepsilon > 0$, then for any $n \in \mathbb{N}$,

$$(1+\varepsilon)^n = \sum_{k=0}^{n} \binom{n}{k} \varepsilon^k = 1 + n\varepsilon + \sum_{k=2}^{n} \varepsilon^k \ge 1 + n\varepsilon,$$

by the binomial theorem and the fact that the terms in the sum are all positive.

From the Archimedean property, it follows that there is an $n_\varepsilon \in \mathbb{N}$ such that $n_\varepsilon \varepsilon > a - 1$ and it follows that

$$(1+\varepsilon)^{n_\varepsilon} \ge 1 + n_\varepsilon \varepsilon > a,$$

and therefore $1 + \varepsilon > a^{1/n_\varepsilon}$. It follows from part (i) that if $n > n_\varepsilon$, then $a^{1/n} < a^{1/n_\varepsilon} < 1 + \varepsilon$, which proves (iv). □

1.6.3. Rational powers

If $a > 0$, $n \in \mathbb{N}$ and $m \in \mathbb{Z}$, then $a^m > 0$ so $\left(a^m\right)^{1/n}$ exists in $\mathbb{R}$ and since $a > 0$, $\left(a^{1/n}\right)^m$ is also well-defined in $\mathbb{R}$. If $\alpha = \left(a^m\right)^{1/n}$, then $\alpha^n = a^m$. On the other hand, since $m, n \in \mathbb{Z}$ it follows from Theorem 1.6.1 that

$$\left[\left(a^{1/n}\right)^m\right]^n = \left(a^{1/n}\right)^{mn} = \left[\left(a^{1/n}\right)^n\right]^m = a^m,$$

so $\left(a^{1/n}\right)^m = \alpha = \left(a^m\right)^{1/n}$.

It follows that for $m, n \in \mathbb{Z}$ with $n > 0$, we can define $a^{m/n}$ unambiguously as $\left(a^m\right)^{1/n} = \left(a^{1/n}\right)^m$.

For $a > 0$ and $r = m/n \in \mathbb{Q}$, we want to define $a^r = a^{m/n}$. However, every $r \in \mathbb{Q}$ can be expressed as a quotient of integers in infinitely many ways. The following proposition shows that defining $a^r = a^{m/n}$ does not depend on the choice of the representative quotient.

Proposition. *Suppose that $k, l, m, n \in \mathbb{Z}$, with $l, n > 0$. If $k/l = m/n$, then $a^{k/l} = a^{m/n}$, for all $a > 0$.*

Proof. By assumption, $kn = lm$. Therefore if $a > 0$, then

$$\left(a^{k/l}\right)^n = \left(\left(a^{1/l}\right)^k\right)^n = \left(a^{1/l}\right)^{kn} = \left(a^{1/l}\right)^{lm} = \left(\left(a^{1/l}\right)^l\right)^m = a^m.$$

It now follows that $a^{k/l} = (a^m)^{1/n} = a^{m/n}$, by the uniqueness of nth roots. □

It is worth emphasizing that $a^r > 0$ for all $a > 0$ and $r \in \mathbb{Q}$. This is immediate, since $a^{1/n}$ is always positive by definition and integer powers of positive numbers are positive. Moreover, if $a > 1$ and $r > 0$, then $a^r > 1$, as follows from similar reasoning.

The algebraic properties of integer exponents (Theorem 1.6.1) extend to rational exponents.

Theorem. *If $a, b > 0$ and $r, s \in \mathbb{Q}$, then $(ab)^r = a^r b^r$, $a^r a^s = a^{r+s}$ and $(a^r)^s = a^{rs}$.*

Throughout the proof, we rely on the fact that these properties hold for integer exponents.

Proof. Let $a, b > 0$ and let $n \in \mathbb{N}$. Then

$$(a^{1/n}b^{1/n})^n = (a^{1/n})^n(b^{1/n})^n = ab,$$

so $a^{1/n}b^{1/n} = (ab)^{1/n}$ by uniqueness of nth roots. If $r = m/n$, then

$$a^r b^r = (a^{1/n})^m(b^{1/n})^m = (a^{1/n}b^{1/n})^m = ((ab)^{1/n})^m = (ab)^r.$$

Next, let $a > 0$ and $r = m/n$ and $s = k/l$ be rational numbers. Then $\left(a^r\right)^{ln} = \left(a^{m/n}\right)^{ln} = a^{lm}$ and $\left(a^s\right)^{ln} = \left(a^{k/l}\right)^{ln} = a^{kn}$. It follows that $\left(a^r a^s\right)^{ln} = \left(a^r\right)^{ln}\left(a^s\right)^{ln} = a^{lm}a^{kn} = a^{lm+kn}$ and therefore

$$a^r a^s = \left(a^{lm+kn}\right)^{1/ln} = a^{(lm+kn)/ln} = a^{r+s},$$

by the uniqueness of $(ln)^{th}$ roots.

Finally, we will prove that $(a^r)^s = a^{rs}$ in the case that $r, s > 0$, and leave the general case as an exercise. If $r = m/n > 0$ and $s = k/l > 0$, then $m, k > 0$ and we have

$$(a^r)^k = \overbrace{a^r \cdots a^r}^{k} = a^{\overbrace{r+\cdots+r}^{k}} = a^{kr}.$$

Hence,

$$((a^r)^s)^{ln} = \left(((a^r)^k)^{1/l}\right)^{ln} = \left(a^{rk}\right)^n = \left(a^{mk/n}\right)^n = a^{mk},$$

and once again it follows from the uniqueness of ln^{th} roots that

$$(a^r)^s = (a^{mk})^{1/ln} = a^{mk/ln} = a^{rs}.$$ □

Corollary. *If $a > 0$, then for all $r \in \mathbb{Q}$, $a^{-r} = 1/a^r$ and if $r \neq 0$, then $(a^r)^{1/r} = a$.*

Proof. From part (i) above, we have $a^r a^{-r} = a^{r-r} = a^0 = 1$, and hence $a^{-r} = 1/a^r$ by the uniqueness of multiplicative inverses.

From part (ii), we have $(a^r)^{1/r} = a^{r \cdot 1/r} = a^1 = a$. □

The analytic (order-related) properties of nth roots extend to all rational exponents.

Lemma. *Let $a, b \in \mathbb{R}$ and $r, s \in \mathbb{Q}$.*

(i) *If $a > 1$ and $r < s$, then $a^r < a^s$.*

(ii) *If $0 < a < b$ and $r > 0$, then $a^r < b^r$.*

(iii) *If $a > 0$ and $r > 0$, then for any $\varepsilon > 0$, there is a $\delta > 0$ such that if $|a - b| < \delta$, then $|a^r - b^r| < \varepsilon$.*

(iv) *For any $s \in \mathbb{Q}$ and any $\varepsilon > 0$, there is an $n \in \mathbb{N}$ such that if $|s - r| < 1/n$, then $|a^s - a^r| < \varepsilon$.*

Proof. If $r = m/n$ and $s = k/l$, then $r < s$ implies that $lm < nk$ because $n, l > 0$. Given that $a > 1$, it follows from Lemma 1.7.1(i) that $a^{lm} < a^{nk}$ and therefore

$$\left(a^{m/n}\right)^l = \left(a^{lm}\right)^{1/n} < \left(a^{nk}\right)^{1/n} = a^k,$$

by Lemma 1.7.2(ii). As a result, we have

$$a^r = \left(\left(a^{m/n}\right)^l\right)^{1/l} < \left(a^k\right)^{1/l} = a^s.$$

This proves (i).

If $0 < a < b$ and $0 < r$, then writing $r = m/n$, we have $a^{1/n} < b^{1/n}$ by Lemma 1.7.2(ii) and therefore

$$a^r = (a^{1/n})^m < (b^{1/n})^m = b^r$$

by Theorem 1.4.3(vii), proving (ii).

Next, writing $r = m/n$, then $r > 0$ implies $m, n > 0$. Given $\varepsilon > 0$, there exists $\delta_1 > 0$ such that if $|a^{1/n} - b^{1/n}| < \delta_1$, then

$$|a^r - b^r| = |(a^{1/n})^m - (b^{1/n})^m| < \varepsilon,$$

by Lemma 1.7.1(ii). Moreover, by Lemma 1.7.2(iii), there is a $\delta > 0$, such that if $|a - b| < \delta$, then $|a^{1/n} - b^{1/n}| < \delta_1$. In other words, if $|a - b| < \delta$, then $|a^r - b^r| < \varepsilon$, proving (iii).

Finally, let $s \in \mathbb{Q}$ and $\varepsilon > 0$ be given. According to Proposition 1.7.2, we can find $n \in \mathbb{N}$ such that

$$1 < a^{1/n} < 1 + \frac{\varepsilon}{a^s}.$$

If $r \in \mathbb{Q}$ with $0 < |s - r| < 1/n$, then there are two possibilities to consider, $r < s < r + 1/n$ and $s < r < s + 1/n$.

If $r < s < r + 1/n$, then

$$a^r < a^s < a^{r+1/n} = a^r a^{1/n} < a^r\left(1 + \frac{\varepsilon}{a^s}\right) < a^r + \varepsilon$$

(because $a^r/a^s < 1$) and if $s < r < s + 1/n$, then

$$a^s < a^r < a^{s+1/n} = a^s a^{1/n} < a^s\left(1 + \frac{\varepsilon}{a^s}\right) = a^s + \varepsilon.$$

In both cases, we have $|a^s - a^r| < \varepsilon$, proving (iv). □

Exercises

ex1.6.1. Prove Theorem 1.6.1.

ex1.6.2. Prove Corollary 1.6.1.

ex1.6.3. Show that for all $a \in \mathbb{R}$, $|a| = \sqrt{a^2}$.

ex1.6.4. Show that if $a > 0$ and $k \in \mathbb{Z}$ with $k < 0$, then for any $\varepsilon > 0$ there is a $\delta > 0$ such that $|a - b| < \delta$ implies $|a^k - b^k| < \varepsilon$. (i.e., extend Proposition 1.7.1 to negative integer powers).

Hint: If $a > 0$ then $1/a > 0$ and for $k < 0$, $a^k - b^k = (1/a)^{-k} - (1/b)^{-k}$, so you can make use of Proposition 1.7.1. Also, $|(1/a) - (1/b)| = |a - b|/ab$ and if $a/2 < b$, then $|(1/a) - (1/b)| < |a - b|/(a^2/2)$.

ex1.6.5. Suppose that $0 < a < 1$. Show that for any $\varepsilon > 0$ there is an $n_\varepsilon \in \mathbb{N}$ such that $1 - \varepsilon < a^{1/n} < 1$ for all $n > n_\varepsilon$.

Hint: First $1 < 1/a$. Second, you may assume that $0 < \varepsilon < 1$ (why?). In this case $0 < 1 - \varepsilon$, so $1/(1 - \varepsilon) = 1 + \delta$ with $\delta > 0$. Now what?

ex1.6.6. Show that $(a^r)^s = a^{rs}$ for $a > 0$ and *all* $r, s \in \mathbb{Q}$ (i.e., complete the proof of Theorem 1.7.3).

*1.7. Real powers of real numbers

Now, while $a^2 = a \cdot a$ makes perfect sense in any field, and as we have seen, we can make sense of a^r algebraically when $r \in \mathbb{Q}$, it makes no sense to think of $a^{\sqrt{2}}$ in terms of repeated multiplication or as the solution of a simple algebraic equation. In the previous section, the nth root was defined algebraically and we used the completeness of $\mathbb{R}$ to prove that $a^{1/n}$ exists in $\mathbb{R}$ if $a > 0$.

In this section we will also rely on the completeness of $\mathbb{R}$ to *define* a^u, when $u \notin \mathbb{Q}$.

1.7.1. Defining a^u

We now want to extend the definition of a^u to all $u \in \mathbb{R}$, for (fixed) $a > 0$. To start, we set $1^u = 1$ for all u. We will focus here on the case that $a > 1$ and leave the case of $0 < a < 1$ for the exercises.

Suppose that $a > 1$ and for any $u \in \mathbb{R}$ we define the set A_u by

$$A_u = \{a^r : r \in \mathbb{Q} \cap (-\infty, u)\}.$$

This set is not empty because there are rational numbers less than u, and by Lemma 1.6.3 (i), if $s \in \mathbb{Q}$ with $u < s$, then $a^r < a^s$ for all $r < u$, and therefore A_u is bounded by a^s. It follows that $\sup A_u$ exists in $\mathbb{R}$.

We want to define $a^u = \sup A_u$, but for this to work, we need to verify that it agrees with the definition of a^u in the case that $u \in \mathbb{Q}$.

Theorem. *If* $s \in \mathbb{Q}$, *then* $\sup A_s = a^s$.

Proof. On the one hand, we have $\sup A_s \le a^s$ because $a^r < a^s$ for all $a^r \in A_s$ by Lemma 1.6.3 (i).

On the other hand, it follows from Lemma 1.7.3(iv) that for any $\varepsilon > 0$ there is an $n_\varepsilon \in \mathbb{N}$ such that if r is rational and $s - 1/n_\varepsilon < r < s$, then $a^s < a^r + \varepsilon$. Thus, given $\varepsilon > 0$, if we set $r_\varepsilon = s - 1/2n_\varepsilon$, then $s - 1/n_\varepsilon < r_\varepsilon < s$ so $a^{r_\varepsilon} \in A_s$ and it follows that $a^s < a^{r_\varepsilon} + \varepsilon < \sup A_s + \varepsilon$. This means that $a^s \le \sup A_s$ and therefore $a^s = \sup A_s$.[12] □

With this out of the way, we can now formally define a^u for $a > 1$ and $u \in \mathbb{R}$.

Definition. If $a > 1$ and $u \in \mathbb{R}$, then

$$a^u = \sup A_u = \sup\{a^r : r \in \mathbb{Q} \quad \text{and} \quad r < u\}.$$

We remark that for rational powers, the definition given in the previous section is the more useful one.

Lemma. *Suppose that $a > 1$ and $u \in \mathbb{R}$.*

(i) *If $r \in \mathbb{Q}$ and $r < u$, then $a^r < a^u$.*

(ii) *If $u > 0$, then $a^u > 1$.*

(iii) *Given $\varepsilon > 0$, there exists $\delta > 0$ such that if $r \in \mathbb{Q} \cap (u - \delta, u)$, then $a^u - \varepsilon < a^r$.*

Proof.

(i) If $r < u$, then there is an $s \in \mathbb{Q}$ such that $r < s < u$, from which it follows that $a^r < a^s \le a^u$.

(ii) If $u > 0$, then there is $r \in \mathbb{Q}$ with $0 < r < u$, so that $1 < a^r < a^u$.

(iii) If $a^u - \varepsilon < a^{r_0}$ for some $r_0 \in \mathbb{Q} \cap (-\infty, u)$, then $a^u - \varepsilon < a^{r_0} < a^r$ for all $r \in (r_0, u)$ and we may take $\delta = u - r_0$. To show that such an r_0 exists, we suppose to the contrary that for some $\varepsilon > 0$, we have $a^r \le a^u - \varepsilon$ for all $a^r \in A_u$. But this implies that $a^u = \sup A_u \le a^u - \varepsilon < a^u$, a contradiction. □

1.7.2. Algebraic properties of real powers

Theorem. *If $a, b > 1$ and $u, v \in \mathbb{R}$, then $(ab)^u = a^u b^u$, $a^{u+v} = a^u a^v$ and if $u, v > 0$, then $(a^u)^v = a^{uv}$.*

Proof. From the definition, we have

$$(ab)^u = \sup\{(ab)^r : r \in \mathbb{Q} \text{ and } r < u\} = \sup\{a^r b^r : r \in \mathbb{Q} \text{ and } r < u\},$$

since $(ab)^r = a^r b^r$ for rational r. Now,

$$\sup\{a^r b^r : r \in \mathbb{Q} \text{ and } r < u\} \le \sup\{a^r b^s : r, s \in \mathbb{Q} \text{ and } r, s < u\}$$

[12] Remember **ex1.4.6**?

because the set on the left is a subset of the set on the right. Writing $A_u = \{a^r : r \in \mathbb{Q} \text{ and } r < u\}$ and $B_u = \{a^s : s \in \mathbb{Q} \text{ and } s < u\}$, it follows that

$$\begin{aligned}(ab)^u &= \sup\{a^r b^r : r \in \mathbb{Q} \text{ and } r < u\}\\ &\le \sup\{a^r b^s : r, s \in \mathbb{Q} \text{ and } r, s < u\}\\ &= \sup\{xy : x \in A_u \text{ and } y \in B_u\}\\ &= (\sup A_u)(\sup B_u) = a^u b^u,\end{aligned}$$

by **ex1.6.5**. On the other hand, for any $r, s \in \mathbb{Q}$ with $r, s < u$, there is a $t \in \mathbb{Q}$ such that $r, s < t < u$ from which it follows that $a^r < a^t$ and $b^s < b^t$, so $a^r b^s < a^t b^t = (ab)^t < (ab)^u$. This implies that

$$a^u b^u = \sup\{a^r b^s : r, s \in \mathbb{Q} \text{ and } r, s < u\} \le (ab)^u,$$

and therefore $(ab)^u = a^u b^u$.

Next, let u and v be arbitrary real numbers and consider the set $A_1 = \{a^{r+s} : r, s \in \mathbb{Q},\ r < u,\ s < v\}$.

We claim that $A_1 = \{a^t : t \in \mathbb{Q},\ t < u + v\} = A_{u+v}$, and we will show this by showing that $A_1 \subset A_{u+v}$ and $A_1 \supset A_{u+v}$. One direction is immediate: if $r < u$ and $s < v$, then $t = r + s < u + v$, so $a^{r+s} \in A_{u+v}$, showing that $A_1 \subset A_{u+v}$.

In the other direction, suppose that $t \in \mathbb{Q}$ and $t < u + v$. Let $\varepsilon = (u + v) - t > 0$ (so $t = u + v - \varepsilon$) and choose $r \in \mathbb{Q}$ such that $u - \varepsilon/2 < r < u$ (such an r exists by the density of $\mathbb{Q}$ in $\mathbb{R}$). Then $t - r \in \mathbb{Q}$ and

$$s = t - r < (u + v - \varepsilon) - (u - \varepsilon/2) = v - \varepsilon/2 < v.$$

We have $r < u$ and $s < v$, so $a^t = a^{r+s} \in A_1$. This shows that $A_1 \supset A_{u+v}$, so the two sets are the same.

If follows that $a^{u+v} = \sup A_{u+v} = \sup A_1$, and since $a^{r+s} = a^r a^s$ for $r, s \in \mathbb{Q}$, we have

$$\begin{aligned}a^{u+v} &= \sup\{a^r a^s : r, s \in \mathbb{Q},\ r < u,\ s < v\}\\ &= \sup\{xy : x \in A_u,\ y \in A_v\}\\ &= (\sup A_u)(\sup A_v) = a^u a^v,\end{aligned}$$

as follows again from **ex1.6.5**.

It remains to prove that $(a^u)^v = a^{uv}$, and we begin by noting that if $u > 0$ then $a^u > 1$ by Lemma 1.7.1, so $(a^u)^v$ is well-defined. We will show that $(a^u)^v = a^{uv}$ indirectly, by showing that $|(a^u)^v - a^{uv}| < \varepsilon$, for any $\varepsilon > 0$, as follows.

First, if $r, s \in \mathbb{Q}$, then $(a^r)^s = a^{rs}$ by Theorem 1.6.3, so

$$|(a^u)^v - a^{uv}| \le |(a^u)^v - (a^u)^s| + |(a^u)^s - (a^r)^s| + |a^{rs} - a^{uv}|, \tag{1.7.1}$$

by the triangle inequality. Thus, given $\varepsilon > 0$, it is enough to find $r, s \in \mathbb{Q}$ such that r is close enough to u, s is close enough to v, and rs is close enough to uv so that each of the three terms on the right-hand side of the inequality (1.7.1) is less that $\varepsilon/3$.

Let $\varepsilon > 0$ be given. By Lemma 1.7.1, there exist $\delta_v > 0$ and $\delta_{uv} > 0$ such that if $v - \delta_v < s < v$ and $uv - \delta_{uv} < rs < uv$, then

$$(a^u)^v - \varepsilon/3 < (a^u)^s < (a^u)^v \quad \text{and} \quad a^{uv} - \varepsilon/3 < a^{rs} < a^{uv}.$$

Now, by Lemma 1.6.3, for fixed rational $s > 0$, there exists $\delta_s > 0$, such that if $|a^u - a^r| < \delta_s$, then $|(a^u)^s - (a^r)^s| < \varepsilon/3$. Invoking Lemma 1.7.1 again, there is a $\delta_{u,s} > 0$ such that if $u - \delta_{u,s} < r < u$, then $a^u - \delta_s < a^r < a^u$, so that

$$|(a^u)^s - (a^r)^s| < \varepsilon/3,$$

as required.

While $r < u$ and $s < v$ together guarantee that $rs < uv$, the bounds $v - \delta_v < s < v$ and $u - \delta_{u,s} < r < u$ do not necessarily ensure that $uv - \delta_{uv} < rs$. To ensure this last condition, we need one more observation. Namely, if $\delta_1, \delta_2 > 0$, $u - \delta_1 < r$ and $v - \delta_2 < s$, then

$$rs > (u - \delta_1)(v - \delta_2) > uv - (u\delta_2 + v\delta_1).$$

Finally, let $\delta_2 = \min(\delta_v, \delta_{uv}/2u, v/2)$ and choose $s_0 \in \mathbb{Q}$ such that $v - \delta_2 < s_0 < v$, ensuring among other things that $s_0 > v/2 > 0$. Having chosen s_0, we set $\delta_1 = \min(\delta_{u,s_0}, \delta_{uv}/2v)$, and choose $r_0 \in \mathbb{Q}$ such that $u - \delta_1 < r_0 < u$. With these choices, we have

- $|(a^u)^v - (a^u)^{s_0}| = (a^u)^v - (a^u)^{s_0} < \varepsilon/3$, because $\delta_2 \le \delta_v$.
- $|(a^u)^{s_0} - (a^{r_0})^{s_0}| < \varepsilon/3$, because $\delta_1 \le \delta_{u,s_0}$.
- $|a^{uv} - a^{rs}| = a^{uv} - a^{rs} < \varepsilon/3$, because $uv - rs < u\delta_2 + v\delta_1 \le \delta_{uv}$.

Using these bounds in (1.7.1) (with $r = r_0$ and $s = s_0$) shows that

$$|(a^u)^v - a^{uv}| < \varepsilon,$$

as required. □

Corollary. *Suppose that $a > 1$ and $u \in \mathbb{R}$. Then $a^{-u} = 1/a^u$ and if $u > 0$, then $(a^u)^{1/u} = a$.*

Proof. Exercise. □

1.7.3. Analytic properties of real powers

Theorem. *Let $a, b, u, v \in \mathbb{R}$, with $a, b > 1$*

(i) *If $u < v$, then $a^u < a^v$.*

(ii) *If $a < b$ and $u > 0$, then $a^u < b^u$.*

(iii) *If u and a are fixed with $u > 0$, then for any $\varepsilon > 0$, there is a $\delta > 0$ such that if $|a - b| < \delta$, then $|a^u - b^u| < \varepsilon$.*

(iv) *If u and a are fixed, then for any $\varepsilon > 0$, there is a $\delta > 0$ such that if $|u - v| < \delta$, then $|a^u - a^v| < \varepsilon$.*

Proof.

(i) Let $u, v \in \mathbb{R}$ with $u < v$. The density of $\mathbb{Q}$ in $\mathbb{R}$ implies that there is an $s \in \mathbb{Q}$ such that $u < s < v$, and from the definition of a^v it follows that $a^s < a^v$. If $r \in \mathbb{Q}$ and $r < u$, then $r < s$ so $a^r < a^s$ by Lemma 1.7.3(i). It follows that $a^u = \sup A_u \leq a^s < a^v$.

(ii) If $a < b$ and $r \in \mathbb{Q} \cap (0, u)$, then $a^r < b^r < b^u$, by Lemma 1.6.3 (ii) and Proposition 1.7.1, and hence $a^u = \sup A_u \leq b^u$. If $a^u = b^u$, then by Corollary 1.7.2, $a = (a^u)^{1/u} = (b^u)^{1/u} = b$, which contradicts the assumption that $a < b$, so $a^u < b^u$.

(iii) Let $\varepsilon > 0$ be given, and set $a_1 = a + (a - 1)$. By Proposition 1.7.1, there exists $r_1 \in \mathbb{Q}, 0 < r_1 < u$, such that $0 < a_1^u - a_1^{r_1} < \varepsilon/3$. If $1 < x < a_1$, then $1 < a_1/x$ and therefore by part (i),

$$0 < (a_1/x)^{r_1} - 1 < (a_1/x)^u - 1.$$

Since $1 < x^{r_1} < x^u$, it also follows that

$$a_1^{r_1} - x^{r_1} = x^{r_1}((a_1/x)^{r_1} - 1) < x^u((a_1/x)^u - 1) = a_1^u - x^u,$$

and adding $x^u - a_1^{r_1}$ to both sides of the inequality, shows that

$$x^u - x^{r_1} < a_1^u - a_1^{r_1} < \varepsilon/3$$

for all $1 < x < a_1$. In particular, this inequality holds for $x = a$ and it holds for $x = b$ if $|a - b| < (a - 1)$.

By Lemma 1.6.3 (iii), there exists a $\delta_1 > 0$ such that if $|a - b| < \delta_1$, then $|a^{r_1} - b^{r_1}| < \varepsilon/3$. Hence if $|a - b| < \delta = \min(\delta_1, (a - 1))$, then

$$\begin{aligned} |a^u - b^u| &\leq |a^u - a^{r_1}| + |a^{r_1} - b^{r_1}| + |b^{r_1} - b^u| \\ &< \varepsilon/3 + \varepsilon/3 + \varepsilon/3 = \varepsilon. \end{aligned}$$

(iv) The proof of Lemma 1.6.3 (iv) can be repeated word for word here, with $\delta = 1/n$ and r, s being replaced by u, v. □

Comment. Getting a little ahead of ourselves, property (i) implies that if $a > 1$, then the *exponential* function $y = a^x$ is *increasing* in the variable x; property (ii) implies that if $u > 0$, then the *power* function $y = x^u$ is likewise increasing in the variable x; property (iii) implies that the power function $y = x^u$ is *continuous* in the variable x; and property (iv) implies that the exponential function $y = a^x$ is continuous in the variable x.

Exercises

ex1.7.1. Prove Corollary 1.7.2.

ex1.7.2. For $0 < a < 1$ and $u \in \mathbb{R}$, define $a^u = (1/a)^{-u}$.

(a) Show that $a^{u+v} = a^u a^v$ for $0 < a < 1$ and $u, v \in \mathbb{R}$.

(b) Show that $(ab)^u = a^u b^u$ for all $u \in \mathbb{R}$ and all $a, b > 0$.

(c) Show that if $u, v > 0$, then $(a^u)^v = a^{uv}$.

ex1.7.3. Show that for any $u \in \mathbb{R}$ and any $\varepsilon > 0$, there is a $\delta > 0$ such that if $|a-1| < \delta$, then $|a^u - 1| < \varepsilon$.

Hint: Show that it's true for $u \in \mathbb{Q}$, use this to show that there is a $\delta_1 > 0$ such that $1 < a < 1 + \delta_1$ implies $1 < a^u < 1 + \varepsilon$. What next?

★1.8. Euclidean space

The focus of this book will be analysis on $\mathbb{R}$, but much of what we do here generalizes in a straightforward way to *Euclidean space* and we point out these generalizations from time to time in the coming chapters.

Definition. For $n \in \mathbb{N}$, n-dimensional *Euclidean space* is the set

$$\mathbb{R}^n = \{(x_1, x_2, \dots, x_n) : x_1, x_2, \dots, x_n \in \mathbb{R}\}$$

of *ordered n-tuples* of real numbers. Elements in $\mathbb{R}^n$ are called *points* (or *vectors*) and denoted with boldfaced letters, e.g., $(x_1, x_2, \dots, x_n) = \mathbf{x}$.

1.8.1. Addition and scalar multiplication in $\mathbb{R}^n$

We define addition in $\mathbb{R}^n$ "coordinate-wise": if $\mathbf{x} = (x_1, x_2, \dots, x_n)$ and $\mathbf{y} = (y_1, y_2 \dots, y_n)$, then $\mathbf{x} + \mathbf{y} = (x_1 + y_1, x_2 + y_2 \dots, x_n + y_n)$.

Setting $\mathbf{0} = (0, 0, \dots, 0)$ and $-\mathbf{x} = (-x_1, -x_2, \dots, -x_n)$, it is a simple exercise to verify that $\mathbb{R}^n$ with addition defined this way, satisfies axioms **add** 1 - 4 (see section 1.4.1), because these axioms are satisfied in $\mathbb{R}$. We will adopt the usual notation with respect to adding the negative, i.e., $\mathbf{x} - \mathbf{y} = \mathbf{x} + (-\mathbf{y})$.

Points in $\mathbb{R}^n$ may be multiplied by real numbers. If $a \in \mathbb{R}$ and $\mathbf{x} \in \mathbb{R}^n$, then

$$a \cdot \mathbf{x} = (ax_1, ax_2, \dots, ax_n).$$

In this context, we often refer to the real number a as a *scalar* and this operation is referred to as *scalar multiplication*. Scalar multiplication has the following properties.

Proposition. *For $a, b \in \mathbb{R}$ and $\mathbf{x}, \mathbf{y} \in \mathbb{R}^n$,*

(i) $a(\mathbf{x} + \mathbf{y}) = a\mathbf{x} + a\mathbf{y}$.

(ii) $(a + b)\mathbf{x} = a\mathbf{x} + b\mathbf{x}$.

(iii) $0\mathbf{x} = \mathbf{0}$.

(iv) $1\mathbf{x} = \mathbf{x}$.

Proof. Exercise. □

Comment. With addition and scalar multiplication defined as above, the Euclidean space $\mathbb{R}^n$ is a *vector space over* $\mathbb{R}$.

1.8.2. The absolute value on $\mathbb{R}^n$

The absolute value in $\mathbb{R}$ generalizes to $\mathbb{R}^n$.

Definition. The *absolute value* of $\mathbf{x} = (x_1, \dots, x_n) \in \mathbb{R}^n$, denoted $||\mathbf{x}||$, is defined by

$$||\mathbf{x}|| = \left(\sum_{i=1}^{n} x_i^2\right)^{1/2}.$$

The absolute value of $\mathbf{x}$ is also called the *norm* of $\mathbf{x}$ and can be thought of as the *length* of the vector $\mathbf{x}$.

We observe that $\left(\sum_{i=1}^{n} x_i^2\right)^{1/2}$ is defined for all $\mathbf{x} \in \mathbb{R}^n$, because sums of squares are always nonnegative.

The absolute value on $\mathbb{R}^n$ has properties analogous to the properties on the absolute value on $\mathbb{R}$, including the triangle inequality. To prove this, we will use the *Cauchy-Schwarz inequality*:

Lemma. *For any real numbers $x_1, x_2, \dots, x_n$ and $y_1, y_2, \dots, y_n$*

$$\left(\sum_{i=1}^{n} x_i y_i\right)^2 \leq \left(\sum_{i=1}^{n} x_i^2\right)\left(\sum_{i=1}^{n} y_i^2\right). \tag{1.8.1}$$

Proof. Exercise. □

Theorem. *For all $\mathbf{x}, \mathbf{y} \in \mathbb{R}^n$ and $a \in \mathbb{R}$:*

(i) $||\mathbf{x}|| \geq 0$ *and* $||\mathbf{x}|| = 0$ *if and only if* $\mathbf{x} = \mathbf{0}$.

(ii) $||a \cdot \mathbf{x}|| = |a| \cdot ||\mathbf{x}||$.

(iii) $||\mathbf{x} + \mathbf{y}|| \leq ||\mathbf{x}|| + ||\mathbf{y}||$.

Proof. We leave the proofs of (i) and (ii) as exercises.

The proof of (iii) follows the pattern of the proof of Theorem 1.5.5. From the definition, we have

$$||\mathbf{x} + \mathbf{y}||^2 = \sum_{i=1}^{n} (x_i + y_i)^2 = \sum_{i=1}^{n} x_i^2 + 2\sum_{i=1}^{n} x_i y_i + \sum_{i=1}^{n} y_i^2. \tag{1.8.2}$$

From the Cauchy-Schwarz inequality (1.8.1) (and **ex1.7.4**) it follows that

$$\sum_{i=1}^{n} x_i y_i \leq \left|\sum_{i=1}^{n} x_i y_i\right| = \sqrt{\left(\sum_{i=1}^{n} x_i y_i\right)^2} \leq \sqrt{\left(\sum_{i=1}^{n} x_i^2\right) \cdot \left(\sum_{i=1}^{n} y_i^2\right)},$$

and inserting this bound in (1.8.2) shows that

$$\begin{aligned} ||\mathbf{x} + \mathbf{y}||^2 &\leq \sum_{i=1}^{n} x_i^2 + 2\left(\sum_{i=1}^{n} x_i^2\right)^{1/2} \cdot \left(\sum_{i=1}^{n} y_i^2\right)^{1/2} + \sum_{i=1}^{n} y_i^2 \\ &= \left[\left(\sum_{i=1}^{n} x_i^2\right)^{1/2} + \left(\sum_{i=1}^{n} y_i^2\right)^{1/2}\right]^2 = \left(||\mathbf{x}|| + ||\mathbf{y}||\right)^2. \end{aligned}$$

Taking square roots of both sides of this last inequality completes the proof. □

Exercises

ex1.8.1. Prove Proposition 1.8.1.

ex1.8.2. Prove Theorem 1.8.2 (i) and (ii).

ex1.8.3. Prove the *Cauchy-Schwarz inequality*: For any real numbers $x_1, x_2, \ldots, x_n$ and $y_1, y_2, \ldots, y_n$

$$\left(\sum_{i=1}^{n} x_i y_i\right)^2 \le \left(\sum_{i=1}^{n} x_i^2\right)\left(\sum_{i=1}^{n} y_i^2\right).$$

Hint: Consider the quadratic function $Q(u) = \sum_{i=1}^{n}(x_i u + y_i)^2$. Simplify $Q(u)$ to the form $Au^2 + Bu + C$ and use the result of **ex1.4.5**.

Chapter 2

Sequences and Series

In Chapter 1, we defined the real numbers as the (unique) complete ordered field containing the rational numbers as a subfield. The completeness of $\mathbb{R}$ makes it much more fertile ground than $\mathbb{Q}$ for the purposes of analysis. This somewhat abstract conception of the real numbers is not meant to undermine readers' intuition of what the real numbers are, but to strengthen this intuition with a solid foundation.

In this chapter, we continue this process by studying sequences and series of real numbers and their limits. Limits are defined in terms of a distance function, and this is where we start.

2.1. Metric spaces

Distance functions are more formally known as metrics, and sets on which a metric is defined are called metric spaces.

Definition. A *metric* on a set X is a function

$$\rho : X \times X \mapsto \{r \in \mathbb{R} : r \geq 0\} = \mathbb{R}^+$$

satisfying the following conditions:

met 1. For all $x, y \in X$, $\rho(x, y) = \rho(y, x) \geq 0$, and $\rho(x, y) = 0$ if and only if $x = y$.

met 2. *Triangle inequality*: For all $x, y, z \in X$,

$$\rho(x, z) \leq \rho(x, y) + \rho(y, z).$$

A set X with a metric ρ is called a *metric space*, and we denote such a space by(X, ρ).

If (X, ρ) is a metric space and $Y \subset X$, then the restriction of ρ from $X \times X$ to $Y \times Y$ is a metric on Y, because ρ satisfies **met** 1 and 2 on all of X, and therefore also on $Y \subset X$. The restriction of ρ to $Y \times Y$ is called the *induced* or *inherited* metric on Y.

2.1.1. Bounded subsets in a metric space

While a general metric space X may not be an ordered set, we can still define *bounded subsets* of X, using the metric on X and the order on $\mathbb{R}$.

Definition. Suppose that (X, ρ) is a metric space. An *open ball of radius r* in X is a set of the form $\{x \in X : \rho(x, x_0) < r\}$, where x_0 is some point in X and r is a positive real number. We denote such open balls as $B(x_0, r)$ or $B_r(x_0)$.

A subset Y of X is *bounded in X* if $Y \subset B(x_0, r)$ for some $x_0 \in X$ and $r > 0$.

2.1.2. The standard metric on $\mathbb{R}$

Theorem. *The function* $\rho(x, y) = |x - y|$ *defines a metric on* $\mathbb{R}$.

This is the standard metric on $\mathbb{R}$.

Proof. Exercise (see §1.5.5). □

Comment. If $x_0 \in \mathbb{R}$ and $r > 0$, then

$$B(x_0, r) = \{x \in \mathbb{R} : |x - x_0| < r\} = (x_0 - r, x_0 + r),$$

so if $E \subset \mathbb{R}$ is bounded in the metric space $\mathbb{R}$, then E is bounded in the ordered set $\mathbb{R}$, as described in Sections 1.5.1 and 1.6.

On the other hand, if E is bounded in (the ordered set) $\mathbb{R}$, then there is an interval (a, b), such that

$$E \subset (a, b) = (x_0 - r, x_0 + r) = B(x_0, r),$$

where $x_0 = (a + b)/2$ and $r = (b - a)/2$. I.e., E is bounded in the metric space $\mathbb{R}$. Thus the two notions of "bounded set" are equivalent in $\mathbb{R}$.

*2.1.3. The standard Euclidean metric

The metric on $\mathbb{R}^n$ is defined in the same way it is defined on $\mathbb{R}$, using the absolute value.

Theorem. *The function* $\rho(\mathbf{u}, \mathbf{v}) = |\mathbf{u} - \mathbf{v}|$ *is a metric on* $\mathbb{R}^n$.

This is the *standard Euclidean metric.*

Proof. Exercise (see §1.8.2). □

Observation. The metrics on $\mathbb{R}$ and $\mathbb{R}^n$ are closely related to each other. If $\mathbf{x}, \mathbf{y} \in \mathbb{R}^n$, with $\mathbf{x} = (x_1, x_2, \ldots, x_n)$ and $\mathbf{y} = (y_1, y_2, \ldots, y_n)$, then

$$|\mathbf{x} - \mathbf{y}|^2 = \sum_{j=1}^{n} (x_j - y_j)^2.$$

From this we derive two useful inequalities.

First, it follows that for each $1 \leq k \leq n$, $|\mathbf{x} - \mathbf{y}|^2 \geq (x_k - y_k)^2$, so

$$|x_k - y_k| \leq |\mathbf{x} - \mathbf{y}|. \tag{2.1.1}$$

On the other hand, it also follows that $|\mathbf{x}-\mathbf{y}|^2 \leq n \cdot \max_{1\leq j\leq n}(x_j - y_j)^2$, so

$$|\mathbf{x}-\mathbf{y}| \leq \sqrt{n} \cdot \max_{1\leq j\leq n} |x_j - y_j| . \tag{2.1.2}$$

Exercises

ex2.1.1. Let X be any nonempty set. Define ρ on X by setting $\rho(x, x) = 0$ and $\rho(x, y) = 1$ if $x \neq y$, for all $x, y \in X$. Show that ρ is a metric on X.

This is called the *discrete metric.*

ex2.1.2. Show that $||x| - |y|| \leq |x - y|$ for all $x, y \in \mathbb{R}$. Is the same true in $\mathbb{R}^n$? If so, prove it. If not, give an example showing that it fails.

ex2.1.3. For $\mathbf{x} = (x_1, \ldots, x_n)$ and $\mathbf{y} = (y_1, \ldots y_n)$ in $\mathbb{R}^n$, define

$$\rho_2(\mathbf{x}, \mathbf{y}) = \max_{1\leq j\leq n} |x_j - y_j| .$$

Show that ρ_2 is also a metric on $\mathbb{R}^n$. What shape does the open "ball" $B(\mathbf{x}_0, \varepsilon)$ have in $\mathbb{R}^3$ with respect to the metric ρ_2, above?

2.2. Sequences

A *sequence* in a set X is a function $x : \mathbb{N} \to X$ whose domain is $\mathbb{N}$ or a subset of $\mathbb{N}$. In this context we use x_n to indicate the image of n instead of $x(n)$, and we typically write sequences in one of the following ways: $\{x_n\}_{n=1}^{\infty}$, $\{x_n\}_{n\in\Lambda}$, where $\Lambda \subset \mathbb{N}$ is the domain of the sequence, or simply $\{x_n\}$, when the domain is known. In this context, we assume that Λ is an infinite subset of $\mathbb{N}$.

The domain, $\Lambda \subset \mathbb{N}$, of a sequence is also called the *index set* of the sequence. If $\Lambda = \{n_1, n_2, n_3, \ldots, n_k, \ldots\}$ (where it is understood that $n_1 < n_2 < n_3 < \ldots$), we also use the notation $\{x_{n_k}\}_{k=1}^{\infty}$ for the sequence $\{x_n\}_{n\in\Lambda}$. If an index set is not specified, then we assume that it is $\mathbb{N}$.

If $\{x_n\}_{n\in\Lambda}$ is a sequence and Λ_1 is an infinite subset of Λ, then the sequence $\{x_n\}_{n\in\Lambda_1}$ is called a *subsequence* of $\{x_n\}_{n\in\Lambda}$.

2.2.1. Convergent sequences

Definition. Let $\{x_n\}$ be a sequence in a metric space (X, ρ). The sequence $\{x_n\}$ *converges to* the point $z \in X$ if and only if for every $\varepsilon > 0$, there is an $N_\varepsilon \in \mathbb{N}$ such that $\rho(x_n, z) < \varepsilon$ for all $n > N_\varepsilon$. Equivalently, the sequence $\{x_n\} \subset X$ converges to $z \in X$ if and only if for every $\varepsilon > 0$ the set $\{n : \rho(x_n, z) > \varepsilon\} \subset \mathbb{N}$ is finite.

If $\{x_n\}$ converges to z, we write $\lim_{n\to\infty} x_n = z$, $\lim x_n = z$ or $x_n \to z$ and we say that z is the *limit* of the sequence $\{x_n\}$.

A sequence that is converging to a limit is called a *convergent sequence.*

Proposition. *If (X, ρ) is a metric space and $\{x_n\} \subset X$ is a convergent sequence then $\{x_n\}$ is bounded in X.*

Proof. Suppose that $\lim x_n = z \in X$. Then there exists an $N \in \mathbb{N}$ such that $\rho(x_n, z) < 1$ for all $n > N$. Let $M = \max\{\rho(x_1, z), \rho(x_2, z), \dots, \rho(x_N, z), 1\}$ then $\rho(x_n, z) < M + 1$ for all $n \in \mathbb{N}$ and therefore $\{x_n\} \subset B(z, M + 1)$. □

2.2.2. Cauchy sequences

The elements of a convergent sequence are getting close to one specific point. This implies that they are also getting close to each other. This intuition gives rise to the concept of a Cauchy sequence.[1]

Definition. A sequence $\{x_n\}_{n=1}^{\infty}$ in a metric space (X, ρ) is called a *Cauchy sequence* if for every $\varepsilon > 0$ there exists an integer $N = N(\varepsilon)$ such that if $n > N$ and $m > N$ then $\rho(x_n, x_m) < \varepsilon$.

A metric space (X, ρ) is *complete* if every Cauchy sequence in X converges to a limit in X.

Theorem. *Every convergent sequence is a Cauchy sequence.*

Proof. Suppose that $x_n \to z$. Given $\varepsilon > 0$, there exists N_ε such that $\rho(x_n, z) < \varepsilon/2$ for all $n > N_\varepsilon$. If both n and m are bigger than N_ε then $\rho(x_n, x_m) \le \rho(x_n, z) + \rho(x_m, z) < \varepsilon$. □

The fact that convergent sequences are always Cauchy sequences is useful because it gives a criterion for convergence that is intrinsic: it depends only on the elements of the sequence (and not the limit point of the sequence). This is only useful in complete metric spaces. In the coming sections we show that $\mathbb{R}$ and $\mathbb{R}^n$, more generally, are complete.

Exercises

ex2.2.1. Suppose that $\{x_n\}_{n=1}^{\infty}$ converges to z in the metric space X. Show that the subsequence $\{x_n\}_{n \in \Lambda}$ converges to z, for every infinite subset Λ of $\mathbb{N}$.

ex2.2.2. Show that if $\{x_n\}$ is a Cauchy sequence in a metric space (X, ρ), then $\{x_n\}$ is bounded in X.

ex2.2.3. Suppose that $\{x_n\}_{n=1}^{\infty}$ converges to z, and that $\rho(x_n, x_m) < \varepsilon$ for $n, m > N$. Show that $\rho(x_n, z) \le \varepsilon$ for $n > N$.

2.3. Sequences of real numbers

The set of real numbers form a metric space with the standard metric, so the general conclusions about convergence in metric spaces are valid in $\mathbb{R}$. On the other hand, $\mathbb{R}$ has the additional structure of a complete ordered field. This structure gives additional information about convergent sequences in $\mathbb{R}$.

[1] Named for Augustin-Louis Cauchy.

2.3.1. Convergent sequences in $\mathbb{R}$

We saw that boundedness in $\mathbb{R}$ in the metric sense is equivalent to boundedness in the order-sense (in Section 2.1.2). As a result, we can rephrase Proposition 2.2.1 for real numbers as follows.

Proposition. *If the sequence $\{x_n\}$ is convergent in $\mathbb{R}$, then there is an $M > 0$ such that $|x_n| < M$ for all n.*

Proof. By Proposition 2.2.1, the sequence $\{x_n\}$ is bounded in $\mathbb{R}$. Hence there exist $a, b \in \mathbb{R}$, such that $a < x_n < b$ for all n. It follows that $|x_n| < \max(|a|, |b|)$ for all n. □

2.3.2. Combining convergent sequences

Convergent sequences of real numbers play nicely with the basic arithmetic operations in $\mathbb{R}$. in particular, sums and products of convergent sequences are convergent.

Theorem. *Suppose that $\{x_n\}$ and $\{y_n\}$ are convergent sequences of real numbers, with $\lim x_n = a$ and $\lim y_n = b$. Then the sequences $\{x_n + y_n\}$ and $\{x_n y_n\}$ are also convergent with $\lim(x_n + y_n) = a + b$ and $\lim(x_n y_n) = ab$.*

Proof. Let $\varepsilon > 0$ be given. From the assumption that $x_n \to a$ and $y_n \to b$, it follows that there are $N_1, N_2 \in \mathbb{N}$ such that $|x_n - a| < \varepsilon/2$ for all $n > N_1$ and $|y_n - b| < \varepsilon/2$ for all $n > N_2$.

If $n > N_\varepsilon = \max(N_1, N_2)$, we then have

$$\begin{aligned}|(x_n + y_n) - (a + b)| &= |(x_n - a) + (y_n - b)| \\ &\leq |x_n - a| + |y_n - b| < \varepsilon/2 + \varepsilon/2 = \varepsilon,\end{aligned}$$

by the triangle inequality. Since $\varepsilon > 0$ was arbitrary,[2] it follows that $\lim(x_n+y_n) = a+b$.

To prove that $x_n y_n \to ab$ we begin with the identity

$$x_n y_n - ab = x_n y_n - a y_n + a y_n - ab = (x_n - a) y_n + a(y_n - b),$$

from which it follows that

$$|x_n y_n - ab| \leq |x_n - a||y_n| + |a||y_n - b| \tag{2.3.1}$$

by the triangle inequality.

By Proposition 2.4, there is a $Y > 0$ such that $|y_n| < Y$ for all n. For any $\varepsilon > 0$, there are $N_1, N_2 \in \mathbb{N}$ such that $|x_n - a| < \varepsilon/2Y$ for all $n > N_1$ and $|y_n - b| < \varepsilon/2(|a| + 1)$ for all $n > N_2$. If $n > \max(N_1, N_2)$, then it follows from the inequality (2.3.1) that

$$|x_n y_n - ab| < |x_n - a|Y + |a||y_n - b| < \frac{\varepsilon}{2Y} \cdot Y + \frac{\varepsilon}{2(|a| + 1)} \cdot |a| < \varepsilon,$$

and therefore $\lim x_n y_n = ab$. □

Comment. We divided ε by $2(|a| + 1)$ instead of dividing by $2|a|$ in the proof above to account for the possibility that $|a| = a = 0$.

[2]The same argument works for any $\varepsilon > 0$ and therefore for all $\varepsilon > 0$.

Corollary. *If* $\lim x_n = \xi$ *and* $\lim y_n = \eta$, *then for any* $a, b \in \mathbb{R}$,

$$\lim_{n\to\infty} (ax_n + by_n) = a\xi + b\eta.$$

Proof. Let $u_n = a$ and $v_n = b$ for all $n \in \mathbb{N}$. Then $\lim u_n = a$ and $\lim v_n = b$, so

$$\lim_{n\to\infty} ax_n = \lim_{n\to\infty} u_n x_n = a\xi \quad \text{and} \quad \lim_{n\to\infty} by_n = \lim_{n\to\infty} v_n y_n = b\eta,$$

by the second part of Theorem 2.3.2. It follows then from the first part of the theorem that

$$\lim_{n\to\infty} ax_n + by_n = \lim_{n\to\infty} ax_n + \lim_{n\to\infty} by_n = a\xi + b\eta. \qquad \square$$

Limits also play nicely with reciprocals.

Proposition. *If* $\lim_{n\to\infty} x_n = a$, *with* $a \neq 0$ *and* $x_n \neq 0$ *for all n, then* $\lim_{n\to\infty} 1/x_n = 1/a$.

Proof. Since $\lim_{n\to\infty} x_n = a$ and $a \neq 0$, there is an $N_1 \in \mathbb{N}$, such that $|x_n - a| < |a|/2$, or what is the same,

$$a - \frac{|a|}{2} < x_n < a + \frac{|a|}{2}$$

for all $n > N_1$. It follows that $|x_n| > |a|/2$, so $|ax_n| > a^2/2$, for all $n > N_1$.

Now for any $\varepsilon > 0$, there is an $N_\varepsilon \in \mathbb{N}$ such that $|x_n - a| < a^2\varepsilon/2$ for all $n > N_\varepsilon$. Therefore for all $n > \max(N_1, N_\varepsilon)$, we have

$$\left|\frac{1}{x_n} - \frac{1}{a}\right| = \left|\frac{a - x_n}{ax_n}\right| = \frac{|x_n - a|}{|ax_n|} < \frac{a^2\varepsilon/2}{a^2/2} = \varepsilon,$$

so $\lim 1/x_n = 1/a$, as claimed. $\square$

Comment. It may be tempting to argue that since $x_n \cdot 1/x_n = 1$ for all n, then

$$1 = \lim_{n\to\infty} 1 = \lim_{n\to\infty} x_n \cdot 1/x_n = \lim_{n\to\infty} x_n \cdot \lim_{n\to\infty} 1/x_n = a \lim_{n\to\infty} 1/x_n,$$

so that $\lim_{n\to\infty} 1/x_n = 1/a$, by uniqueness of the inverse. The problem with this argument is that it assumes the existence of $\lim_{n\to\infty} 1/x_n$, which is part of what needs to be proved.

2.3.3. The squeeze law

We can often show that a sequence converges by comparing it to other sequences. The theorem below is referred to as "the squeeze law".[3]

Theorem. *Suppose that* $\{x_n\}_{n=1}^\infty$, $\{y_n\}_{n=1}^\infty$ *and* $\{z_n\}_{n=1}^\infty$ *are sequences of real numbers. If* $\lim_{n\to\infty} x_n = \lim_{n\to\infty} z_n = a$ *and there is an* $N_0 \in \mathbb{N}$ *such that* $x_n \leq y_n \leq z_n$ *for all* $n > N_0$, *then* $\lim_{n\to\infty} y_n = a$.

Proof. We first observe that for all $n > N_0$, then $x_n \leq y_n \leq z_n$. In this case, if $a \leq y_n$, then

$$|y_n - a| = y_n - a \leq z_n - a = |z_n - a|$$

[3]Because the sequence $\{y_n\}$ is squeezed between the other two.

and if $y_n \leq a$, then

$$|y_n - a| = a - y_n \leq a - x_n = |x_n - a|.$$

In other words, $|y_n - a| \leq \max(|x_n - a|, |z_n - a|)$ for all $n > N_0$.

Next, given any $\varepsilon > 0$, there are $N_1(\varepsilon) \in \mathbb{N}$ and $N_2(\varepsilon) \in \mathbb{N}$ such that $|x_n - a| < \varepsilon$ for all $n > N_1(\varepsilon)$ and $|z_n - a| < \varepsilon$ for all $n > N_2(\varepsilon)$.

Hence, for any $\varepsilon > 0$, if $n > N = \max(N_0, N_1(\varepsilon), N_2(\varepsilon))$, then we have $|x_n - a| < \varepsilon$, $|z_n - a| < \varepsilon$, and $x_n \leq y_n \leq z_n$, and therefore

$$|y_n - a| \leq \max(|x_n - a|, |z_n - a|) < \varepsilon,$$

It follows that $\lim_{n\to\infty} y_n = a$, as claimed. □

Corollary. *If* $z_n \to 0$ *and for some* $N \in \mathbb{N}$, $z_n \geq |y_n| \geq 0$ *for all* $n > N$, *then* $\lim y_n = 0$.

Proof. Exercise. □

2.3.4. Sequences in $\mathbb{R}$ that diverge to $\pm\infty$

If a sequence $\{x_n\} \subset \mathbb{R}$ fails to converge, it is said to *diverge*, and sequences can diverge in many ways. We will come back to this topic in Section 2.6, but here we will briefly discuss two special types of divergence.

Definition. If the sequence $\{x_n\}$ has the property that for any $B > 0$, there is an $N \in \mathbb{N}$ such that $x_n > B$ for all $n > N$, then we say that the sequence *diverges to infinity* and write $\lim x_n = \infty$ or $x_n \to \infty$.

Likewise, if for every $b < 0$, there is an $N \in \mathbb{N}$ such that $x_n < b$ for all $n > N$, then we say that the sequence *diverges to negative infinity* and write $\lim x_n = -\infty$ or $x_n \to -\infty$.

If $\lim |x_n| = \infty$, but neither $\lim x_n = \infty$ nor $\lim x_n = -\infty$ (e.g., if $x_n = (-1)^n \cdot n$), we say that $\lim x_n = \pm\infty$.

It is clear for example that $\lim n = \infty$, and this statement can be generalized to all positive powers.

Theorem. *If* $a > 0$, *then* $\lim\limits_{n\to\infty} n^a = \infty$.

Proof. We prove the statement first for $a = 1/k$, with $k \in \mathbb{N}$ and then for all $a > 0$.

So suppose that $k \in \mathbb{N}$. Given $B \in \mathbb{R}$, there is an $N \in \mathbb{N}$ such that $N > |B|^k$, by the Archimedean principle. Therefore $n^{1/k} > N^{1/k} > B$ for all $n > N$, and it follows that $\lim\limits_{n\to\infty} n^{1/k} = \infty$.

Now, if $a > 0$, then there is a $k \in \mathbb{N}$ such that $0 < 1/k < a$ (also by the Archimedean property), from which it follows that $n^a > n^{1/k}$ for all $n \in \mathbb{N}$. From this and the preceding paragraph, given $B \in \mathbb{R}$, we may conclude that there is an $N \in \mathbb{N}$ such that $n^a > n^{1/k} > B$ for all $n > N$, and therefore $\lim\limits_{n\to\infty} n^a = \infty$. □

Sequences that diverge to $\pm\infty$ are closely related to sequences that converge to 0.

Proposition. *Suppose that* $x_n \neq 0$ *for all* n *then* $\lim_{n\to\infty} |x_n| = \infty$ *if and only if* $\lim_{n\to\infty} 1/x_n = 0$.

Proof. Exercise. □

2.3.5. Examples

It is time for some concrete examples. We will find the limits of several sequences that crop up from time to time in what follows. The squeeze law plays an important role in the derivations, along with the Archimedean property and the binomial theorem.

Theorem.

(i) *If* $|a| < 1$, *then* $\lim_{n\to\infty} a^n = 0$.

(ii) *If* $a > 0$, *then* $\lim_{n\to\infty} \frac{1}{n^a} = 0$.

(iii) *If* $a > 0$, *then* $\lim_{n\to\infty} a^{1/n} = 1$.

(iv) *If* $a > 1$, *then for any* $b \in \mathbb{R}$, $\lim_{n\to\infty} n^b/a^n = 0$.

(v) *For any* $b \in \mathbb{R}$, $\lim_{n\to\infty} n^{b/n} = 1$.

Proof.

(i) We first observe that $|a^n - 0| = |a^n| = |a|^n$, and therefore if the claim is true for $0 < a < 1$, then it is also true for $-1 < a < 0$. If $0 < a < 1$, then $1/a > 1$ so $1/a = 1 + \delta$ and $a^n = 1/(1+\delta)^n$, for some $\delta > 0$.

From the binomial theorem it follows that for all $n \in \mathbb{N}$, we have

$$(1+\delta)^n = \sum_{k=0}^{n} \binom{n}{k} \delta^k \geq 1 + n\delta > n\delta, \tag{2.3.2}$$

and consequently, $0 < a^n = 1/(1+\delta)^n < 1/n\delta$ for all $n \in \mathbb{N}$.[4]

Finally, given any $\varepsilon > 0$, there is an $n_\varepsilon \in \mathbb{N}$ such that $n_\varepsilon > 1/(\delta \cdot \varepsilon)$ (by the Archimedean property). It follows that if $n > n_\varepsilon$, then $n\delta > n_\varepsilon \delta > 1/\varepsilon$, and therefore $0 < 1/n\delta < \varepsilon$ for all $n > n_\varepsilon$. This means that $\lim_{n\to\infty} 1/n\delta = 0$, and hence $\lim_{n\to\infty} a^n = 0$ by the (corollary to the) squeeze law.

(ii) This follows from Theorem 2.4.3 and Lemma 2.4.3.

(iii) First, we have $1^{1/n} = 1$ for all n, so the claim is true for $a = 1$. Next, suppose that the claim is true for $a > 1$. If $0 < a < 1$, then $1/a > 1$ so $\lim(1/a)^{1/n} = 1$ and

$$\lim_{n\to\infty} a^{1/n} = \lim_{n\to\infty} \frac{1}{(1/a)^{1/n}} = \frac{1}{\lim_{n\to\infty} (1/a)^{1/n}} = 1,$$

by Proposition 2.4.1. Thus it remains to prove the claim for $a > 1$.

If $a > 1$, then $a^{1/n} > 1$ for all n, so $a^{1/n} = 1 + \delta_n$, with $\delta_n > 0$. It follows from the inequality (2.3.2) that $a = (1+\delta_n)^n > n\delta_n$, which implies that $\delta_n < a/n$ for all n.

[4]The inequality $(1+\delta)^n > 1 + n\delta$, for $\delta > 0$, is known as Bernoulli's inequality.

Thus we have $1 < a^{1/n} < 1 + \frac{a}{n}$ for all n, and therefore $a^{1/n} \to 1$ by the squeeze law since

$$\lim_{n\to\infty}\left(1+\frac{a}{n}\right) = 1 + a\lim_{n\to\infty}\frac{1}{n} = 1,$$

by Corollary 2.3.2. and part (ii), above.

(iv) If $b < 0$, then $n^b = n^{-|b|} = 1/n^{|b|} \to 0$ by (ii), so $n^b/a^n \to 0$ by Corollary 2.3.3, because $0 < n^b/a^n < n^b$ (since $a > 1$).

If $b = 0$, then $n^b/a^n = 1/a^n = (1/a)^n \to 0$, by (i). Thus it remains to consider the case that $b > 0$.

Once again we write $a = 1 + \delta$ for some $\delta > 0$ and consider the binomial formula

$$a^n = (1+\delta)^n = \sum_{k=0}^{n}\binom{n}{k}\delta^k,$$

from which it follows that $a^n > \binom{n}{k}\delta^k$ for all $0 \le k \le n$, so that

$$\frac{n^b}{a^n} < \frac{n^b}{\binom{n}{k}\delta^k}.$$

The idea is to find a fixed $k > 0$ such that the expression the right tends to 0 as $n \to \infty$.

To this end, let $k_0 > b$ be a natural number. If $n > 2k_0$, then

$$n > (n-1) > \cdots > (n-k_0+1) > n/2,$$

so

$$\binom{n}{k_0} = \frac{n(n-1)\cdots(n-k_0+1)}{k_0!} > \frac{\overbrace{(n/2)\cdots(n/2)}^{k_0}}{k_0!} = \frac{n^{k_0}}{2^{k_0}k_0!}$$

and therefore

$$\binom{n}{k_0}\delta^{k_0} > \frac{n^{k_0}\delta^{k_0}}{2^{k_0}k_0!},$$

for all $n > 2k_0$.

It follows that $0 < n^b/a^n < cn^{b-k_0}$ for all $n > 2k_0$, where $c = (2/\delta)^{k_0}k_0!$ is a constant. Since $b - k_0 < 0$, we have $cn^{b-k_0} \to 0$ by (ii), and hence $\lim(n^b/a^n) = 0$ by the squeeze law.

(v) For all $n > 1$, we have $n^{1/n} > 1$ so similarly to the observation in the proof of (iii), we have $n^{1/n} = 1 + \delta_n$, for some $\delta_n > 0$.

Now, like the argument in the proof of (iv), we note that if $n > 2$, then $n - 1 > n/2$ and hence

$$n = (1+\delta_n)^n > \binom{n}{2}\delta_n^2 = \frac{n(n-1)}{2}\delta_n^2 > \frac{n^2}{4}\delta_n^2.$$

Dividing the inequality by $n^2/4$ shows that $\delta_n^2 < 4/n$ and thus $\delta_n < 2/n^{1/2}$, when $n > 2$. As a consequence, we have

$$1 < n^{1/n} = 1 + \delta_n < 1 + \frac{2}{n^{1/2}},$$

for all $n > 2$, and therefore $\lim_{n\to\infty} n^{1/n} = 1$ by part (ii) and the squeeze law. □

2.3.6. Monotone sequences in $\mathbb{R}$

Previously we saw that convergent sequences are always bounded (Proposition 2.4). For certain types of sequences, the reverse is also true.

Definition. A sequence $\{x_n\}$ of real numbers is *monotone nondecreasing* if $x_n \geq x_m$ when $n > m$. If $x_n > x_m$ for $n > m$, then the sequence is *monotone increasing*.

Likewise, the sequence is *monotone nonincreasing* if $x_n \leq x_m$ when $n > m$, and if $x_n < x_m$ for $n > m$, then the sequence is *monotone decreasing*.

Theorem. *If $\{x_n\}$ is a monotone nondecreasing sequence of real numbers that is bounded above, then $\{x_n\}$ is convergent, with*

$$\lim_{n\to\infty} x_n = \sup\{x_n : n \in \mathbb{N}\}.$$

Likewise, if $\{x_n\}$ is a monotone nonincreasing sequence of real numbers that is bounded below, then $\{x_n\}$ is convergent, with

$$\lim_{n\to\infty} x_n = \inf\{x_n : n \in \mathbb{N}\}.$$

Proof. We give the proof for a bounded nondecreasing sequence. The proof for nonincreasing sequences is analogous and we leave it as an exercise.

The order-completeness of $\mathbb{R}$ guarantees that $z = \sup\{x_n\}$ exists in $\mathbb{R}$. Now if $\varepsilon > 0$, then $z - \varepsilon$ is *not* an upper bound for $\{x_n\}$ and therefore there exists n_ε such that $z - \varepsilon < x_{n_\varepsilon} \leq z$. The assumption that the sequence is monotone nondecreasing implies that if $n > n_\varepsilon$, then

$$z - \varepsilon < x_{n_\varepsilon} \leq x_n \leq z,$$

and therefore $|x_n - z| < \varepsilon$ for all $n > n_\varepsilon$. □

Exercises

ex2.3.1. Suppose that $\lim x_n = a$. Prove that $\lim x_n^k = a^k$ for all $k \in \mathbb{N}$.

Hint: Use induction on k.

ex2.3.2. Prove Corollary 2.3.3.

Hint: Consider the sequence $x_n = 0$ for all n.

ex2.3.3. Suppose that $\lim_{n\to\infty} x_n = a$, with $a > 0$ and $x_n > 0$ for all $n \in \mathbb{N}$. Show that $x_n^u \to a^u$ for all $u \in \mathbb{R}$.

ex2.3.4. Prove the second half of Theorem 2.4.3: *If* $\{x_n\}$ *is monotone nonincreasing and bounded below, then* $\lim x_n = \inf\{x_n\}$.

ex2.3.5. Suppose that $j, k \in \mathbb{N}$ and $a_0, a_1, \ldots, a_j, b_0, b_1, \ldots, b_k$ are real numbers with $a_j \neq 0$ and $b_k \neq 0$. Further suppose that $b_k n^k + \cdots + b_1 n + b_0 \neq 0$ for all $n \in \mathbb{N}$. Prove:

(i) If $j < k$, then

$$\lim_{n\to\infty} \frac{a_j n^j + \cdots + a_1 n + a_0}{b_k n^k + \cdots + b_1 n + b_0} = 0.$$

(ii) If $j = k$, then

$$\lim_{n\to\infty} \frac{a_k n^k + \cdots + a_1 n + a_0}{b_k n^k + \cdots + b_1 n + b_0} = \frac{a_k}{b_k}.$$

(iii) If $j > k$, then

$$\lim_{n\to\infty} \frac{a_j n^j + \cdots + a_1 n + a_0}{b_k n^k + \cdots + b_1 n + b_0} = \begin{cases} \infty & : a_j/b_k > 0 \\ -\infty & : a_j/b_k < 0. \end{cases}$$

Hint: $a_j n^j + \cdots + a_1 n + a_0 = n^j \left(a_j + a_{j-1} n^{-1} + a_{j-2} n^{-2} + \cdots + a_0 n^{-j}\right)$.

2.4. The density of the rational numbers in the real numbers

In Chapter 1, we saw that $\mathbb{Q}$ is *dense* in $\mathbb{R}$ in the sense that there is a rational number between every pair of real numbers (Theorem 1.6.2). We can call this "order-density". In this section we will discuss another sense in which $\mathbb{Q}$ is dense in $\mathbb{R}$.

2.4.1. The sequential density of $\mathbb{Q}$ in $\mathbb{R}$

Theorem. *For every* $\alpha \in \mathbb{R}$, *there is a sequence* $\{a_n\} \subset \mathbb{Q}$ *such that* $\lim a_n = \alpha$

We say that $\mathbb{Q}$ is *sequentially dense* in $\mathbb{R}$.

Proof. Let $\alpha \in \mathbb{R}$. For every $n \in \mathbb{N}$, there is an $a_n \in \mathbb{Q}$, such that $\alpha < a_n < \alpha + 1/n$, by the order density of $\mathbb{Q}$ in $\mathbb{R}$. It follows that for every $n \in \mathbb{Q}$, $|a_n - \alpha| = a_n - \alpha < 1/n$.

Given $\varepsilon > 0$, there is an $N_\varepsilon \in \mathbb{N}$ such that $N_\varepsilon > 1/\varepsilon$, because $\mathbb{R}$ is an Archimedean field. For all $n > N_\varepsilon$,

$$|a_n - \alpha| < \frac{1}{n} < \frac{1}{N_\varepsilon} < \varepsilon,$$

from which it follows that $\lim a_n = \alpha$. □

The conclusion of this theorem is simple and the proof is straightforward. On the other hand, the result may be a little unsatisfying because it proves the existence of a sequence converging to α without specifying what the elements of the sequence are. In the next section, we take a more concrete approach to finding a sequence converging to $\sqrt{2}$.

2.4.2. A rational sequence converging to $\sqrt{2}$

To construct an explicit sequence of rational numbers that converge to $\sqrt{2}$, we need to rely on the defining properties of $\sqrt{2}$: $\sqrt{2} > 0$ and $(\sqrt{2})^2 = 2$. Hence, to find a sequence $\{a_n\} \subset \mathbb{Q}$ converging to $\sqrt{2}$, we will construct a sequence $\{a_n\} \subset \mathbb{Q}$ such that $a_n > 0$ for all n and $a_n^2 \to 2$. It seems intuitive that this will imply that $a_n \to \sqrt{2}$, but this intuition still requires justification.

Proposition. *Let $y > 0$. If $x_n \geq 0$ for all $n \in \mathbb{N}$ and $\lim_{n\to\infty} x_n^2 = y$, then $\lim_{n\to\infty} x_n = \sqrt{y}$.*

Proof. We first note that $x_n + \sqrt{y} \geq \sqrt{y} > 0$ for all $n \in \mathbb{N}$, because $x_n \geq 0$. From the identity $x_n^2 - y = (x_n - \sqrt{y})(x_n + \sqrt{y})$, it follows that

$$|x_n - \sqrt{y}| = \left|\frac{x_n^2 - y}{x_n + \sqrt{y}}\right| \leq \frac{|x_n^2 - y|}{\sqrt{y}}. \tag{2.4.1}$$

From the assumption that $x_n^2 \to y$, we know that for any $\varepsilon > 0$, there is an N such that $|x_n^2 - y| < \sqrt{y} \cdot \varepsilon$ for all $n > N$. It follows from this and the inequality (2.4.1) that $|x_n - \sqrt{y}| < \varepsilon$ for all $n > N$ and therefore $x_n \to \sqrt{y}$. □

We begin the construction of the sequence $\{a_n\}$ with the observation that $1 < 2 < 4$, so $1 < \sqrt{2} < 2$, and we set $a_1 = 2$ (setting $a_1 = 1$ is another natural choice). Since $a_1 > \sqrt{2}$, it follows that $2/a_1 < \sqrt{2}$, and it is reasonable to guess that

$$a_2 = \frac{1}{2}\left(a_1 + \frac{2}{a_1}\right) = \frac{3}{2},$$

is closer to $\sqrt{2}$ than $a_1 = 2$. Repeating our guess, we define

$$a_3 = \frac{1}{2}\left(a_2 + \frac{2}{a_2}\right) = \frac{17}{12}, \quad a_4 = \frac{1}{2}\left(a_3 + \frac{2}{a_3}\right) = \frac{577}{408},$$

and so on. In other words, we define the sequence $\{a_n\}$ *recursively* by setting $a_1 = 2$ and for $n \geq 1$, we define

$$a_{n+1} = \frac{1}{2}\left(a_n + \frac{2}{a_n}\right) = \frac{a_n^2 + 2}{2a_n}.$$

Clearly, $a_1, a_2, a_3, a_4 \in \mathbb{Q}$, and if $a_n \in \mathbb{Q}$, then $a_n^2 \in \mathbb{Q}$, so $a_{n+1} \in \mathbb{Q}$ as a quotient of rational numbers, so $a_n \in \mathbb{Q}$ for all $n \in \mathbb{N}$, as follows by induction. We claim that $\lim a_n^2 = 2$.

To prove this, we first note that $a_1^2 = 4 > 2$ and if $n > 1$, then

$$a_n^2 - 2 = \frac{(a_{n-1}^2 + 2)^2}{4a_{n-1}^2} - 2 = \frac{a_{n-1}^4 + 4a_{n-1}^2 + 4 - 8a_{n-1}^2}{4a_{n-1}^2} = \frac{(a_{n-1}^2 - 2)^2}{4a_{n-1}^2} > 0,$$

so $a_n^2 > 2$ for all $n \geq 1$. It further follows that

$$0 < a_n^2 - 2 = \frac{(a_{n-1}^2 - 2)^2}{4a_{n-1}^2} < \frac{(a_{n-1}^2 - 2)^2}{8}, \tag{2.4.2}$$

for all $n > 1$, because $4a_{n-1}^2 > 8$.

Claim. $a_n^2 - 2 < \left(\frac{1}{2}\right)^{5n-8}$ for all $n \geq 1$.

We leave the proof (by induction) of the claim as an exercise, but it follows from the claim that $0 < a_n^2 - 2 < 256/32^n$ for all $n \geq 1$. Therefore $(a_n^2 - 2) \to 0$, by Corollary 2.3.3, so $\lim a_n^2 = 2$.

Observation. Since $a_n^2 > 2$, it follows that $a_n > \sqrt{2}$ for all n, and therefore

$$a_n - \sqrt{2} = \frac{a_n^2 - 2}{a_n + \sqrt{2}} < \frac{(1/2)^{5n-8}}{2} = \left(\frac{1}{2}\right)^{5n-7}$$

for all n, because $a_n + \sqrt{2} > 2\sqrt{2} > 2$. In other words, we can estimate how closely a_n approximates $\sqrt{2}$ even though we don't have a precise value for $\sqrt{2}$.

Comment. The intuition (guess) that led to the construction of the sequence above works equally well to find approximate square roots of any natural number (that is not a perfect square). This approach to approximating square roots has been known since antiquity, and is known variously as "Heron's formula" or the "Babylonian method".

Exercises

ex2.4.1. Use induction and the inequality (2.4.2) to prove the claim that $a_n^2 - 2 < \left(\frac{1}{2}\right)^{5n-8}$ for all $n \geq 1$.

ex2.4.2. Show that the sequence $\{a_n\}$ constructed above is monotone decreasing.

ex2.4.3. Follow the steps of the example above to find a sequence $\{b_n\}$ that converges to $\sqrt{3}$.

ex2.4.4. Let $y > 0$ and $k \in \mathbb{N}$. Show that if $x_n \geq 0$ for all n and $\lim x_n^k \to y$, then $\lim x_n \to y^{1/k}$.

Hint: Follow the pattern of the proof of Proposition 2.4.2 and use the identity (1.1.1).

2.5. Subsequences

As we discussed briefly in Section 2.3.4, a general sequence of real numbers need not converge, and sequences can diverge in various ways. On the other hand, divergent sequences may have *convergent subsequences*.

For example, the sequence $y_n = (-1)^n$, i.e., $\{-1, 1, -1, 1, \ldots\}$ is divergent, but the subsequence $\{y_{2n}\}_{n=1}^{\infty} = \{y_2, y_4, \ldots\} = \{1, 1, \ldots\}$ is convergent with $\lim y_{2n} = 1$. Likewise the subsequence $\{y_1, y_3, y_5, \ldots\} = \{y_{2k+1}\}_{k\in\mathbb{N}}$ converges to -1.

2.5.1. Convergent subsequences

Recall that if $\{x_n\}_{n=1}^{\infty} \subset \mathbb{R}$ is a sequence and $\Lambda = \{n_j\}_{j=1}^{\infty}$ is a strictly increasing sequence of natural numbers, then $\{x_{n_j}\}_{j=1}^{\infty} = \{x_n\}_{n\in\Lambda}$ is a *subsequence* of $\{x_n\}_{n=1}^{\infty}$.

Definition. If $\lim_{j\to\infty} x_{n_j} = a$, then $\{x_{n_j}\}_{j=1}^{\infty}$ is a *convergent subsequence* of $\{x_n\}_{n=1}^{\infty}$ and a is called a *limit point*, or a *subsequential limit* of the sequence $\{x_n\}_{n=1}^{\infty}$. In this case, we us the notation $\lim_{j\to\infty} x_{n_j} = a$ or $\lim_{n\in\Lambda} x_n = a$ interchangeably.

The following theorem is useful in that it allows us to identify a limit point of a sequence without identifying a particular subsequence that converges to it.

Theorem. *The real number ξ is a limit point of $\{x_n\}$ if and only if for every $\varepsilon > 0$ the set $\{n \in \mathbb{N} : |x_n - \xi| < \varepsilon\}$ is infinite.*

In other words, ξ is a limit point of $\{x_n\}$ if and only if every interval that ξ also contains *infinitely many* of the terms of the sequence $\{x_n\}$.[5]

Proof. Let ξ be a limit point of $\{x_n\}$ and let $\{x_{n_k}\}_{k=1}^{\infty}$ be a subsequence that is converging to ξ. Then for any $\varepsilon > 0$, there is $K \in \mathbb{N}$ such that $|x_{n_k} - \xi| < \varepsilon$ for all $k > K$, so the set $\{n \in \mathbb{N} : |x_n - \xi| < \varepsilon\}$ is infinite, since it contains the infinite subset $\{n_k \in \mathbb{N} : k > K\}$.

In the other direction, let $\xi \in \mathbb{R}$ satisfy the condition in the statement. We will construct a subsequence of $\{x_n\}$ that converges to ξ as follows.

Let $A_1 = \{n \in \mathbb{N} : |x_n - \xi| < 1\}$. By assumption, A_1 is infinite and therefore not empty and we set n_1 to be the least element in A_1. In particular, $|x_{n_1} - \xi| < 1$.

Let $A_2 = \{n \in \mathbb{N} : n_1 < n \text{ and } |x_n - \xi| < 1/2\}$. Since $\{n \in \mathbb{N} : |x_n - \xi| < 1/2\}$ is infinite by assumption, it follows that A_2 is not empty and we set n_2 to be the least element in A_2. It follows that $n_1 < n_2$ and $|x_{n_2} - \xi| < 1/2$.

We continue to define the indices of the subsequence recursively in this way: having found $n_1, n_2, \ldots, n_k \in \mathbb{N}$ satisfying $n_1 < n_2 < \cdots < n_k$ and $|x_{n_j} - \xi| < 1/j$ for all $1 \le j \le k$, we set n_{k+1} to be the least element of the (nonempty) set $A_{k+1} = \{n \in \mathbb{N} : n_k < n \text{ and } |x_n - \xi| < 1/(k+1)\}$. With this choice, $n_k < n_{k+1}$ and $|x_{n_{k+1}} - \xi| < 1/(k+1)$.

Now, given $\varepsilon > 0$, there is an $m \in \mathbb{N}$ such that $1/m < \varepsilon$ (by Lemma 1.6.2) and for all $k > m$ we have $|x_{n_k} - \xi| < 1/k < 1/m < \varepsilon$. Hence $\lim_{k\to\infty} x_{n_k} = \xi$, so ξ is a subsequential limit of $\{x_n\}$. □

2.5.2. $\limsup$ and $\liminf$

A general sequence need not have any limit points. For example, if $x_n \to \infty$, then $\{x_n\}$ has no convergent subsequences: if $x_n \to \infty$ and $\xi \in \mathbb{R}$, then $x_n > \xi + 1$ for all sufficiently large $n \in \mathbb{N}$, so ξ cannot be a limit point of $\{x_n\}$.

Let $\{x_n\}_{n\in\mathbb{N}} \subset \mathbb{R}$ and let $y_m = \sup\{x_n : n > m\}$ for all $m \in \mathbb{N}$. then the sequence $\{y_m\}$ is monotone nonincreasing, because if $m_1 < m_2$, then

$$\sup\{x_n : n > m_2\} \le \sup\{x_n : n > m_1\}.$$

If the sequence $\{y_m\}$ is bounded below, then it converges. Likewise, if $z_m = \inf\{x_n : n > m\}$, then $\{z_m\}$ is monotone nondecreasing, and it converges if it is bounded above.

[5]In contrast, $\lim x_n = \xi$ if and only if every interval containing ξ, contains *all but finitely many* terms of $\{x_n\}$.

Definition. The *limit superior* (lim sup) of the sequence $\{x_n\}$ is defined

$$\limsup_{n\to\infty} x_n = \lim_{m\to\infty}\left(\sup_{n>m} x_n\right), \tag{2.5.1}$$

The *limit inferior* (lim inf) of the sequence is

$$\liminf_{n\to\infty} x_n = \lim_{m\to\infty}\left(\inf_{n>m} x_n\right). \tag{2.5.2}$$

Proposition. *If* $S = \limsup_{n\to\infty} x_n$ *and* $s = \liminf_{n\to\infty} x_n$ *both exist in* $\mathbb{R}$, *then* $S \geq s$.

Proof. Exercise. □

The next theorem gives practical characterizations of $\limsup x_n$ and $\liminf x_n$.

Theorem. *Suppose that* $\{x_n\}_{n\in\mathbb{N}}$ *is a sequence of real numbers.*

(i) $\limsup x_n = S \in \mathbb{R}$ *if and only if for every* $\varepsilon > 0$, $\{n \in \mathbb{N} : x_n > S + \varepsilon\}$ *is finite, and* $\{n \in \mathbb{N} : x_n > S - \varepsilon\}$ *is infinite.*

(ii) $\liminf x_n = s \in \mathbb{R}$ *if and only if for every* $\varepsilon > 0$, $\{n \in \mathbb{N} : x_n < s - \varepsilon\}$ *is finite, and* $\{n \in \mathbb{N} : x_n < s + \varepsilon\}$ *is infinite.*

We will prove part (a) and leave part (b) as an exercise.

Proof. First suppose that $S = \limsup x_n = \lim_{m\to\infty}(\sup_{n>m} x_n)$. Then for any $\varepsilon > 0$,

$$S + \varepsilon > \lim_{m\to\infty}\left(\sup_{n>m} x_n\right) > S - \varepsilon. \tag{2.5.3}$$

The left-hand inequality implies that there exists $M \in \mathbb{N}$ such that $S + \varepsilon > \sup_{n>m} x_n$ for all $m \geq M$. Hence if $n > M$, then $x_n < S + \varepsilon$, so the set $\{n : x_n > S + \varepsilon\}$ contains at most M indices. From the right-hand inequality of (2.5.3), it follows that there is an n_0 such that $\sup_{n>m} x_n > S - \varepsilon$ for all $m \geq n_0$. In particular, there is an $n_1 > n_0$ such that $x_{n_1} > S - \varepsilon$. Likewise, since $n_1 > n_0$, there is an $n_2 > n_1 > n_0$, such that $x_{n_2} > S - \varepsilon$. And so on: having found indices $n_k > \cdots > n_1 > n_0$ such that $x_{n_j} > S - \varepsilon$, it follows that there is an $n_{k+1} > n_k$ such that $x_{n_{k+1}} > S - \varepsilon$. Thus we can recursively construct an infinite sequence $n_1 < n_2 < \cdots < n_k < \cdots$ in $\mathbb{N}$ such that $x_{n_j} > S - \varepsilon$ for all $j \in \mathbb{N}$, so $\{n \in \mathbb{N} : x_n > S - \varepsilon\}$ is infinite.

In the other direction, suppose that S has property described in (i) and let $\varepsilon > 0$. If M_1 is the maximum of the *finite* set $\{n :\} x_n > S + \varepsilon$ then $\sup_{n>m} x_n \leq S + \varepsilon$ for all $m > M_1$, so $\lim_{m\to\infty}(\sup_{n>m} x_n) \leq S + \varepsilon$.

On the other hand, the fact that $\{n :\} x_n > S - \varepsilon$ is infinite implies $\sup_{n>m} x_n > S - \varepsilon$ for all m, so $\lim_{m\to\infty}(\sup_{n>m} x_n) \geq S - \varepsilon$. Combining the two inequalities, we see that for any $\varepsilon > 0$,

$$S - \varepsilon \leq \lim_{m\to\infty}\left(\sup_{n>m} x_n\right) \leq S + \varepsilon,$$

and therefore $S = \lim_{m\to\infty}\left(\sup_{n>m} x_n\right) = \limsup_{n\to\infty} x_n$. □

To illustrate the usefulness of these characterizations, we use them to prove the following

Corollary. *Suppose that* $\{x_n\}$ *is a sequence of real numbers. Then, if they exist,* $\limsup x_n$ *and* $\liminf x_n$ *are limit points of* $\{x_n\}$.

Proof. If $S = \limsup x_n$, then for any $\varepsilon > 0$ the set $\{n : x_n > S - \varepsilon\}$ is infinite and the set $\{n : x_n > S+\varepsilon/2\}$ is finite, by part (a) of the theorem above. It follows that there is an $N \in \mathbb{N}$ such that $x_n \leq S+\varepsilon/2 < S+\varepsilon$ for all $n > N$ and therefore the set $\{n : |x_n - S| < \varepsilon\}$ is infinite because it contains the infinite set $\{n : n > N \text{ and } x_n > S - \varepsilon\}$ as a subset. This implies that S is a limit point of $\{x_n\}$ by Theorem 2.5.2. The proof that $\liminf x_n$ is a limit point of $\{x_n\}$ is analogous and we leave the details as an exercise. □

2.5.3. The Bolzano-Weierstrass theorem

As noted, an unbounded sequence need not have any limit points. Bounded sequences in $\mathbb{R}$, on the other hand, always have limit points in $\mathbb{R}$.

Theorem (Bolzano-Weierstrass). *If the sequence* $\{x_n\} \subset \mathbb{R}$ *is bounded then it has a convergent subsequence.*

Proof. If $\{x_n\}$ is bounded, then $a \leq x_n \leq b$ for some $a, b \in \mathbb{R}$. It follows that the sequence $y_m = \sup\{x_n : n > m\}$ is bounded below, so $\limsup x_n = \lim y_m$ exists because $\{y_m\}$ is bounded and monotone. Hence $\{x_n\}$ has limit points by Corollary 2.5.2. □

2.5.4. The metric completeness of $\mathbb{R}$

As we mentioned in §2.2.2, $\mathbb{R}$ is a *complete metric space*. We prove this following the lemma below.

Lemma. *The sequence* $\{x_n\} \subset \mathbb{R}$ *converges if and only if it is bounded and*

$$\limsup x_n = \liminf x_n.$$

Proof. First suppose that the sequence converges and $\lim x_n = y$. Then the sequence is bounded (by Proposition 2.2.1) and for any ε, there is an N such that $y-\varepsilon < x_n < y+\varepsilon$ for all $n > N$.

This implies that the sets $\{n : x_n < y - \varepsilon\}$ and $\{n : x_n > y + \varepsilon\}$ are both finite with at most N elements, and the sets $\{n : x_n > y - \varepsilon\}$ and $\{n : x_n < y + \varepsilon\}$ are both infinite, since both contain the (infinite) set $\{n : y - \varepsilon < x_n < y + \varepsilon\}$ as a subset. It follows from Theorem 2.5.2 that $\limsup x_n = y = \liminf x_n$.

Now suppose that $\{x_n\}$ is bounded and $\limsup x_n = \liminf x_n = y$. Given $\varepsilon > 0$, the sets $\{n : x_n < y - \varepsilon\}$ and $\{n : x_n > y + \varepsilon\}$ are both finite, by Theorem 2.5.2. Thus, there exists an $N_\varepsilon \in \mathbb{N}$ such that if $x_n < y - \varepsilon$ or $x_n > y + \varepsilon$, then $n \leq N_\varepsilon$. Hence, $y - \varepsilon \leq x_n \leq y + \varepsilon$ for all $n > N_\varepsilon$, and since $\varepsilon > 0$ was arbitrary, it follows that $\lim x_n = y$. □

Theorem. *If* $\{x_n\} \subset \mathbb{R}$ *is a Cauchy sequence, then it converges to a limit in* $\mathbb{R}$.

Proof. Let $\{x_n\} \subset \mathbb{R}$ be a Cauchy sequence. We will show that it converges by showing that it is bounded and that $\limsup x_n = \liminf x_n$.

The Cauchy condition implies that there is an N such that if $n, m \geq N$, then $|x_n - x_m| < 1$. In particular, $|x_n - x_N| < 1$ and so $x_N - 1 < x_n < x_N + 1$ for all $n > N$. Setting $m = \min\{x_1, x_2, \ldots, x_{N-1}, x_N - 1\}$ and $M = \max\{x_1, x_2, \ldots, x_{N-1}, x_N + 1\}$, it follows that $m \leq x_n \leq M$ for all n, so the sequence is bounded.

As in the proof of the Bolzano-Weierstrasse theorem, it follows that both $S = \limsup x_n$ and $s = \liminf x_n$ exist. We have $S \geq s$, by Proposition 2.5.2 and we claim that in fact $S = s$ so that $\lim x_n = S$. To show this, we prove that the assumption $S > s$ is not consistent with the Cauchy condition on $\{x_n\}$.

Suppose that $S - s > 0$ and let $\varepsilon = \frac{S-s}{3} > 0$. Then it follows from Theorem 2.5.2 that the sets $\{n \in \mathbb{N} : x_n > S - \varepsilon\}$ and $\{n \in \mathbb{N} : x_n < s + \varepsilon\}$ are both infinite. This implies that for any $N \in \mathbb{N}$, there exist $n > N$ and $m > N$ such that $x_n < s + \varepsilon$ and $x_m > S - \varepsilon$. But then for any $N \in \mathbb{N}$, there are $n, m > N$ such that

$$x_m - x_n > (S - \varepsilon) - (s + \varepsilon) = (S - s) - 2\varepsilon = 3\varepsilon - 2\varepsilon = \varepsilon,$$

which contradicts the assumption that $\{x_n\}$ is a Cauchy sequence. □

Comment. The sequence $\{a_n\} \subset \mathbb{Q}$ that we constructed in Example 2.4.2 is monotone decreasing because

$$a_n - a_{n+1} = a_n - \frac{a_n^2 + 2}{2a_n} = \frac{a_n^2 - 2}{2a_n} > 0$$

for all n (since $a_n^2 > 2$ for all n).

It follows that if $n > m > N$, then $\sqrt{2} < a_n < a_m < a_N$, and therefore

$$|a_n - a_m| < a_N - \sqrt{2} < \frac{1}{2^{5N-7}}.$$

This means that $\{a_n\}$ is a Cauchy sequence of rational numbers that does *not converge* in $\mathbb{Q}$, because $a_n \to \sqrt{2} \notin \mathbb{Q}$.

In other words, $\mathbb{Q}$ is *not* a complete metric space. Another standard approach to constructing the real numbers is by *completing* the rational numbers.[6] In principle, this means adding points to $\mathbb{Q}$—one limiting value for each Cauchy sequence of rational numbers that does not have a rational limit. In practice, the technical details are a bit more complicated, and we provide them in the appendix.

⋆2.5.5. The completeness of $\mathbb{R}^n$

To show that $\mathbb{R}^n$ is a complete metric space, we first need to establish some notation and prove a lemma. For the elements of a sequence $\{\mathbf{x}_m\}$ in $\mathbb{R}^n$, we write

$$\mathbf{x}_m = (x_{m,1}, x_{m,2}, \ldots, x_{m,n}).$$

This means that a sequence $\{\mathbf{x}_m\}_{m=1}^{\infty}$ in $\mathbb{R}^n$ gives rise to n sequences in $\mathbb{R}$: $\{x_{m,1}\}_{m=1}^{\infty}$, $\{x_{m,2}\}_{m=1}^{\infty}, \ldots, \{x_{m,n}\}_{m=1}^{\infty}$.

Lemma. *The sequence $\{\mathbf{x}_m\}_{m=1}^{\infty}$ is a Cauchy sequence in $\mathbb{R}^n$ if and only if for each $1 \leq k \leq n$, the sequence $\{x_{m,k}\}_{m=1}^{\infty}$ is a Cauchy sequence in $\mathbb{R}$.*

[6] As opposed to the "Dedekind cut" approach.

Proof. First suppose that $\{\mathbf{x}_m\}_{m=1}^{\infty}$ is a Cauchy sequence in $\mathbb{R}^n$ and let $\varepsilon > 0$. It follows that there is an $M \in \mathbb{N}$ such that $|\mathbf{x}_t - \mathbf{x}_s| < \varepsilon$ for all $t > s > M$. It thus follows from the inequality (2.1.1) that

$$|x_{t,k} - x_{s,k}| \leq |\mathbf{x}_t - \mathbf{x}_s| < \varepsilon$$

for all $t > s > M$, so $\{x_{m,k}\}$ is a Cauchy sequence in $\mathbb{R}$ for each $1 \leq k \leq n$.

Now suppose that $\{x_{m,k}\}_{m=1}^{\infty}$ is a Cauchy sequence for each $1 \leq k \leq n$, and let $\varepsilon > 0$. In this case, there are constants $M_1, M_2, \ldots, M_n$ such that for each k, if $t > s > M_k$, then $|x_{t,k} - x_{s,k}| < \varepsilon/\sqrt{n}$. Setting $M = \max(M_1, \ldots, M_n)$, if follows that if $t > s > M$, then

$$|x_{t,k} - x_{s,k}| < \varepsilon/\sqrt{n}$$

for each $1 \leq k \leq n$ and therefore by (2.1.2)

$$|\mathbf{x}_t - \mathbf{x}_s| \leq \sqrt{n} \cdot \max_{1 \leq k \leq n} |x_{t,k} - x_{s,k}| < \varepsilon.$$

I.e., $\{\mathbf{x}_m\}$ is a Cauchy sequence in $\mathbb{R}^n$. □

Theorem. *If $\{\mathbf{x}_m\}_{m=1}^{\infty}$ is a Cauchy sequence in $\mathbb{R}^n$, then it converges to a limit in $\mathbb{R}^n$.*

Proof. If $\{\mathbf{x}_m\}_{m=1}^{\infty}$ is a Cauchy sequence in $\mathbb{R}^n$, then by the lemma above, $\{x_{m,k}\}_{m=1}^{\infty}$ is a Cauchy sequence in $\mathbb{R}$ for each $1 \leq k \leq n$. Because of the completeness of $\mathbb{R}$, it follows that there are n real numbers, $a_1, a_2, \ldots, a_n$, such that $\lim_{m\to\infty} x_{m,k} = a_k$ for each $1 \leq k \leq n$. We claim that $\lim_{m\to\infty} \mathbf{x}_m = \mathbf{a} = (a_1, a_2, \ldots, a_n)$.

Indeed, if $\varepsilon > 0$, then for each $1 \leq k \leq n$, we can find $M_k \in \mathbb{N}$ such that $|x_{m,k} - a_k| < \varepsilon/\sqrt{n}$, for all $m > M_k$. Setting $M = \max(M_1, \ldots, M_n)$, if $m > M$, then $|x_{m,k} - a_k| < \varepsilon/\sqrt{n}$ for all $1 \leq k \leq m$ and therefore

$$|\mathbf{x}_m - \mathbf{a}| \leq \sqrt{n} \cdot \max_{1 \leq k \leq n} |x_{m,k} - a_k| < \varepsilon,$$

by the inequality (2.1.2). □

*2.5.6. The Bolzano-Weierstrass theorem in $\mathbb{R}^n$

The Bolzano-Weierstrass theorem is equally valid in $\mathbb{R}^n$. We can think of Theorem 2.5.3 as a special case of the one below. On the other hand, we will quote the 1-dimensional case in our proof of the general case.

Before proving the theorem, we prove an analog of Lemma*2.5.5.

Lemma. *The sequence $\{\mathbf{x}_m\}_{m=1}^{\infty}$ is bounded in $\mathbb{R}^n$ if and only if $\{x_{m,k}\}_{m=1}^{\infty}$ is bounded in $\mathbb{R}$ for $1 \leq k \leq n$.*

Proof. If $\{x_{m,k}\}_{m=1}^{\infty}$ is bounded for $1 \leq k \leq n$, then there exist real numbers $y_1, \ldots, y_n \in \mathbb{R}$ and positive real numbers $r_1, \ldots, r_n$ such that

$$|x_{m,k} - y_k| < r_k \tag{2.5.4}$$

for all $m \in \mathbb{N}$ and $1 \leq k \leq n$.

Setting $r = \max(r_1, r_2, \ldots, r_n)$ and $\mathbf{y} = (y_1, y_2, \ldots, y_n)$, it follows from (2.5.4) and (2.1.2) that $|\mathbf{x}_m - \mathbf{y}| < \sqrt{n} \cdot r$ for all $m \in \mathbb{N}$, so $\{\mathbf{x}_m\}_{m=1}^{\infty}$ is bounded in $\mathbb{R}^n$.

If $\{\mathbf{x}_m\}_{m=1}^{\infty}$ is bounded, then there exists $\mathbf{y} = (y_1, y_2, \ldots, y_n) \in \mathbb{R}^n$ and $r > 0$, such that $|\mathbf{x}_m - \mathbf{y}| < r$ for all $m \in \mathbb{N}$ and it follows from this and (2.1.1) that $|x_{m,k} - y_k| < r$ for all $m \in \mathbb{N}$ and $1 \leq k \leq n$. Hence, $\{x_{m,k}\}_{m=1}^{\infty}$ is bounded in $\mathbb{R}$ for $1 \leq k \leq n$. □

Theorem (Bolzano-Weierstrass). *If $\{\mathbf{x}_m\}_{m=1}^{\infty}$ is a bounded sequence in $\mathbb{R}^n$, then it has a convergent subsequence.*

Proof. Suppose that $\{\mathbf{x}_m\}_{m=1}^{\infty}$ is a bounded sequence in $\mathbb{R}^n$. By the lemma above, the sequence $\{x_{m,1}\}_{m=1}^{\infty}$ is bounded in $\mathbb{R}$ and so has a convergent subsequence. We denote this subsequence $\{x_{m,1}\}_{m\in\Lambda_1}$, where $\Lambda_1 = \{m_j\}_{j=1}^{\infty}$ is a strictly increasing sequence in $\mathbb{N}$. We denote the limit of this subsequence by y_1.

The sequence $\{x_{m,2}\}_{m=1}^{\infty}$ is also bounded, and so by the lemma, so is the subsequence $\{x_{m,2}\}_{m\in\Lambda_1} \subset \{x_{m,2}\}_{m=1}^{\infty}$. Once again, by Theorem 2.5.3, there is a subsequence of natural numbers, $\Lambda_2 \subset \Lambda_1$, such that $\{x_{m,2}\}_{m\in\Lambda_2}$ converges to a limit $y_2 \in \mathbb{R}$. Furthermore, since $\Lambda_2 \subset \Lambda_1$ and $\lim_{m\in\Lambda_1} x_{m,1} = y_1$, we also have $\lim_{m\in\Lambda_2} x_{m,1} = y_1$.

Thus, after these first two steps, we have found a strictly increasing sequence $\Lambda_2 \subset \mathbb{N}$ such that $\lim_{m\in\Lambda_2} x_{m,1} = y_1$ and $\lim_{m\in\Lambda_2} x_{m,2} = y_2$.

Likewise, since $\{x_{m,3}\}_{m\in\Lambda_2} \subset \{x_{m,3}\}_{m=1}^{\infty}$ is bounded, there is a subsequence Λ_3 of Λ_2 and $y_3 \in \mathbb{R}$ such that $\lim_{m\in\Lambda_3} x_{m,3} = y_3$. And as before, $\lim_{m\in\Lambda_3} x_{m,1} = y_1$ and $\lim_{m\in\Lambda_3} x_{m,2} = y_2$.

Continuing in this way, we can find n strictly increasing sequences of natural numbers, $\Lambda_n \subset \Lambda_{n-1} \subset \cdots \subset \Lambda_2 \subset \Lambda_1 \subset \mathbb{N}$, and real numbers $y_1, y_2, \ldots, y_n$ such that for $1 \leq k \leq n$ and $1 \leq j \leq k$, $\lim_{m\in\Lambda_k} x_{m,j} = y_j$. In particular, for $j = 1 \ldots n$, we have $\lim_{m\in\Lambda_n} x_{m,j} = y_j$, and therefore $\lim_{m\in\Lambda_n} \mathbf{x}_m = \mathbf{y} = (y_1, \ldots, y_n)$. □

Exercises

ex2.5.1. Show that if $x_n \to \infty$, then $\{x_n\}$ has no subsequential limits.

ex2.5.2. Prove Proposition 2.5.2.

ex2.5.3. Prove part (b) of Theorem 2.5.2.

ex2.5.4. Show that a sequence $\{x_n\} \subset \mathbb{R}$ is bounded if and only if $\limsup x_n$ and $\liminf x_n$ both exist.

ex2.5.5. Prove the second half of Corollary 2.5.2: If $\{x_n\} \subset \mathbb{R}$ and $\liminf x_n$ exists, then it is a limit point of $\{x_n\}$.

ex2.5.6. Given a bounded sequence $\{x_n\}_{n\in\mathbb{N}}$ of real numbers, show that

$$\liminf x_n = -\limsup(-x_n).$$

ex2.5.7. Construct a sequence $\{x_n\} \subset \mathbb{R}$ that is not bounded below such that $\limsup x_n = 1$.

ex2.5.8. Given a bounded sequence $\{x_n\}_{n\in\mathbb{N}}$ of real numbers. Show that

$$\limsup x_n = \inf\{a : \text{the set } \{n : x_n \geq a\} \text{ is finite}\}.$$

ex2.5.9. Let $\{x_n\}$ be a bounded sequence in $\mathbb{R}$ and let A be the set of limit points of $\{x_n\}$. Show that $\limsup x_n = \sup A$ and $\liminf x_n = \inf A$.

ex2.5.10. Given bounded sequences $\{a_n\}$ and $\{b_n\}$ of real numbers, prove or disprove (by giving a counterexample) each of the following statements:

(i) $\limsup(a_n + b_n) \leq \limsup a_n + \limsup b_n$.

(ii) $\limsup a_n b_n \leq \limsup a_n \limsup b_n$.

(iii) $\liminf(a_n + b_n) \leq \liminf a_n + \liminf b_n$.

(iv) $\liminf(a_n + b_n) \geq \liminf a_n + \liminf b_n$.

2.6. Series

Calculating the sum of finitely many terms can be done in any field, relying only on the field axioms. To extend addition in $\mathbb{R}$ to an infinite sequence of terms, we make use of the metric space structure and define the sum of a series to be the limit of a sequence, if the limit exists.

2.6.1. Convergence

Definition. Given a sequence of real numbers $\{x_n\}_{n=1}^{\infty}$, the sum of the *series* $\sum_{n=1}^{\infty} x_n$ is defined to be the limit of the sequence of *partial sums*

$$S_n = \sum_{j=1}^{n} x_j,$$

if $\lim S_n = S$ exists. We say in this case that $\sum x_n$ converges to S, or has sum S, and write $\sum_{n=1}^{\infty} x_n = S$. If the limit does not exist, we say that the series *diverges*, or that it is a *divergent series.*

The *tails* of a series $S = \sum_{n=1}^{\infty} x_n$, are the series $T_n = \sum_{j>n} x_j$ (one for each $n \in \mathbb{N}$).

Proposition. *If* $\sum_{n=1}^{\infty} x_n = S$, *then the tail* T_n *converges to* $S - S_n$ *for each* n.

Proof. Let $n > 0$, and for $m > n$, set $\tau_m = \sum_{j=n+1}^{m} x_j$. Then $\{\tau_m\}_{m>n}$ is the sequence of partial sums of T_n, so $T_n = \lim_{m\to\infty} \tau_m = \lim_{m\to\infty} (S_m - S_n) = S - S_n$ □

Corollary. *The tails of a convergent series tend to* 0.

2.6.2. The Cauchy criterion

A sequence of real numbers converges if and only if it is a Cauchy sequence. For series, this leads to the following criterion.

Theorem. *The series* $\sum_{n=1}^{\infty} x_n$ *converges in* $\mathbb{R}$ *if and only if for every* $\varepsilon > 0$, *there is an* $N_\varepsilon \in \mathbb{N}$ *such that*

$$\left| \sum_{j=m+1}^{n} x_j \right| < \varepsilon$$

for all $n > m \geq N_\varepsilon$.

Proof. The series converges if and only if the sequence of partial sums, $S_n = \sum_{j=1}^n x_j$, converges, and the sequence $\{S_n\}$ converges if and only if it is a Cauchy sequence. In other words, $\sum x_n$ converges if and only if for every $\varepsilon > 0$, there is an $N_\varepsilon \in \mathbb{N}$ such that $|S_n - S_m| < \varepsilon$ for all $n > m \geq N_\varepsilon$. The theorem now follows from the observation that for any $n > m \geq 1$,

$$|S_n - S_m| = \left|\sum_{j=1}^n x_j - \sum_{j=1}^m x_j\right| = \left|\sum_{j=m+1}^n x_j\right|. \qquad \square$$

Corollary. *If the series $\sum_{n=1}^\infty x_n$ converges, then $\lim\limits_{n\to\infty} x_n = 0$.*

Proof. If the series converges then for any $\varepsilon > 0$ there is an N_ε such that $\left|\sum_{j=m+1}^n x_j\right| < \varepsilon$ for all $n > m > N_\varepsilon$. In particular, $|x_n| < \varepsilon$ for all $n > N_\varepsilon+1$, because $\sum_{j=n}^n x_j = x_n$. $\square$

The converse of this corollary is false in general. The fact that $\lim x_n = 0$, is *not sufficient* to guarantee the convergence of the series $\sum x_n$.

Example. The series $\sum \frac{1}{n}$ is called the *harmonic series*. We have $1/n \to 0$, but for any $N \geq 1$, if $N < j \leq 2N$, then

$$\sum_{j=N+1}^{2N} \frac{1}{j} \geq \sum_{j=N+1}^{2N} \frac{1}{2N} = N \cdot \frac{1}{2N} = \frac{1}{2}.$$

The harmonic series fails the Cauchy criterion with $\varepsilon = 1/2$, and so does not converge.

2.6.3. Geometric series

Determining whether a sequence converges or not can be tricky, and in this context it is useful to have sequences whose convergence or divergence is well understood.

Definition. For $c, q \in \mathbb{R}$, we call the sequence $\{cq^n\}_{n\in\mathbb{N}}$ a *geometric progression* and the series $\sum cq^n$ is called a *geometric series.*

Theorem. *If $c \neq 0$, then $\sum\limits_{n=0}^{\infty} cq^n$ converges if and only if $|q| < 1$.*

Proof. If $q = 1$, then $|S_n| = \left|\sum_{k=0}^n c1^n\right| = (n+1)|c| \to \infty$ so $\{S_n\}$ diverges. If $q = -1$, then $S_n = \sum_{k=0}^n c(-1)^n = (1 + (-1)^n)c$, so for $k \geq 1$, $S_{2k} = 2c$ and $S_{2k+1} = 0$ and once again, the sequence $\{S_n\}$ fails to converge. I.e., if $q = \pm 1$, then $\sum cq^n$ diverges.

Next, for any $q \in \mathbb{R}$ and $n \in \mathbb{N}$, the identity (1.1.1) gives

$$1 - q^{n+1} = (1-q)(1 + q + \cdots + q^n) = (1-q)\sum_{j=0}^n q^n,$$

and if $q \neq 1$, it follows that

$$\sum_{j=0}^n cq^n = \frac{c - cq^{n+1}}{1-q}.$$

If $|q| < 1$, then $\lim_{n\to\infty} q^n = 0$ (by Theorem 2.4.4 (i)) and in this case, $\sum cq^n$ converges because

$$\sum_{n=0}^{\infty} cq^n = \lim_{n\to\infty}\left(\sum_{j=0}^{n} cq^n\right) = \lim_{n\to\infty}\frac{c - cq^{n+1}}{1-q} = \frac{c}{1-q}.$$

If $|q| > 1$, then $|q|^n \to \infty$ and

$$|S_n| = \left|\sum_{j=0}^{n} cq^n\right| = |c|\left|\frac{1-q^{n+1}}{1-q}\right| = |c|\frac{|q^{n+1}-1|}{|q-1|} \geq \frac{(|q|^{n+1}-1)|c|}{|q-1|},$$

so $|S_n| \to \infty$ and $\sum q^n$ diverges. □

Comment. Geometric series provide a useful yardstick with which to judge the convergence or divergence of other series, as described in Sections 2.7.1 and 2.7.3, for example.

2.6.4. Sums and multiples of series

Convergent series can be treated in certain respects like finite sums.[7]

Theorem. *If the series $\sum_{n=1}^{\infty} x_n$ and $\sum_{n=1}^{\infty} y_n$ converge to the limits X and Y respectively, then for any $a, b \in \mathbb{R}$, the series $\sum_{n=1}^{\infty}(ax_n + by_n)$ converges and*

$$\sum_{n=1}^{\infty}(ax_n + by_n) = aX + bY.$$

Proof. Let $S_{x,n} = \sum_{k=1}^{n} x_n$ and $S_{y,n} = \sum_{k=1}^{n} y_n$. Then $\lim_n S_{x,n} = X$ and $\lim_n S_{y,n} = Y$, by definition, and therefore

$$\sum_{n=1}^{\infty} ax_n + by_n = \lim_{n\to\infty}(aS_{x,n} + bS_{y,n}) = aX + bY,$$

as follows from Theorem 2.3.2. □

This property can be extended to any finite collection of convergent sums.

Corollary. *Let $m \in \mathbb{N}$ and suppose that the series $\sum_{n=1}^{\infty} x_{j,n}$ converges to X_j, for $1 \leq j \leq m$. Then for any $a_1, \ldots, a_m \in \mathbb{R}$*

$$\sum_{n=1}^{\infty}(a_1x_{1,n} + a_2x_{2,n} + \cdots + a_mx_{m,n}) = a_1X_1 + a_2X_2 + \cdots a_mX_m.$$

Proof. By induction on m. Exercise. □

[7] But not in all respects, as we see in Section 2.8 for example.

2.6.5. Absolute convergence

Definition. If the series $\sum_{n=1}^{\infty} |x_n|$ converges, then we say that the series $\sum_{n=1}^{\infty} x_n$ converges *absolutely*, or that it is *absolutely convergent*.

If $\sum_{n=1}^{\infty} x_n$ converges, but does not converge absolutely, we say that the series converges *conditionally*, or that it is *conditionally convergent*.

Theorem 2.6.3 says that the geometric series $\sum cq^n$ converges if and only if $|q| < 1$. We can now amend that statement to say that the geometric series converges *absolutely* if $|q| < 1$ and diverges if $|q| \geq 1$.

Proposition. *If $\sum |x_n|$ converges, then $\sum x_n$ converges.*

Proof. Suppose that $\sum |x_n|$ converges. By the Cauchy criterion, given $\varepsilon > 0$, there is an $N \in \mathbb{N}$ such that

$$\left| \sum_{j=m}^{n} x_j \right| \leq \sum_{j=m}^{n} |x_j| < \varepsilon$$

for all $n \geq m > N$, where the left-hand inequality is the triangle inequality. □

Absolutely convergent series behave very much like finite sums. For example, the terms in an absolutely convergent series can be rearranged without affecting the limit of the series (see Section 2.10.1).

2.6.6. The Cauchy product

Definition. Given two series, $\sum a_n$ and $\sum b_n$, their *Cauchy product* is the series $\sum c_n$, where for all $n \geq 0$,

$$c_n = \sum_{k=0}^{n} a_{n-k} b_k.$$

Theorem (Mertens). *If $\sum a_n$ and $\sum b_n$ both converge, and one of the two series converges absolutely, then their Cauchy product converges to the product of their sums.*

If neither of the two series converges absolutely, the conclusion may fail. See **ex2.7.7**.

Proof. We may suppose without loss of generality that $\sum a_n$ converges absolutely. Set $A = \sum a_n$ and $B = \sum b_n$, and denote the partial sums $\sum_{j=0}^{n} a_j$, $\sum_{j=0}^{n} b_j$ and $\sum_{j=0}^{n} c_j$, by A_n, B_n and C_n, respectively. We want to show that $C_n \to AB$.

Given $\varepsilon > 0$, there exists an N_0 such that $|AB - A_n B_n| < \varepsilon/2$ for all $n > N_0$ because $A_n B_n \to AB$, by Theorem 2.3.2. Hence

$$|AB - C_n| \leq |AB - A_n B_n| + |A_n B_n - C_n| < \frac{\varepsilon}{2} + |A_n B_n - C_n|, \tag{2.6.1}$$

for all $n > N_0$. To bound $|A_n B_n - C_n|$, we rearrange the terms in C_n to obtain the identity

$$C_n = \sum_{j=0}^{n} \sum_{k=0}^{j} a_k b_{j-k} = \sum_{k=0}^{n} a_k \sum_{j=0}^{n-k} b_j = \sum_{k=0}^{n} a_k B_{n-k},$$

from which it follows that

$$|A_nB_n - C_n| = \left|\sum_{k=0}^{n} a_k(B_n - B_{n-k})\right| \le \sum_{k=0}^{n} |a_k||B_n - B_{n-k}|. \tag{2.6.2}$$

Since $\{B_n\}$ converges it is a Cauchy sequence. Hence, there is an N_1 such that $|B_n - B_m| < \varepsilon/4A_1$ for all $n \ge m > N_1$, and if $n > 2N_1$, then $n - k > N_1$ for $0 \le k \le n/2$, so

$$\sum_{k=0}^{n/2} |a_k||B_n - B_{n-k}| < \frac{\varepsilon}{4A_1} \sum_{k=0}^{n/2} |a_k| < \frac{\varepsilon}{4}. \tag{2.6.3}$$

This bounds the "bottom half" of the sum on the right of (2.6.2).

For the "top half", we note that the convergent sequence $\{B_n\}$ is necessarily bounded, so there is an M such that $|B_n - B_m| \le M$ for all n, m. Also, since the series $\sum |a_n|$ converges, its tails tend to 0, so there is an N_2 such that for all $m > N_2$, we have $\sum_{k\ge m} |a_k| < \varepsilon/4M$. It follows that if $n > 2N_2$, then

$$\sum_{k=n/2+1}^{n} |a_k||B_n - B_{n-k}| \le M \sum_{k>N_2} |a_k| < \frac{\varepsilon}{4}. \tag{2.6.4}$$

Finally, if $n > \max(N_0, 2N_1, 2N_2)$, then it follows from (2.6.1), (2.6.2), (2.6.3), and (2.6.4) that

$$|AB - C_n| < \frac{\varepsilon}{2} + \sum_{k=0}^{n/2} |a_k||B_n - B_{n-k}| + \sum_{k=n/2+1}^{n} |a_k||B_n - B_{n-k}| < \varepsilon. \qquad \square$$

Exercises

ex2.6.1. Show that $\sum \frac{1}{n} \to \infty$.

Hint: Use the idea in Example 2.6.2 to show that $\sum_{n\le 2^k} > k/2$ for all $k \ge 1$.

ex2.6.2. Prove Corollary 2.6.4.

ex2.6.3. A sequence $\{x_n\} \subset \mathbb{R}$ has *bounded variation* if $\sum_{n=1}^{\infty} |x_{n+1} - x_n| < \infty$. Show that a sequence of bounded variation converges.

2.7. Convergence criteria

In this section, we introduce some useful tools for determining convergence or divergence of a series.

2.7.1. The comparison test

In many cases, we can determine the convergence or divergence of a series based on comparison to a series whose behavior is known. Most commonly, the series to which we are comparing is a geometric series.

Theorem. *Suppose that $\{a_n\}, \{b_n\}$ and $\{c_n\}$ are sequences of real numbers, such that $b_n \geq 0$ and $c_n \geq 0$ for all n, $\sum b_n$ converges and $\sum c_n$ diverges.*

(i) *If there exists $N_1 \in \mathbb{N}$ such that $|a_n| \leq b_n$ for all $n > N_1$, then $\sum a_n$ converges absolutely.*

(ii) *If there exists $N_2 \in \mathbb{N}$ such that $a_n \geq c_n$ for all $n > N_2$, then $\sum a_n$ diverges.*

Proof.

(i) Let $\varepsilon > 0$ be given. Since $\sum b_n$ converges, then by the Cauchy criterion, there is an $N \in \mathbb{N}$ such that for all $n \geq m > N$,

$$\sum_{j=m}^{n} b_n = \left|\sum_{j=m}^{n} b_n\right| < \varepsilon,$$

because $b_n \geq 0$ for all n. It follows that if $n \geq m > N_\varepsilon = \max(N, N_1)$, then

$$\left|\sum_{j=m}^{n} |a_n|\right| = \sum_{j=m}^{n} |a_n| \leq \sum_{j=m}^{n} b_n < \varepsilon,$$

so $\sum |a_n|$ converges by the Cauchy criterion.

(ii) Since $\sum c_n$ diverges, there exists $\varepsilon_0 > 0$, such that for all $N \in \mathbb{N}$, there are $n \geq m > N$ such that

$$\left|\sum_{j=m}^{n} c_n\right| = \sum_{j=m}^{n} c_n > \varepsilon_0.$$

It follows that for any $N \in \mathbb{N}$, there exist $n \geq m > \max(N, N_2)$ (so in particular, $n \geq m > N$), such that

$$\left|\sum_{j=m}^{n} a_n\right| = \sum_{j=m}^{n} a_n \geq \sum_{j=m}^{n} c_n > \varepsilon_0.$$

Thus, $\sum a_n$ diverges because it fails the Cauchy criterion for convergence. □

Corollary. *If there exist $q, c \in \mathbb{R}$, with $0 < q < 1$, $0 < c$, and $N \in \mathbb{N}$ such that $|a_n| \leq cq^n$ for all $n \geq N$, then $\sum a_n$ converges absolutely.*

Proof. This follows by comparison to the convergent geometric series $\sum cq^n$. □

Example. Consider the series

$$\sum_{n=1}^{\infty} \left(\frac{n^2 + 10n + 7}{2n^2 + 1}\right)^{n/2}.$$

We will show that it converges using the corollary to the comparison test. To do so, we need to show that the (nonnegative) terms $\left(\frac{n^2+10n+7}{2n^2+1}\right)^{n/2}$ are eventually bounded by q^n for some $0 < q < 1$.[8]

If $n > 100$, then $n^2 > 100n$, and hence

$$\frac{1}{2}n^2 > 50n = 15n + 35n > 15n + 35.$$

It follows that

$$2n^2 + 1 > \frac{3}{2}n^2 + \frac{1}{2}n^2 > \frac{3}{2}n^2 + 15n + 35 > \frac{3}{2}\left(n^2 + 10n + 7\right),$$

and therefore

$$\frac{n^2 + 10n + 7}{2n^2 + 1} < \frac{n^2 + 10 + 7}{\frac{3}{2}(n^2 + 10n + 7)} = \frac{2}{3}$$

for all $n > 100$.

It therefore follows that

$$\left(\frac{n^2 + 10n + 7}{2n^2 + 1}\right)^{n/2} < \left(\frac{2}{3}\right)^{n/2} = \left(\sqrt{\frac{2}{3}}\right)^n,$$

for all $n > 100$, so the series converges by Corollary 2.7.1.

2.7.2. The number e

The number e is defined as the sum of a series:

$$e = \sum_{n=0}^{\infty} \frac{1}{n!},$$

where $n!$ is the factorial function. The fact that this series converges is an immediate consequence of the theorem below, and boils down to the fact that $n!$ grows very rapidly and therefore $1/n!$ goes to 0 very rapidly.

Lemma. *For all $n > 1$, $n! > (n/2)^{n/2} = \left(\sqrt{n/2}\right)^n$.*

Proof. If $n = 2k$ is an even integer, then $k + 1 > k = n - k = n/2$ and if $n = 2k + 1$ is an odd integer, then $n - k = k + 1 > k + 1/2 = n/2$. Now, for any $n \in \mathbb{N}$, let $k = n/2$ if n is even and $k = (n - 1)/2$ if n is odd. It then follows from the observation above that

$$n! = \overbrace{n(n-1)\cdots(n-k)}^{k+1 \text{ factors}} \cdot (n-k-1)\cdots 2 \cdot 1 > (n/2)^{k+1} > (n/2)^{n/2},$$

because $n > (n-1) > \cdots (n-k+1) > (n-k) \geq n/2$ and all the other factors in the product are greater than or equal to 1. □

Comment. The correct order of magnitude of $n!$ is significantly bigger. There are constants $0 < a < b$ such that $an^{n+\frac{1}{2}} \leq n! \leq bn^{n+\frac{1}{2}}$ for all $n \in \mathbb{N}$.[9]

[8] "Eventually bounded" means that the bound is valid for all $n > N$, for some $N \in \mathbb{N}$.
[9] This follows from *Stirling's formula*.

Theorem. *The series*

$$\sum_{n=0}^{\infty} \frac{x^n}{n!}$$

converges absolutely for every $x \in \mathbb{R}$.

In particular, the series converges for $x = 1$, showing that the series defining e converges.

Proof. For any $x \in \mathbb{R}$, it follows from the Archimedean principle that there is an $N_x \in \mathbb{N}$ such that $N_x > 2(2|x| + 1)^2$. For $n > N_x$, we have $\sqrt{n/2} > \sqrt{N_x/2} > 2|x| + 1$, and hence by the lemma above

$$\frac{|x|^n}{n!} < \frac{|x|^n}{\left(\sqrt{n/2}\right)^n} < \frac{|x|^n}{(2|x|+1)^n} = \left(\frac{|x|}{(2|x|+1)}\right)^n < \left(\frac{1}{2}\right)^n, \tag{2.7.1}$$

for all $n > N_x$. It follows from Corollary 2.7.1 that the series converges absolutely. □

2.7.3. The root and ratio tests

In Example 2.7.4, we show in effect that

$$\sqrt[n]{\left(\frac{n^2 + 10n + 7}{2n^2 + 1}\right)^{n/2}} < \sqrt{\frac{2}{3}}$$

for all $n > 100$, which implies that

$$\limsup_{n\to\infty} \sqrt[n]{\left(\frac{n^2 + 10n + 7}{2n^2 + 1}\right)^{n/2}} \le \sqrt{\frac{2}{3}}.$$

Likewise, from the inequality (2.7.1) in the proof of Theorem 2.7.2, it follows that for all $x \in \mathbb{R}$

$$\limsup_{n\to\infty} \sqrt[n]{\frac{|x|^n}{n!}} < \frac{1}{2}.$$

These examples might lead you to guess that

If $\limsup_{n\to\infty} \sqrt[n]{|a_n|} < 1$, *then* $\sum a_n$ *converges absolutely.*

This guess is correct and comprises part of the "root test". There is also a "ratio test", that is similar in nature.

Theorem. *Let* $\{a_n\}$ *be a bounded sequence of real numbers.*

(i) *The series* $\sum a_n$ *converges absolutely if* $\limsup \sqrt[n]{|a_n|} < 1$, *and the series diverges if* $\limsup \sqrt[n]{|a_n|} > 1$.

(ii) *The series* $\sum a_n$ *converges absolutely if* $\limsup\left|\frac{a_{n+1}}{a_n}\right| < 1$ *and the series diverges if* $\liminf\left|\frac{a_{n+1}}{a_n}\right| > 1$.

Comments. Part (i) is called the 'root test' and part (ii) is called the 'ratio test'. If $\limsup \sqrt[n]{|a_n|} = 1$, then the root test is inconclusive. Likewise, if $\limsup|a_{n+1}/a_n| = 1$ (or $\liminf|a_{n+1}/a_n| = 1$), the ratio test is inconclusive.

Proof.

(i) If $\limsup \sqrt[n]{|a_n|} = \rho_1 < 1$, then $\varepsilon = (1-\rho_1)/2 > 0$ and hence $0 < \rho_1 + \varepsilon < 1$. By Theorem 2.5.2(a), the set $\{n : \sqrt[n]{|a_n|} > \rho_1 + \varepsilon\}$ is finite, so there exists $n_0 \in \mathbb{N}$ such that $\sqrt[n]{|a_n|} \leq \rho_1 + \varepsilon$ for all $n \geq n_0$. It follows that $|a_n| \leq (\rho_1 + \varepsilon)^n$ for all $n \geq n_0$, and the series $\sum a_n$ therefore converges absolutely by Corollary 2.7.1 (comparison to a geometric series).

If $\limsup \sqrt[n]{|a_n|} = \rho_2 > 1$, then $\varepsilon = (\rho_2 - 1)/2 > 0$ so $\sqrt[n]{|a_n|} > \rho_2 - \varepsilon > 1$ for infinitely many $n \in \mathbb{N}$, by Theorem 2.5.2(a). Thus $|a_n| > (1-\varepsilon)^n > 1$ for infinitely many $n \in \mathbb{N}$ so $a_n \not\to 0$, and $\sum a_n$ diverges.

(ii) We may assume that $a_n \neq 0$ for all n, because any 0 terms may be removed from the series without affecting its convergence/divergence properties. In other words, we assume that a_{n+1}/a_n is well defined for all n.

Now, if $\limsup |a_{n+1}/a_n| = \eta < 1$, then $\varepsilon = (1-\eta)/2 > 0$. As in the proof of (i), there is an $n_0 \in \mathbb{N}$ such that $|a_{n+1}/a_n| < \eta + \varepsilon = q < 1$ for all $n \geq n_0$. This implies that $|a_{n_0+1}| < q|a_{n_0}|$, and thus for $n_0 + 2$ and $n_0 + 3$ we have

$$|a_{n_0+2}| < q|a_{n_0+1}| < q^2|a_{n_0}| \quad \text{and} \quad |a_{n_0+3}| < q|a_{n_0+2}| < q^3|a_{n_0}|,$$

and so on. More generally, it follows by induction on m that $|a_{n_0+m}| < q^m|a_{n_0}|$, for all $m \geq 1$. Setting $n = n_0 + m$, we can restate this last inequality to say that

$$|a_n| < q^{n-n_0}|a_{n_0}| = cq^n$$

for all $n > n_0$, where $c = |a_{n_0}|q^{-n_0}$. The absolute convergence of $\sum a_n$ now follows from Corollary 2.7.1.

For the second half of (ii), we first observe that if there is an $N \in \mathbb{N}$ such that $|a_{n+1}/a_n| \geq 1$ for all $n \geq N$, then

$$|a_n| \geq |a_{n-1}| \geq \cdots \geq |a_{N+1}| \geq |a_N| > 0$$

for all $n > N$. This implies that $a_n \not\to 0$, so $\sum a_n$ diverges.

If $\liminf |a_{n+1}/a_n| = \kappa > 1$, then $\varepsilon = (\kappa - 1)/2 > 0$ and by Theorem 2.6.2(b), the set $\{n : |a_{n+1}/a_n| < \kappa - \varepsilon\}$ is finite. I.e., there exists an $N \in \mathbb{N}$ such that $|a_{n+1}/a_n| \geq \kappa - \varepsilon > 1$ for all $n > N$ and $\sum a_n$ diverges, as follows from the paragraph above. □

2.7.4. Series with nonnegative terms

If $a_n \geq 0$ for all n, then $S_{n+1} - S_n = a_{n+1} \geq 0$, so the sequence of partial sums $\{S_n\}$ is monotone nonincreasing. As a consequence, a series with nonnegative terms converges if and only if its sequence of partial sums is bounded, as follows from Theorem 2.4.5.

Proposition. *The series* $\sum_{n=1}^{\infty} \frac{1}{n^2}$ *converges.*

Proof. Since $1/n^2 > 0$ for all $n \in \mathbb{N}$, it follows from the preceding observation that to prove convergence of the series, it is enough to show that the sequence of partial sums is bounded.

To start, let $n \in \mathbb{N}$ and choose $m \in \mathbb{N}$ such that $n \leq 2^m$, so that $S_n \leq S_{2^m} < S_{2^{m+1}-1}$. The plan is to find a bound for the partial sum $S_{2^{m+1}-1}$ by breaking it up into subsums of size 2^k, for $0 \le k \le m$ and bounding each subsum. Specifically, we have

$$\begin{aligned} S_{2^{m+1}-1} = \sum_{j=1}^{2^{m+1}-1} \frac{1}{j^2} &= 1 + \left(\frac{1}{4} + \frac{1}{9}\right) + \left(\frac{1}{16} + \frac{1}{25} + \frac{1}{36} + \frac{1}{49}\right) + \cdots \\ &\cdots + \left(\frac{1}{(2^k)^2} + \cdots + \frac{1}{(2^{k+1}-1)^2}\right) + \cdots \\ &\cdots + \left(\frac{1}{(2^m)^2} + \cdots + \frac{1}{(2^{m+1}-1)^2}\right) \\ &= \sum_{k=0}^{m} \left(\sum_{j=2^k}^{2^{k+1}-1} \frac{1}{j^2} \right) \end{aligned}$$

In the kth subsum, the terms are all bounded by the first term, $\frac{1}{(2^k)^2}$ (because the terms are monotonically decreasing), and since there are $2^{k+1} - 1 - (2^k - 1) = 2^k$ terms in this sum, we have

$$\sum_{j=2^k}^{2^{k+1}-1} \frac{1}{j^2} < 2^k \cdot \frac{1}{(2^k)^2} = \frac{1}{2^k}$$

and therefore

$$S_n < S_{2^{m+1}-1} < \sum_{k=0}^{m} \frac{1}{2^k} < 1. \qquad \square$$

Comment. By Theorem 2.3.5(v), $\lim_{n\to\infty} n^{-2/n} = \lim_{n\to\infty} n^{-1/n} = 1$, and therefore

$$\limsup_{n\to\infty} \sqrt[n]{n^{-2}} = \limsup_{n\to\infty} \sqrt[n]{n^{-1}} = 1.$$

These examples show that the root test is inconclusive if $\limsup \sqrt[n]{|a_n|} = 1$, because $\sum \frac{1}{n}$ diverges while $\sum \frac{1}{n^2}$ converges.

The following theorem of Cauchy generalizes the result above and we prove it in much the same way.

Theorem. *If $a_n \ge 0$ for all $n \ge 0$ and the sequence $\{a_n\}$ is monotone nonincreasing, then the series $\sum_{n=1}^{\infty} a_n$ converges if and only if the series $\sum_{m=0}^{\infty} 2^m a_{2^m}$ converges.*

Proof. We set $S_n = \sum_{j=1}^{n} a_j$ and $\sigma_m = \sum_{k=0}^{m} 2^k a_{2^k}$. We will show that the sequence $\{S_n\}$ is bounded if and only if the sequence $\{\sigma_m\}$ is bounded, from which it follows that either both series converge or both series diverge, since both series have nonnegative terms.

We begin by choosing $n, m \in \mathbb{N}$ such that $n \le 2^m$, which implies that $S_n \le S_{2^{m+1}-1}$. Next we break up the partial sum $S_{2^{m+1}-1}$ into subsums and bound each subsum, as we did in the proof of Proposition 2.7.4. This gives

$$S_n \le S_{2^{m+1}-1} = \sum_{j=1}^{2^{m+1}-1} a_j = \sum_{k=0}^{m} \left(\sum_{j=2^k}^{2^{k+1}-1} a_j \right) \le \sum_{k=0}^{m} 2^k a_{2^k} = \sigma_m,$$

because the subsum $\sum_{j=2^k}^{2^{k+1}-1} a_j$ has 2^k terms, all of which are bounded by the first term, a_{2^k}. It follows that if the sequence $\{\sigma_m\}$ is bounded, then so is the sequence $\{S_n\}$.

On the other hand, if $m \in \mathbb{N}$ and $n \geq 2^m$, then breaking the partial sum S_{2^m} into subsums in a different way, we can show that $S_n \geq \frac{1}{2}\sigma_m$. Specifically, we have

$$
\begin{aligned}
S_n \geq S_{2^m} &= \sum_{j=1}^{2^m} a_j = a_1 + a_2 + (a_3 + a_4) + \cdots + (a_{2^{m-1}+1} + \cdots + a_{2^m}) \\
&= a_1 + \sum_{k=1}^{m} \left(\sum_{j=2^{k-1}+1}^{2^k} a_j \right) \geq a_1 + \sum_{k=1}^{m} 2^{k-1} a_{2^k} \\
&\geq \frac{1}{2}\left(a_1 + \sum_{k=1}^{m} 2^k a_{2^k} \right) = \frac{1}{2}\sigma_m,
\end{aligned}
$$

where the inequality in the second line above follows from the fact that the kth subsum, $\sum_{j=2^{k-1}+1}^{2^k} a_j$, contains 2^{k-1} terms, each of which is $\geq a_{2^k}$.

In other words, if $n \geq 2^m$, then $S_n \geq \frac{1}{2}\sigma_m$ and therefore if the sequence $\{S_n\}$ is bounded, then so is the sequence $\{\sigma_m\}$. □

As an immediate application, we have the following result that includes the proposition in Proposition 2.7.4 as a special case.

Corollary. *The series* $\sum_{n=1}^{\infty} \frac{1}{n^s}$ *converges if* $s > 1$ *and diverges if* $s \leq 1$.

Proof. Writing $a_n = 1/n^s$ we have

$$
2^n a_{2^n} = \frac{2^n}{(2^n)^s} = 2^{n-ns} = \left(2^{1-s}\right)^n,
$$

and therefore $\sqrt[n]{2^n a_{2^n}} = 2^{1-s}$ for all n, so $\limsup_{n\to\infty} \sqrt[n]{|a_n|} = 2^{1-s}$.

By Cauchy's theorem and the root test, it follows that the series converges if $s > 1$, since $2^{1-s} < 1$ in this case, and the series diverges if $s < 1$, since $2^{1-s} > 1$ in that case. If $s = 1$, the series is the harmonic series and therefore also diverges. □

2.7.5. Alternating series

If $a_n \geq 0$ for all $n \geq 0$, then the terms in the series $\sum_{n=0}^{\infty}(-1)^n a_n$ have *alternating* signs, and such a series is called an *alternating series*. The *Leibniz criterion* says that if the sequence $\{a_n\}$ is also decreasing monotonically to 0, then the alternating series above converges. This criterion is a corollary of the following, more general theorem of Dirichlet.

Theorem. *Suppose that* $a_n \geq a_{n+1} \geq 0$ *for all* $n \in \mathbb{N}$ *and* $\lim a_n = 0$. *If the partial sums of the series* $\sum e_n$ *are bounded, then the series* $\sum_{n=1}^{\infty} e_n a_n$ *converges.*

Proof. We set $E_n = \sum_{j=1}^{n} e_n$, so by assumption, there is an $E > 0$, such that $|E_n| < E$ for all $n \in \mathbb{N}$. For $n > m > 1$ we have the identity (called summation by parts)

$$\begin{aligned}\sum_{k=m}^{n} e_k a_k &= \sum_{k=m}^{n} (E_k - E_{k-1}) a_k \\ &= \sum_{k=m}^{n} E_k a_k - \sum_{k=m}^{n} E_{k-1} a_k \\ &= \sum_{k=m}^{n} E_k a_k - \sum_{k=m-1}^{n-1} E_k a_{k+1} \\ &= E_n a_n - E_{m-1} a_m + \sum_{k=m}^{n-1} E_k (a_k - a_{k+1}),\end{aligned}$$

from which it follows that

$$\left|\sum_{k=m}^{n} e_k a_k\right| \leq |E_n a_n| + |E_{m-1} a_m| + \sum_{k=m}^{n-1} |E_k (a_k - a_{k+1})|, \tag{2.7.2}$$

by the triangle inequality.

By assumption, we have $a_k - a_{k+1} \geq 0$ and $|E_k| < E$ for all k and therefore

$$\sum_{k=m}^{n-1} |E_k (a_k - a_{k+1})| \leq E \sum_{k=m}^{n-1} (a_k - a_{k+1}) = E(a_m - a_n) \leq E a_m.$$

Inserting this into (2.7.2), together with the bounds $|E_n a_n| \leq E a_m$ and $|E_{m-1} a_m| \leq E a_m$ shows that

$$\left|\sum_{k=m}^{n} e_k a_k\right| \leq 3E a_m.$$

Finally, given $\varepsilon > 0$ there is an n_0 such that $a_n < \varepsilon / 3E$ for all $n > n_0$, because $\lim a_n = 0$. Therefore if $n > m > n_0$, then

$$\left|\sum_{k=m}^{n} e_k a_k\right| \leq 3E a_m < \varepsilon,$$

so the series $\sum e_n a_n$ converges by the Cauchy criterion. □

Corollary. *If $a_n \geq a_{n+1} \geq 0$ for all $n \geq 0$ and $\lim a_n = 0$, then the alternating series $\sum_{n=0}^{\infty} (-1)^n a_n$ converges.*

Proof. Apply the previous theorem with $e_n = (-1)^n$ in which case, $|E_n| \leq 1$ for all n. □

Comment. The comparison, root and ratio tests all provide criteria for absolute convergence, whereas the Leibniz criterion and Theorem 2.7.5, more generally, only guarantee conditional convergence. For example, though the harmonic series diverges, it follows from the Leibniz criterion that the alternating harmonic series

$$\sum_{n=1}^{\infty} \frac{(-1)^{n-1}}{n} = 1 - \frac{1}{2} + \frac{1}{3} - \frac{1}{4} + \cdots$$

does converge, but conditionally.

Exercises

ex2.7.1. Prove Corollary 2.6.4.

ex2.7.2. Show that the series

$$\sum_{n=0}^{\infty} \left(\frac{n^2+2}{5n^2+1}\right)^{n/3}$$

converges.

ex2.7.3. Show that for any $a > 0$, there is an $N \in \mathbb{N}$ such that $n! > a^n$ for all $n > N$.

ex2.7.4. Show that

$$\limsup_{n\to\infty} \sqrt[n]{\frac{|u|^n}{n!}} = 0$$

for all $u \in \mathbb{R}$.

ex2.7.5. Suppose that the sequence $\{a_n\}$ is bounded and that $a_n > 0$ for all $n \geq 0$.

(i) Show that

$$\lim_{n\to\infty} \sqrt[n]{a_n} = \lim\left(\frac{a_n}{a_{n-1}} \cdot \frac{a_{n-1}}{a_{n-2}} \cdots \frac{a_1}{a_0}\right)^{1/n},$$

and that the same is true if "lim" is replaced by "lim inf" or "lim sup".

Hint: Theorem 2.3.5(v) may be useful here.

(ii) Use (i) to deduce the ratio test from the root test.

Hint: The expression $(b_1 \cdots b_m)^{1/m}$ is the *geometric mean* of the positive numbers $b_1, \ldots, b_m$. Show that it is bounded between the smallest and largest of these numbers.

Hint: Given $u \in \mathbb{R}$ and $\varepsilon > 0$ let $a = |u|/\varepsilon$ in **ex2.7.2.**

ex2.7.6. Prove Corollary 2.7.5 directly from the Cauchy criterion.

Hint: Show that $\left|\sum_{j=m}^{n}(-1)^n a_j\right| \leq a_m$.

ex2.7.7. Find series $\sum a_n$ and $\sum b_n$ that both converge conditionally, such that their Cauchy product does not converge.

Hint: Find a convergent alternating series $\sum a_n$ with $|c_n| = \left|\sum_{k=0}^{n} a_k a_{n-k}\right| 6\rightarrow 0$.

2.8. The decimal expansion of a real number

The familiar decimal representation of a real number x is the expression of x as a series of (negative) powers of 10, each multiplied by a "decimal digit". We use the shorthand $x = d_0 . d_1 d_2 d_3 \dots d_n \dots$ for the series

$$x = d_0(x) \pm \sum_{n=1}^{\infty} \frac{d_n(x)}{10^n},$$

where $d_0 \in \mathbb{Z}$ and $d_n(x) \in D = \{0, 1, 2, 3, 4, 5, 6, 7, 8, 9\}$ for each $n \in \mathbb{N}$. We choose the "−" sign above if $d_0 < 0$ and the "+" sign otherwise. We refer to this series as a *decimal expansion* or *decimal representation* of x.

Comment. If $0 \leq d_n(x) \leq 9$ for all $n \in \mathbb{N}$, then the series $\sum_{n=1}^{\infty} \frac{d_n(x)}{10^n}$ converges absolutely by comparison to the geometric series $\sum_{n=1}^{\infty} \frac{10}{10^n}$.

2.8.1. "Uniqueness" of the decimal expansion

Proposition. *For all $N \in \mathbb{N}$,*

$$\sum_{n=N+1}^{\infty} \frac{9}{10^n} = \frac{1}{10^N}.$$

Proof. Exercise. □

Theorem. *For every $x \in [0, 1)$, there is a unique set of integers $\{d_n(x)\}$ with $0 \leq d_n(x) \leq 9$ for all $n \in \mathbb{N}$ for which $\{n \in \mathbb{N} : d_n(x) < 9\}$ is infinite, such that*

$$x = \sum_{n=1}^{\infty} \frac{d_n(x)}{10^n}.$$

The condition that the set $\{n \in \mathbb{N} : d_n(x) < 9\}$ is infinite is in place to assure uniqueness.

Proof. We begin by proving existence. Given $0 \leq x < 1$, there is a unique $d \in D$ such that $\frac{d}{10} \leq x < \frac{d+1}{10}$. Denoting this integer by $d_1(x)$, we have

$$0 \leq x - \frac{d_1(x)}{10} < \frac{1}{10}.$$

Setting $x_1 = x - d_1(x)/10$, there is a unique $d \in D$ such that $\frac{d}{100} \leq x_1 < \frac{d+1}{100}$. Denoting this integer by $d_2(x)$, we have $0 \leq x_1 - \frac{d_2(x)}{100} < \frac{1}{100}$, and therefore

$$0 \leq x - \left(\frac{d_1(x)}{10} + \frac{d_2(x)}{100}\right) = x_1 - \frac{d_2(x)}{100} < \frac{1}{100}.$$

We continue to define the sequence $\{d_n(x)\}$ recursively. Suppose that $n \in \mathbb{N}$ and that we have found integers $d_1(x), \dots, d_n(x)$ satisfying $0 \leq d_j(x) \leq 9$ for $1 \leq j \leq n$ and

$$0 \leq x - \sum_{j=1}^{n} \frac{d_j(x)}{10^j} < \frac{1}{10^n}. \tag{2.8.1}$$

Writing $x_n = x - \sum_{j=1}^{n} \frac{d_j(x)}{10^j}$, we have $0 \le x_n < 1/10^n$ so there is a unique $d \in D$ such that $d/10^{n+1} \le x_n < (d+1)/10^{n+1}$. Denoting this d by $d_{n+1}(x)$, we have

$$x - \sum_{j=1}^{n+1} \frac{d_j(x)}{10^j} = x_n - \frac{d_{n+1}(x)}{10^{n+1}}$$

and therefore

$$0 \le x - \sum_{j=1}^{n+1} \frac{d_j(x)}{10^j} < \frac{1}{10^{n+1}}.$$

With the sequence $\{d_n(x)\}_{n\in\mathbb{N}}$ defined this way, it follows that the bounds in (2.8.1) are valid for all $n \in \mathbb{N}$ and therefore the series $\sum_{n=1}^{\infty} \frac{d_n(x)}{10^n}$ converges to x. I.e., if $x \in [0,1)$, then x has a decimal expansion.

If $x \in [0,1)$ has a decimal expansion for which the set $\{n \in \mathbb{N} : d_n(x) < 9\}$ is finite, let $n_0 > 0$ be the greatest integer for which $d_{n_0} < 9$, then it follows from Proposition 2.8.1 that

$$x = \sum_{n=1}^{n_0} \frac{d_n(x)}{10^n} + \sum_{n=n_0+1}^{\infty} \frac{9}{10^n} = \sum_{n=1}^{n_0} \frac{d_n}{10^n} + \frac{1}{10^{n_0}} = \sum_{n=1}^{n_0-1} \frac{d_n(x)}{10^n} + \frac{d_{n_0}(x)+1}{10^{n_0}}.$$

Hence in this case, x has a second expansion for which $d_n(x) = 0$ for all $n > n_0$, for which the set $\{n \in \mathbb{N} : d_n(x) < 9\}$ is infinite.

It remains to show that if $x \in [0,1)$, then x has only one decimal representation for which the set $\{n \in \mathbb{N} : d_n(x) < 9\}$ is infinite. We do this by showing that given two distinct sets of decimal digits satisfying this condition, then the corresponding decimal expansions represent different numbers.

Suppose that $\{d_n\}_{n=1}^{\infty}$ and $\{c_n\}_{n=1}^{\infty}$ are two *different* sequences of integers satisfying $c_n, d_n \in D$ for all n, for which the sets $\{n \in \mathbb{N} : d_n < 9\}$ and $\{n \in \mathbb{N} : c_n < 9\}$ are both infinite, and let $n_0 \ge 1$ be the smallest integer such that $c_{n_0} \ne d_{n_0}$. Without loss of generality, we may assume that $d_{n_0} < c_{n_0}$. Writing $a = \sum_{n=1}^{\infty} \frac{c_n}{10^n}$ and $b = \sum_{n=1}^{\infty} \frac{d_n}{10^n}$, then by Theorem 2.6.4

$$a - b = \sum_{n=1}^{\infty} \frac{c_n - d_n}{10^n} = \sum_{n=n_0}^{\infty} \frac{c_n - d_n}{10^n} \ge \frac{1}{10^{n_0}} + \sum_{n=n_0+1}^{\infty} \frac{c_n - d_n}{10^n}. \tag{2.8.2}$$

From the assumption that the sets $\{n \in \mathbb{N} : d_n < 9\}$ and $\{n \in \mathbb{N} : c_n < 9\}$ are both infinite, it follows that the set $\{n > n_0 : |c_n - d_n| < 9\}$ is infinite, and hence

$$\sum_{n=n_0+1}^{\infty} \frac{|c_n - d_n|}{10^n} < \sum_{n=n_0+1}^{\infty} \frac{9}{10^n} = \frac{1}{10^{n_0}},$$

by Proposition 2.8.1 (again). It follows from this and (2.8.2) that

$$a - b \ge \frac{1}{10^{n_0}} + \sum_{n=n_0+1}^{\infty} \frac{c_n - d_n}{10^n} \ge \frac{1}{10^{n_0}} - \sum_{n=n_0+1}^{\infty} \frac{|c_n - d_n|}{10^n} > 0. \qquad \square$$

Definition. The *integer part* of $x \in \mathbb{R}$, denoted $[x]$, is the greatest $m \in \mathbb{Z}$ such that $m \le x$.

Corollary. *For every $x \in \mathbb{R}$, there is a unique set of integers $\{d_n(x)\}_{n=0}^{\infty}$ with $d_n \in \mathbb{Z}$ for $n \geq 0$, $0 \leq d_n(x) \leq 9$ for $n \geq 1$ and for which $\{n \in \mathbb{N} : d_n(x) < 9\}$ is infinite, such that*

$$x = d_0 \pm \sum_{n=1}^{\infty} \frac{d_n(x)}{10^n},$$

where we choose the $-$ sign if $x < 0$ and the $+$ sign otherwise.

Proof. We have proved this for $x \in [0, 1)$ (with $d_0 = 0$). If $x \geq 1$, then $x - [x] \in [0, 1)$, so the claim is true in this case (with $d_0 = [x]$). Finally, the claim follows for $x < 0$, because it is true for $|x|$, and $x = -|x|$. □

Exercises

ex2.8.1. Prove Proposition 2.8.1.

ex2.8.2. Show that $x \in [0, 1)$ has a finite decimal expansion (one for which $\{n \in \mathbb{N} : d_n(x) > 0\}$ is finite) if and only if $x = l/k \in \mathbb{Q}$ where k divides 10^n for some $n \in \mathbb{N}$.

ex2.8.3. Show that for $x \in [0, 1)$, the construction in the proof of Theorem 2.8.1 yields a set of decimal digits for which the set $\{n \in \mathbb{N} : d_n(x) < 9\}$ is infinite.

*2.9. Binary and ternary representations

There is nothing particularly special about the number 10, other than the fact that most humans have 10 fingers (digits). Everything that we did in the last section to develop the decimal expansion (base 10), can be done for any natural number $m > 1$. In this section we'll briefly describe the binary (base 2) and ternary (base 3) expansions of real numbers in the interval $[0, 1)$. The descriptions are sparse and are only meant to help with the exercises that follow.

2.9.1. Binary expansions

The *binary expansion* of a number $x \in [0, 1)$ is a series of the form

$$x = \sum_{n=1}^{\infty} \frac{b_n(x)}{2^n},$$

where $b_n(x) = 0$ or 1 for all n. The numbers $\{b_n(x)\}$ are the "binary digits" of x. In the same way that we proved Theorem 2.8.1, it can be shown that every $x \in [0, 1)$ has a binary expansion and that this expansion is unique if we stipulate that the set $\{n \in \mathbb{N} : b_n(x) = 0\}$ is infinite. This condition is necessary if we want to claim uniqueness because if $b_n(x) = 0$ and $b_k(x) = 1$ for all $k > n$, then

$$x = \sum_{k=1}^{n-1} \frac{b_k(x)}{2^k} + \sum_{k=n+1}^{\infty} \frac{1}{2^k} = \sum_{k=1}^{n-1} \frac{b_k(x)}{2^k} + \frac{1}{2^n}.$$

2.9.2. Ternary expansions

The *ternary expansion* of a number $x \in [0, 1)$ is a series of the form

$$x = \sum_{n=1}^{\infty} \frac{t_n(x)}{3^n},$$

where in this case $t_n(x) \in \{0, 1, 2\}$ for all n. The set $\{t_n(x)\}$ are the "ternary digits" of x, and it can be shown that every $x \in [0, 1)$ has a unique ternary expansion if we add the condition that the set $\{n \in \mathbb{N} : t_n(x) < 2\}$ is infinite. This condition on the ternary digits is necessary for uniqueness because

$$\sum_{k=n+1}^{\infty} \frac{2}{3^k} = \frac{1}{3^n},$$

for all $n \geq 1$. Thus if $r = m/3^n$ with $0 < m < 3^n$ has two ternary expansions:

$$r = \sum_{k=1}^{n} \frac{t_k(r)}{3^k} = \sum_{k=1}^{n-1} \frac{t_k(r)}{3^k} + \frac{t_n(r) - 1}{3^n} + \sum_{k=n+1}^{\infty} \frac{2}{3^k}.$$

Exercises

ex2.9.1. State and prove an analog of Theorem 2.8.1 for binary representations.

ex2.9.2. Show that $x \in [0, 1)$ has a finite binary expansion if and only if $x = m/2^n$, for some integers $m, n \geq 0$, with $m < 2^n$.

ex2.9.3. State and prove an analog of Theorem 2.8.1 for ternary representations.

ex2.9.4. Show that $x \in [0, 1)$ has a finite ternary expansion if and only if $x = m/2^n$, for some integers $m, n \geq 0$, with $m < 3^n$.

ex2.9.5. State and prove an analog of Theorem 2.8.1 for n-ary representations, where $n \in \mathbb{N}$, $n > 1$, and where the n-ary representation of $x \in [0, 1)$ is a series of the form

$$\sum_{k=1}^{\infty} \frac{l_k(x)}{n^k}$$

with n-ary digits $l_k(x) \in \{0, 1, 2, \ldots, n-1\}$.

ex2.9.6. Show that $x \in [0, 1)$ has a finite n-ary expansion if and only if $x = a/b \in \mathbb{Q}$, where b divides n^k for some $k \in \mathbb{N}$.

⋆2.10. Rearrangements

In a finite sum, the terms may be rearranged in any way we want without affecting the value of the sum, as follows from the commutative and associative properties of addition. The rearrangement of a finite sum $a_1 + a_2 + \cdots + a_n$ can be expressed in terms of a *permutation* of the set of indices $\{1, 2, \ldots, n\}$. A permutation is a bijection $\sigma : \{1, 2, \ldots, n\} \to \{1, 2, \ldots, n\}$, and we have

$$a_1 + a_2 + \cdots + a_n = a_{\sigma(1)} + a_{\sigma(2)} + \cdots + a_{\sigma(n)},$$

for any such permutation.

A rearrangement of an infinite series $\sum a_n$ involves a permutation of the natural numbers, i.e., a bijection $\sigma : \mathbb{N} \to \mathbb{N}$. The sum of a convergent infinite series is defined as the limit of its partial sums and rearranging the terms in the series changes the sequence of its partial sums. Hence it is reasonable to ask when or whether it is true that

$$\sum_{n=1}^{\infty} a_n = \sum_{n=1}^{\infty} a_{\sigma(n)}.$$

Definition. A series $\sum_{n=1}^{\infty} a_n$ converges *unconditionally*, or is *unconditionally convergent* if the series converges and every rearrangement of the series converges to the same sum.

2.10.1. Absolute convergence implies unconditional convergence

The following theorem was originally proven by Dirichlet in 1837.

Theorem. *If $\sum_{n=1}^{\infty} a_n$ converges absolutely then*

$$\sum_{n=1}^{\infty} a_{\sigma(n)} = \sum_{n=1}^{\infty} a_n,$$

for any permutation $\sigma : \mathbb{N} \to \mathbb{N}$.

Proof. Suppose that $\sum_{n=1}^{\infty} a_n = a$ and that the convergence is absolute. Let σ be a permutation of $\mathbb{N}$ and let $\varepsilon > 0$. It follows from the absolute convergence of the series and the Cauchy criterion that there is an $N_1 \in \mathbb{N}$ such that $\sum_{n=N}^{M} |a_n| < \varepsilon/2$, for all $M > N > N_1$. Likewise, since $\sum a_n = a$, there exists an $N_2 \in \mathbb{N}$ such that $\left|a - \sum_{n=1}^{N} a_n\right| < \varepsilon/2$ for all $N \geq N_2$. Set $N_\varepsilon = \max(N_1, N_2)$.

If $M > N_\varepsilon$ and A is any subset of $\mathbb{N}$ satisfying $\{1, 2, \ldots, N_\varepsilon\} \subset A \subset \{1, 2, \ldots, M\}$, then setting $A^+ = \{n \in A : n > N_\varepsilon\}$, we have

$$\left|a - \sum_{n \in A} a_n\right| = \left|a - \sum_{n=1}^{N_\varepsilon} a_n - \sum_{n \in A^+} a_n\right| \leq \left|a - \sum_{n=1}^{N_\varepsilon} a_n\right| + \left|\sum_{n \in A^+} a_n\right|$$

$$\leq \left|a - \sum_{n=1}^{N_\varepsilon} a_n\right| + \sum_{n \in A^+} |a_n| \leq \left|a - \sum_{n=1}^{N_\varepsilon} a_n\right| + \sum_{N=N_\varepsilon+1}^{M} |a_n|,$$

and therefore for such $A \subset \mathbb{N}$,

$$\left|a - \sum_{n \in A} a_n\right| < \varepsilon/2 + \varepsilon/2 = \varepsilon. \tag{2.10.1}$$

Let $B_\varepsilon = \sigma^{-1}(\{1,2,\ldots,N_\varepsilon\}) = \{n \in \mathbb{N} : 1 \le \sigma(n) \le N_\varepsilon\}$ and set $L_\varepsilon = \max B_\varepsilon$. If $L > L_\varepsilon$, then B_ε is a proper subset of $\{1,\ldots,L\}$ so $\{1,\ldots,N_\varepsilon\} = \sigma(B_\varepsilon) \subset \sigma(\{1,\ldots,L\} = A_L$, and $M_L = \max A_L > N_\varepsilon$. It follows that $\{1,2,\ldots,N_\varepsilon\} \subset A_L \subset \{1,2,\ldots,M_L\}$ and therefore

$$\left|a - \sum_{n=1}^{L} a_{\sigma(n)}\right| = \left|a - \sum_{n \in A_L} a_n\right| < \varepsilon,$$

as follows from (2.10.1). Hence $\sum a_\sigma(n) = a$, since ε was arbitrary. □

2.10.2. Riemann's rearrangement theorem

The following theorem of Riemann shows that nonabsolutely convergent series are as far from unconditionally convergent as possible.

Theorem. *If $\sum_{n=1} a_n$ converges but not absolutely, then for any $u \in \mathbb{R}$, there is a permutation σ of $\mathbb{N}$ such that*

$$\sum_{n=1}^{\infty} a_{\sigma(n)} = u.$$

Comment. It follows from this and Theorem 2.10.1 that absolute convergence and unconditional convergence are equivalent in $\mathbb{R}$.[10]

Proof. We will sketch the main ideas and leave the technical details to the motivated reader (and to the exercises). Also, to keep things simple, we will assume that $a_n \neq 0$ for all $n \in \mathbb{N}$ and that $u > 0$.

The observations that drive the proof are the following. If $\sum_{n=1}^{\infty} a_n$ converges, but $\sum_{n=1}^{\infty} |a_n|$ diverges then

(i) The sets $\{n \in \mathbb{N} : a_n > 0\}$ and $\{n \in \mathbb{N} : a_n < 0\}$ are both infinite.

(ii) If we list the positive terms in the sequence $\{a_n\}$ as the subsequence $I^+ = \{b_{n_j}\}_{j=1}^{\infty}$ and the negative terms as the subsequence $I^- = \{c_{n_k}\}_{k=1}^{\infty}$, then

$$\sum_{j=1}^{\infty} b_{n_j} = \infty \quad \text{and} \quad \sum_{k=1}^{\infty} c_{n_k} = -\infty.$$

(iii) Finally, $\lim_{n\to\infty} a_n = 0$ (by Corollary 2.6.2).

To obtain a rearrangement of $\sum a_n$ that converges to u, we choose elements successively from either I^+ or I^-, in the order in which they appear, as follows.

We begin by choosing elements $b_{n_1}, \ldots, b_{n_{j_1}}$ from I^+ such that their sum is greater than u, i.e., $T_1 = \sum_{j=1}^{j_1} b_{n_j} > u$, and where j_1 is the first index for which this holds. Next, we choose elements $c_{n_1}, \ldots, c_{n_{k_1}}$ from I^- so that $T_2 = T_1 + \sum_{k=1}^{k_1} c_{n_k} < u$, where k_1 is the first index for which this happens.

Once again, we choose elements $b_{n_{j_1+1}}, \ldots, b_{n_{j_2}}$ from I^+ so that $T_3 = T_2 + \sum_{j=j_1+1}^{j_2} b_{n_j} > u$, where as before, j_2 is the first index for which this happens. Then we choose elements $c_{n_{k_1+1}}, \ldots, c_{n_{k_2}}$ from I^- so that $T_4 = T_3 + \sum_{k=k_1+1}^{k_2} c_{n_k} < u$, (where k_2 is the

[10] This is not the case in all metric spaces.

first index for which this happens). From observation (ii), it follows that this process can be continued indefinitely, resulting in a (sub)sequence of partial sums $\{T_m\}$. These partial sums are characterized by the following two features.

(iv) $T_{2l} < u < T_{2l-1}$ for all $l \in \mathbb{N}$.

(v) For $l \in \mathbb{N}$, $|T_{2l-1} - u| \le b_{n_l}$ and $|T_{2l} - u| \le |c_{n_l}|$.

Denoting by σ_u the permutation of $\mathbb{N}$ that arises from this rearrangement, and setting

$$S_N = \sum_{n=1}^{N} a_{\sigma_u(n)},$$

then for every $N \in \mathbb{N}$, there is an $l \in \mathbb{N}$ such that $T_{2l} \le S_N \le T_{2l-1}$.

Finally, since $a_n \to 0$, it follows that for any $\varepsilon > 0$, there is an $L \in \mathbb{N}$ such that $|c_{n_l}| < \varepsilon$ and $|b_{n_l}| < \varepsilon$ for all $l > L$, and therefore $S_N \to u$. □

The following exercises all refer to the proof of Riemann's rearrangement theorem.

Exercises

ex2.10.1. Prove the claims (i) and (ii).

ex2.10.2. Justify/explain the statement "*Because of observation* (*ii*), *above, this process can be continued indefinitely.*"

ex2.10.3. Describe the permutation σ_u as explicitly as you can.

Hint: For $1 \le n \le j_1$, $\sigma(n) = j_n$; for $j_1 < n < j_1 + k_1$, $\sigma(n) = k_{n-j_1}$; for $j_1 + k_1 < n \le j_1 + j_2 + k_1$, $\sigma(n) = j_{n-(j_1+k_1)}$ ..., etc. The challenge is to give a clear, well-notated recursive definition for all $n \in \mathbb{N}$.

ex2.10.4. Justify the statement "... *for every* $N \in \mathbb{N}$, *there is an* $m \in \mathbb{N}$ *such that* $T_{2m} \le S_N \le T_{2m-1}$."

ex2.10.5. Prove "features" (iv) and (v).

ex2.10.6. Justify the statement "... *for every* $N \in \mathbb{N}$, *there is an* $l \in \mathbb{N}$ *such that* $T_{2l} \le S_N \le T_{2l-1}$."

ex2.10.7. Justify the statement "... *since* $a_n \to 0$, *it follows that for any* $\varepsilon > 0$, *there is an* $L \in \mathbb{N}$ *such that* $|c_{n_l}| < \varepsilon$ *and* $|b_{n_l}| < \varepsilon$ *for all* $l > L$, *and therefore* $S_N \to u$."

ex2.10.8. Construct a rearrangement of $\sum a_n$ that diverges to ∞.

ex2.10.9. Given $M > 0$, Construct a rearrangement of $\sum a_n$ such that the sets $\{N \in \mathbb{N} : \sum_{n=1}^{N} a_n > M\}$ and $\{N \in \mathbb{N} : \sum_{n=1}^{N} a_n < -M\}$ are both infinite.

Chapter 3

The Topology of the Real Line

A *topology* on a set X is a family of distinguished subsets of X called *open sets*. A space with a topology is called a *topological space*. A topology is in a certain sense a generalization of a metric, in that it allows us to carry over many of the ideas and methods metric spaces to topological spaces, without having a distance function.

In a metric space (X, ρ), we usually define a topology in terms of the metric ρ, called the *metric topology*. Everything that can be said or proven in terms of the metric topology can be proven directly in terms of ρ without reference to the topology, but the topological language can often make the discussion more transparent.

We focus primarily on $\mathbb{R}$ and its standard metric topology in an effort to keep the discussion as concrete as possible for students who may need more time to feel comfortable with a higher level of abstraction. On the other hand, much of what we discuss in this chapter is true in general metric spaces, and we state (and prove) such results in general when appropriate.

3.1. Closed sets and open sets in $\mathbb{R}$

To define the topology on $\mathbb{R}$, we need to describe which sets are open. Equivalently, we can define which sets are *closed*. A set F in a topological space X is closed if it is the complement of an open set $O \subset X$. Both concepts are defined in terms of open balls, which in $\mathbb{R}$ are open intervals.

3.1.1. Limit points and closed sets

A set F is closed if any point that "touches" F is contained in F. By "touches", we mean that the point is a limit point of F.

Definition. A point $x \in \mathbb{R}$ is a *limit point* of a set $E \subset \mathbb{R}$ if for all $\varepsilon > 0$, the set $(x-\varepsilon, x+\varepsilon) \cap E$ is infinite. A set $E \subset \mathbb{R}$ is *closed* if and only if E contains all of its limit points.

The set of limit points of a set $E \subset \mathbb{R}$ is denoted by E'.

More generally, if X is a metric space, then x is a limit point of the set $E \subset X$ if for every $\varepsilon > 0$, the set $B(x, \varepsilon) \cap E$ is infinite. A set $E \subset X$ is closed if and only if it contains all of its limit points.

We observe that the condition "for all $\varepsilon > 0$ the set $B(x, \varepsilon) \cap E$ is infinite" is equivalent to the seemingly weaker condition "for all $\varepsilon > 0$, the set $B(x, \varepsilon) \cap E$ contains at least one point other than x." See **ex3.1.2**.

Proposition. *The (closed) interval $[a, b]$ is closed for every $a, b \in \mathbb{R}$, with $a < b$.*

Proof. Suppose that $a < b$. We will show that if $x \notin [a, b]$, then x is not a limit point of $[a, b]$. Hence, $[a, b]$ must contain all of its limit points.

If $x \notin [a, b]$, then either $x < a$ or $b < x$. In the first case, setting $\varepsilon = (a-x)/2 > 0$, we have $(x-\varepsilon, x+\varepsilon) \cap [a, b] = \emptyset$, and in the second case, setting $\varepsilon = (x-b)/2 > 0$, we have $(x-\varepsilon, x+\varepsilon) \cap [a, b] = \emptyset$. Thus, if $x \notin [a, b]$, there is an $\varepsilon > 0$ such that $(x-\varepsilon, x+\varepsilon) \cap [a, b] = \emptyset$, so x cannot be a limit point of $[a, b]$. □

Comment. The limits of convergent subsequences of a sequence $\{x_n\}_{n=1}^{\infty} \subset \mathbb{R}$ are called the "limit points of the sequence". It is important to distinguish between the limit points of a sequence and the limit points of the *range* of the sequence.

A sequence $\{x_n\}_{n=1}^{\infty} \subset \mathbb{R}$ represents a function $f : \mathbb{N} \to \mathbb{R}$, where $x_n = f(n)$. The range E of a sequence is the same as the range of the function that it represents:

$$E = f(\mathbb{N}) = \{x \in \mathbb{R} : x = f(n) \text{ for some } n \in \mathbb{N}\}.$$

The point a is a limit point of the sequence $\{x_n\}$ (i.e., subsequential limit) if the set $I_\varepsilon = \{n \in \mathbb{N} : |x_n - a| < \varepsilon\}$ is infinite for every $\varepsilon > 0$. On the other hand, the point a is a limit point of the range E of $\{x_n\}$ if the set $E_\varepsilon = \{x \in E : |x-a| < \varepsilon\}$ is infinite for every $\varepsilon > 0$.

Theorem. *If $F \subset \mathbb{R}$ is closed, $x_n \in F$ for all $n \in \mathbb{N}$ and $\lim x_n = x^*$, then $x^* \in F$.*

Proof. If $x^* = x_n$ for some $n \in \mathbb{N}$, then $x^* \in F$ and we are done. Otherwise $x^* \neq x_n$ for all $n \in \mathbb{N}$, and we show that in this case, x^* is a limit point of F and hence contained in F.

Let $\varepsilon > 0$, then since $x_n \to x^*$, there is an $N_\varepsilon \in \mathbb{N}$ such that $x_n \in (x^*-\varepsilon, x^*+\varepsilon)$ for all $n \geq N_\varepsilon$. If $x_n = x_m$ for all $n, m \geq N_\varepsilon$, then $x^* = x_N$ contradicting our assumption. Hence there are $n > m \geq N_\varepsilon$ such that $x_n \neq x_m$, and therefore the interval $(x^*-\varepsilon, x^*+\varepsilon) \cap F$ contains at least two points, from which it follows that x^* is a limit point of F. □

3.1.2. Open sets

Open sets may be defined as the complements of closed sets, which they are, but they are naturally defined as follows.

Definition. A set $E \subset \mathbb{R}$ is *open* if for every $x \in E$ there is a positive ε such that the interval $(x-\varepsilon, x+\varepsilon)$ is contained in E.

More generally, if (X, ρ) is a metric space, then $A \subset X$ is open if for any $x \in A$, there is an $\varepsilon > 0$ such that $B(x, \varepsilon) \subset A$.

We observe that in any metric space X, the empty set satisfies the condition of being open, but "vacuously", because the condition "if $x \in E$, then..." is satisfied for $E = \varnothing$ since there is no $x \in \varnothing$.

Proposition. *If (X, ρ) is a metric space then for every $x \in X$ and $\varepsilon > 0$, the sets $B(x, \varepsilon)$ and $R(x, \varepsilon) = \{y \in X : \rho(x, y) > \varepsilon\}$ are both open.*

Proof. Let $x \in X$ and $\varepsilon > 0$. If $y \in B(x, \varepsilon)$, then $\rho(y, x) < \varepsilon$, and $\delta_y = \varepsilon - \rho(y, x) > 0$. If $z \in B(y, \delta_y)$, then

$$\rho(z, x) \leq \rho(z, y) + \rho(y, x) < \delta + \rho(y, x) = \varepsilon,$$

so $z \in B(x, \varepsilon)$ and hence $B(y, \delta_y) \subset B(x, \varepsilon)$.

If $y \in R(x, \varepsilon)$, then $\eta_y = \rho(x, y) - \varepsilon > 0$. If $z \in B(y, \eta_y)$, then

$$\rho(x, y) \leq \rho(x, z) + \rho(z, y) < \rho(x, z) + \eta_y = \rho(x, z) + \rho(x, y) - \varepsilon,$$

from which it follows that $\varepsilon < \rho(x, z)$. Hence if $y \in R(x, \varepsilon)$, then $B(y, \eta_y) \subset R(x, \varepsilon)$, so $R(x, \varepsilon)$ is open. □

Theorem. *The set E in the metric space (X, ρ) is open if and only if E^c is closed.*

Proof. Suppose that E^c is closed and $x \in E$. Then x is not a limit point of E^c and it follows that there exists a positive ε such that the set $A = B(x, \varepsilon) \cap E^c$ is *finite.* This means that $\{\rho(x, y) : y \in A\}$ is a finite set of (strictly) positive numbers, so $\min\{\rho(x, y) : y \in A\} = \varepsilon_A > 0$. Hence, $B(x, \varepsilon_A) \cap E^c = \varnothing$ so $B(x, \varepsilon_A) \subset E$ and E is open.

Conversely, if E is an open set and $x \in E$, then there is a positive ε such that $B(x, \varepsilon) \subset E$. It follows that $B(x, \varepsilon) \cap E^c$ is empty, so x is not a limit point of E^c. Hence, E^c contains all of its limit points and is therefore closed. □

Corollary. *If X is a metric space, then for any $x \in X$ and $\varepsilon > 0$, the set $\overline{B}(x, \varepsilon) = \{y \in X : \rho(x, y) \leq \varepsilon\}$ is closed.*

Proof. We have $\overline{B}(x, \varepsilon) = (R(x, \varepsilon))^c$ and by Proposition 3.1.2, the set $R(x, \varepsilon)$ is open. □

Comment. We call $\overline{B}(x, \varepsilon)$ the *closed* ball of radius ε around x. Note that in $\mathbb{R}$, $[a, b] = \overline{B}(c, \varepsilon)$, where $c = (a+b)/2$ and $\varepsilon = (b-a)/2$, so that the corollary above includes Proposition 3.1.1 as a special case.

3.1.3. Unions and intersections of open sets and closed sets

Certain combinations of open sets are themselves open, and the corresponding combinations of closed sets are closed.

Theorem. *The union of any family of open sets is open and the intersection of any finite collection of open sets is open.*

Proof. Suppose that E_a is an open set for every $a \in A$. If $x \in \bigcup_{a\in A} E_a$, then there is an $a_1 \in A$ such that $x \in E_{a_1}$. Since E_{a_1} is open by assumption, there is a positive ε such that $B(x,\varepsilon) \subset E_{a_1}$, and hence $B(x,\varepsilon) \subset \bigcup_{a\in A} E_a$. Since x was arbitrary, it follows that $\bigcup_{a\in A} E_a$ is open.

Now suppose that E_j is open for $1 \le j \le n$ and let $x \in \bigcap_{j=1}^{n} E_j$, so that $x \in E_j$ for $1 \le j \le n$. Since E_j is open for each j, there is an $\varepsilon_j > 0$ such that $B(x,\varepsilon_j) \subset E_j$. If $\varepsilon = \min(\varepsilon_1, \varepsilon_2, \ldots, \varepsilon_n)$, then $\varepsilon > 0$ and $B(x,\varepsilon) \subset B(x,\varepsilon_j) \subset E_j$ for each $1 \le j \le n$. Hence $B(x,\varepsilon) \subset \bigcap_{j=1}^{n} E_j$, and since x was arbitrary, it follows that $\bigcap_{j=1}^{n} E_j$ is open. □

Corollary. *The intersection of any family of closed sets is closed and the union of any finite collection of closed sets is closed.*

Proof. Exercise □

We observe that there is no mention of $\mathbb{R}$ in the statement or proof of Theorem 3.1.3. This theorem and its corollary are true in any metric space with the metric topology, and are in fact characteristic of topological spaces in general.

3.1.4. Closures and interiors

Given a nonempty subset $E \subset \mathbb{R}$, we might think of "closing it" by adding all of its limits points. Likewise, by removing points from E that are limit points of its complement in $\mathbb{R}$, we might construct the 'open part' of E — the largest open set contained in E. These intuitions are justified after the following definitions.

Definition. Suppose that $E \subset \mathbb{R}$ is nonempty. Let $\mathcal{F}$ be the collection of all closed subsets of $\mathbb{R}$ that contain E and let $\mathcal{O}$ be the collection of all open subsets of $\mathbb{R}$ that are contained in E. The *closure* of E is the set

$$\overline{E} = \bigcap_{F\in\mathcal{F}} F$$

and the *interior* of E is the set

$$E^o = \bigcup_{O\in\mathcal{O}} O.$$

We observe that $E \subset \mathbb{R}$ may contain no open subsets, so the interior of E may be empty.[1] On the other hand, every subset of $\mathbb{R}$ is contained in the closed set $\mathbb{R}$, so the closure of a nonempty set is never empty.

Theorem. *For any $E \subset \mathbb{R}$,*

(i) *$\overline{E}$ is closed, $E \subset \overline{E}$ and if F is closed and contains E, then $\overline{E} \subset F$.*

(ii) *E^o is open, $E^o \subset E$, and if O is open and contained in E, then $O \subset E^o$.*

I.e., the closure of E is the smallest closed set containing E and the interior of E is the largest open set contained in E.

Proof. Let $E \subset \mathbb{R}$ be nonempty, and let $\mathcal{F}$ and $\mathcal{O}$ be the collections of sets described in the definitions of closure and interior, above.

[1] For example, the interior of $\mathbb{Q}$ is empty.

(i) $\overline{E}$ is closed by Corollary 3.1.3 as the intersection of a family of closed sets. If $x \in E$, then $x \in F$ for all $F \in \mathcal{F}$ and therefore $x \in \bigcap_{F\in\mathcal{F}} F = \overline{E}$, so $E \subset \overline{E}$. If F_0 is closed and $E \subset F_0$, then $F \in \mathcal{F}$, so $\overline{E} = \bigcap_{F\in\mathcal{F}} F \subset F_0$.

(ii) E^o is open by Theorem 3.1.3 as the union of a family of open sets. If $x \in E^o = \bigcup_{O\in\mathcal{O}} O$, then $x \in O \subset E$ for some $O \in \mathcal{O}$, hence $E^o \subset E$. If O is an open set contained in E, then $O \in \mathcal{O}$, so $O \subset \bigcup_{O\in\mathcal{O}} O = E^o$. □

Proposition. *Let $E \subset \mathbb{R}$. Then $\overline{E} = E \cup E'$ and $E^o = E \setminus (E^c)'$.*

Recall that E' denotes the set of limit points of E. Also, for any two sets A and B, $A \setminus B$ is the set $A \cap B^c$, of points in A that are not in B.

Proof. First, if x is a limit point of E, then x is a limit point of $\overline{E}$, so $x \in \overline{E}$, because $\overline{E}$ is closed. It follows that $F = E \cup E' \subset \overline{E}$. To show that $F = \overline{E}$, it is enough to show that F is closed, because $E \subset F$.

To that end, Let x be a limit point of F. Then for any $\varepsilon > 0$, the set $B(x, \varepsilon/2) \cap F$ contains infinitely many points, and so contains a point $y \neq x$. If $y \in E$, we set $x_\varepsilon = y$. If $y \notin E$, then y is a limit point of E, and the set $B(y, \varepsilon/2) \cap E$ contains infinitely many points, and therefore contains a point $z \neq x$. In this case, we set $x_\varepsilon = z$ and observe that

$$|x - x_\varepsilon| \leq |x - y| + |y - \varepsilon| < \varepsilon/2 + \varepsilon/2 = \varepsilon.$$

In either case, for any $\varepsilon > 0$, the set $B(x, \varepsilon) \cap E$ contains a point $x_\varepsilon \neq x$, and it follows that x is a limit point of E, by the observation following Definition 3.1.1.

Second, if $x \in E$ is not a limit point of E^c, then there is a $\varepsilon > 0$ such that $B(x, \varepsilon) \cap E^c$ is empty, so $B(x, \varepsilon) \subset E$. It follows that $E \setminus (E^c)'$ is open. Moreover, if $O \subset E$ is open and $x \in O$, then $x \notin (E^c)'$, so $x \in E \setminus (E^c)'$, and it follows that $E \setminus (E^c)'$ contains every open subset of E. □

3.1.5. Topology, more generally

At the beginning of the chapter we wrote that a topology on a set X is a collection of special subsets of X called open sets. However not every collection of subsets can be a topology. Certain conditions have to be met.

Definition. A topology on a set X is a family $\mathcal{O}$ of subsets of X, called open sets, with the following properties.

top1. $\varnothing \in \mathcal{O}$ and $X \in \mathcal{O}$.

top2. If $E_a \in \mathcal{O}$ for all $a \in A$, then $\bigcup_{a\in A} E_a \in \mathcal{O}$.

top3. If $E_j \in \mathcal{O}$ for $1 \leq j \leq n$, then $\bigcap_{j=1}^n E_j \in \mathcal{O}$.

Properties **top**2. and **top**3. are often summarized by saying "$\mathcal{O}$ is closed under arbitrary unions and finite intersections".

A set $F \subset X$ is *closed* if its complement in X is open, and we note that the set X itself is both open and closed, as is the empty set.

A general topological space X with topology $\mathcal{O} \subseteq \mathcal{P}(X)$ is denoted by $(X, \mathcal{O})$.

Theorem 3.1.3 shows that the collection of sets in $\mathbb{R}$ that we defined to be open in Definition 3.1.2 satisfies the definition of a topology. I.e., the metric topology is a topology on $\mathbb{R}$ and the same is true for the metric topology of any metric space.

There are always two topologies that can be defined on any nonempty set X. The first is the trivial topology that includes only the sets X and $\varnothing$ as open sets. At the other extreme is the *discrete topology* that includes all subsets of X as open sets. In the discrete topology for example, each singleton $\{x\} \subset X$ is open.

3.1.6. The relative topology

Nonempty subsets of topological spaces are naturally endowed with a topologies that they "inherit" from the larger space.

Theorem. *If $(X, \mathcal{O})$ is a topological space and Y is a nonempty subset of X, then the family of sets $\mathcal{O}_Y = \{U \cap Y : U \in \mathcal{O}\}$ is a topology on Y.*

The topology $\mathcal{O}_Y$ is called the *relative topology* on $Y \subset X$.

Proof. We need to show that $\mathcal{O}_Y$ satisfies properties **top**1.–**top**3., and we leave this as an exercise. □

If $Y \subset X$ is open and $U \in \mathcal{O}$, then $U \cap Y \in \mathcal{O}$. In this case $\mathcal{O}_Y = \{U \in \mathcal{O} : U \subset Y\}$ is a subset of $\mathcal{O}$.

If Y is not an open subset of X and U is open, then $Y \cap U$ need not be an open set in X (though it can be). For example, if $I = [0, 1]$, then the interval $[0, 1/2) = I \cap (-1/2, 1/2)$ is open in the relative topology on I even though it is not open in $\mathbb{R}$.

3.1.7. Connected sets

One of the nice properties of intervals in $\mathbb{R}$ is that they do not have any gaps. More formally, we say that intervals in $\mathbb{R}$ are *connected.*

Definition. A nonempty subset E of a metric space X is *disconnected* if there are disjoint, open sets $U, V \subset X$ such that $E \cap U \neq \varnothing$, $E \cap V \neq \varnothing$ and $E \subset U \cup V$. A set is *connected* if it is not disconnected.

We observe that in any topological space X, a singleton $\{x\}$ is always connected, however in general, connected sets can have many forms. In $\mathbb{R}$, the only connected sets are intervals (and singletons).

Theorem. *Let $E \subset \mathbb{R}$ contain at least two points. Then E is connected if and only if E is an interval.*

Proof. If E is not an interval, then it follows from Proposition 1.5.1 that there exist points $x, y \in E$ and $z \notin E$ such that $x < z < y$. It follows that $E \subset (-\infty, z) \cup (z, \infty)$, so E is disconnected.

To prove that all intervals are connected, we first prove that bounded closed intervals are connected. Suppose that $a < b$ and that, by way of contradiction, $[a, b] \subset U \cup V$, where U and V are disjoint open sets, such that $U \cap [a, b]$ and $V \cap [a, b]$ are both nonempty. Without loss of generality, we may assume that $a \in U$ and since

U is open, there is an $\varepsilon > 0$, such that $(a - \varepsilon, a + \varepsilon) \subset U$. It follows that the set $\{x \in \mathbb{R} : [a, x] \subset U\}$ is not empty, and we denote by c its supremum. If $c \geq b$, then $[a, b] \subset U$, so $[a, b] \cap V = \emptyset$, which contradicts one of our assumptions. If $c < b$, then $c \in U$ and $(c, b] \subset V$. Since $c \in U$ and U is open, there is an $\delta > 0$, such that $(c - \delta, c + \delta) \subset U$, but $(c - \delta, c + \delta) \cap (c, b] \neq \emptyset$ which implies that $U \cap V \neq \emptyset$. This contradicts our other assumption, so $[a, b]$ must be connected.

Finally, suppose that I is any other type of interval, open, half-open or unbounded, and let $I \subset U \cup V$, where U and V are disjoint open sets. If $[a, b] \subset U$ for all $[a, b] \subset I$ or $[a, b] \subset V$ for all $[a, b] \subset I$, then either $I \subset U$ or $I \subset V$, and U is not disconnected by U and V. Otherwise, there exist closed bounded, intervals $[a, b], [c, d] \subset I$ such that $[a, b] \subset U$ and $[c, d] \subset V$.

If $x \in [a, b] \cap [c, d]$, then $U \cap V \neq \emptyset$, so these intervals are not disjoint, and we may assume that $b < c$. Since $b, c \in I$ and I is an interval, then I must contain the interval $[b, c]$. But $b \in [b, c] \cap U$ and $c \in [b, c] \cap V$, which contradicts the fact that closed bounded intervals are connected. □

3.1.8. Separability and complete separability

We have seen that $\mathbb{Q}$ is dense in $\mathbb{R}$ in two different senses: (i) between every two real numbers, there is a rational number (order density) and (ii) for every real number there sequence of rational numbers converging to it (sequential density). To these two notions, we now add a third, topological sense.

Definition. A subset A in a topological space X is (topologically) *dense* if every non-empty open set $O \subset X$ contains an element from A. If the topological space X has a *countable* dense subset, then X is said to be *separable*. If there is a countable collection $\mathcal{O} = \{O_n\}_{n \in \mathbb{N}}$ of open sets in X such that every open set in X is a union of sets from $\mathcal{O}$, then X is said to be *completely separable*.

Lemma. *Every nonempty open subset of $\mathbb{R}$ contains a rational number.*

I.e., $\mathbb{Q}$ is (topologically) dense in $\mathbb{R}$, and since $\mathbb{Q}$ is countable, $\mathbb{R}$ is a separable metric space.

Proof. If $O \subset \mathbb{R}$ is nonempty and open, then there is an $x \in O$ and an $\varepsilon > 0$ such that $(x - \varepsilon, x + \varepsilon) \subset O$. By the order density of $\mathbb{Q}$ in $\mathbb{R}$, it follows that there is an $r \in \mathbb{Q}$ such that $x - \varepsilon < r < x + \varepsilon$, and hence, $r \in O$. □

The collection of open intervals with rational endpoints, $\mathcal{Q} = \{(a, b) : a, b \in \mathbb{Q}\}$, is countable because $\mathbb{Q}$ is countable, so it follows from the next theorem that $\mathbb{R}$ is completely separable.

Theorem. *Every nonempty open set $O \in \mathbb{R}$ is a union of open intervals with rational endpoints.*

Proof. If $O \subset \mathbb{R}$ is open and $x \in O$, then there is an $\varepsilon_x > 0$ such that $(x - \varepsilon_x, x + \varepsilon_x) \subset O$ and it follows that $O = \bigcup_{x \in O} (x - \varepsilon_x, x + \varepsilon_x)$. To prove the theorem, it is therefore enough to prove that every bounded open interval is a union of intervals with rational endpoints.

Suppose then that $(a, b) \subset \mathbb{R}$. From the density of $\mathbb{Q}$ in $\mathbb{R}$, it follows that for every $n \in \mathbb{N}$, there are $a_n, b_n \in \mathbb{Q}$ such that $a_n \in (a, a + \delta_n)$ and $b_n \in (b - \delta_n)$, where

$$\delta_n = \frac{b-a}{3n}.$$

We have $b_n - a_n > \frac{b-a}{3n} > 0$ and therefore $a < a_n < b_n < b$ for all $n \in \mathbb{N}$ and furthermore, $\delta_n \to 0$, so $\lim a_n = a$ and $\lim b_n = b$.

Let $x \in (a, b)$, and let $\varepsilon = \min(x - a, b - x)$, so that $\varepsilon > 0$. For $n \in \mathbb{N}$ that is sufficiently large, we have $\delta_n < \varepsilon$ and for such n, we have $a_n < x < b_n$. In other words, for every $x \in (a, b)$, there is an $n \in \mathbb{N}$ such that $x \in (a_n, b_n)$, and since $(a_n, b_n) \subset (a, b)$ for all $n \in \mathbb{N}$ this means that

$$(a, b) = \bigcup_{n=1}^{\infty}(a_n, b_n).$$ □

The intervals with rational endpoints guaranteed by the theorem are not necessarily pairwise disjoint. On the other hand, if we drop the condition that the endpoints are rational, we have the following fact.

Proposition. *If $G \subseteq \mathbb{R}$ is open, then there is a countable set of pairwise disjoint open intervals $\{I_n\}$ such that $G = \bigcup_n I_n$.*

Proof. Exercise. □

Exercises

ex3.1.1. Show that any finite subset F of $\mathbb{R}$ is closed.

ex3.1.2. Prove that x is a limit point of a set E if and only if for every $\varepsilon > 0$ the set $(x - \varepsilon, x + \varepsilon)$ contains a point $y \in E$, $y \neq x$.

ex3.1.3. Prove that x is a limit point of a set E if, and only if, there exists a sequence $\{x_n\}_{n \in \mathbb{N}}$ of distinct elements of E such that $x = \lim_{n\to\infty} x_n$.

ex3.1.4. Show that if a is a limit point of the range of the sequence $\{x_n\}$, then it is a limit point of the sequence.

Hint: You may rely on the previous exercise, but there is a little more work to be done beyond quoting it.

ex3.1.5. Suppose that X is a metric space and $\{x_n\}_{n=1}^{\infty} \subset X$ is a sequence with no convergent subsequences. Show that the set $\{x_n\}_{n \in \mathbb{N}}$ is closed in X.

ex3.1.6. Prove that the intersection of any collection of closed sets is closed, and that the union of finitely many closed sets is closed.

Hint: Theorem 1.1 may be useful here, and in the next exercise.

ex3.1.7. Complete the proof of Theorem 3.1.6.

ex3.1.8. Prove that for any set E, the set E' of limit points of E is closed.

Hint: Show that every limit point of E' is also a limit point of E. Adapt (part of) the proof of Proposition 3.1.4 to do this.

ex3.1.9. Prove Proposition 3.1.8.

Hint: If two open intervals overlap, then their union is a single open interval. If $\{I_a\}_{a\in A}$ is a collection of disjoint open intervals then each one of them contains a rational number not contained in any other.

3.2. Compactness

In introductory calculus classes, we learn that a continuous function f defined on a closed interval $[a, b]$ always attains both a minimum and maximum value in the interval. This is an important theorem, with many applications.

The proof of this theorem relies on the continuity of the function f, properly defined, and (at least implicitly) the fact that closed intervals in $\mathbb{R}$ are *compact.* We will return to continuous functions in Chapter 4, and in this section explore the concept of compactness in $\mathbb{R}$ and in metric spaces in general.

3.2.1. Open covers

Compact subsets of $\mathbb{R}$ (and compact sets in general) behave in certain respects like finite sets. For example, if E is a finite set and $x_n \in E$ for all $n \in \mathbb{N}$, then the sequence $\{x_n\}$ has a limit point in E. As we shall see below, the same is true for compact sets.

Compactness can be defined in several ways, and we begin with a definition that incorporates finiteness in a crucial way.

Definition. Let X be a topological space.[2]

(i) An *open cover* of a subset E of X is a family $\{O_a\}_{a\in A}$ of open sets in X such that $E \subset \bigcup_{a\in A} O_a$.

(ii) A *subcover* of an open cover $\{O_a\}_{a\in A}$ of E is a family of open sets $\{O_b\}_{b\in B}$, where $B \subset A$ and $E \subset \bigcup_{b\in B} O_b$. The subcover is a *finite subcover* if the index set B is a finite set. If $\{O_a : a \in A\}$ contains a finite subcover of E, we also say that E *admits a finite subcover from* $\mathcal{O}$.

(iii) A set E in X is *compact* if *every* open cover of E admits a finite subcover of E.

(iv) A metric space X is a *compact metric space* if X is a compact set in itself.

In line with the remarks before the definition, we observe that finite sets are always compact (in any topological space) and leave the proof as an exercise. Moreover, compact sets have a variety of useful characteristics, all of which can be seen as generalizations of corresponding properties of finite sets.

Theorem. *If K is a compact subset of the metric space (X, ρ), then K is closed and bounded in X.*

[2] You can think of X as a metric space with the metric topology or even more specifically as $\mathbb{R}$.

Proof. Choose a point $x_0 \in X$, then $K \subset X = \bigcup_{n\in\mathbb{N}} B(x_0, n)$, so $\{B(x_0, n)\}_{n\in\mathbb{N}}$ is an open cover of K. It follows from the compactness of K that there is a finite set of open balls $\{B(x_0, n_1), \ldots, B(x_0, n_k)\}$, with $n_1 < \cdots < n_k$ in $\mathbb{N}$, such that

$$K \subset \bigcup_{j=1}^{k} B(x_0, n_j) = B(x_0, n_k),$$

so K is bounded.

We prove that K is closed by showing that K^c is open. Let $y \in K^c$ and note that if $x \in K$, then $\rho(x, y) > 0$, so there exists $n \in \mathbb{N}$ such that $\rho(x, y) > 1/n$. It follows that $\{R(y, 1/n)\}_{n\in\mathbb{N}}$ is an open cover of K, using the notation and conclusion of Proposition 3.1.2.[3] Since K is compact, there exist $n_1 < n_2 < \ldots < n_k$ in $\mathbb{N}$ such that

$$K \subset \bigcup_{j=1}^{k} R(y, 1/n_j) = R(y, 1/n_k),$$

and it follows that $\rho(x, y) > 1/n_k$ for all $x \in K$ so that $B(y, 1/n_k) \subset K^c$. Hence, K^c is open and K is closed. □

The proof that a compact subset of a metric space is bounded can be adjusted to show a bit more.

Corollary. *If X is a metric space and $K \subset X$ is compact, then for every $\varepsilon > 0$ there is a finite set $\{x_1, \ldots, x_n\} \subset K$ such that*

$$K \subset \bigcup_{j=1}^{n} B(x_j, \varepsilon).$$

We say that K is *totally bounded* in X in this case.

Proof. Suppose that $K \subset X$ is compact and let $\varepsilon > 0$. The set $\{B(x, \varepsilon) : x \in K\}$ is an open cover of K, so it admits a finite subcover, $\{B(x_1, \varepsilon), \ldots, B(x_n, \varepsilon)\}$. □

If E and F are subsets of a metric space X, then the distance between them is defined by

$$d(E, F) = \inf\{d(x, y) : x \in E,\ y \in F\}.$$

If E and F are disjoint, finite sets, then $d(E, F) > 0$. The same is true of disjoint compact sets.

Proposition. *If K_1 and K_2 are compact subsets of a metric space X, and $K_1 \cap K_2 = \varnothing$, then $d(K_1, K_2) > 0$.*

Proof. For every $x \in K_1$, let $\delta_x = \inf\{d(x, y) : y \in K_2\}$. If $\delta_x = 0$ for some $x \in K_1$, then there is a sequence of points $\{y_n\} \subset K_2$ such that $d(x, y_n) < 1/n$, whence $x \in K_2$ since K_2 is closed. This is impossible because $K_1 \cap K_2 = \varnothing$, so $\delta_x > 0$ for all $x \in K_1$.

[3] Where we defined $R(y, \varepsilon)$ to be the interior of $B(y, \varepsilon)^c$.

The set $\{B(x, \delta_x/2) : x \in K_1\}$ is an open cover of K_1, so there is a finite set $\{x_1, \ldots, x_n\} \subseteq K_1$ such that

$$K_1 \subseteq \bigcup_{j=1}^{n} B(x_j, \delta_{x_j}/2).$$

If $x \in K_1$ and $y \in K_2$, then $x \in B(x_j, \delta_{x_j}/2)$, for some $1 \leq j \leq n$, so $d(x, y) > \delta_{x_j}/2$, by the triangle inequality. It follows that $d(x, y) \geq \frac{1}{2} \min(\delta_{x_j} : 1 \leq j \leq n) > 0$, for all $x \in K_1$ and $y \in K_2$. □

3.2.2. The Heine-Borel Theorem

The converse of Theorem 3.2.1 is not true in all metric spaces—there are metric spaces containing closed and bounded sets that are not compact.

Example. Let $X = (0, 1)$ with the standard metric inherited from $\mathbb{R}$. The set $F = \{1/n : n \geq 2\}$ is bounded because $F \subset B(1/2, 1)$. Furthermore, F doesn't have any limit points in X because $0 \notin X$, so F is closed in X.

For each $n \in \mathbb{N}$, we have $\frac{1}{n} - \frac{1}{n+1} = \frac{1}{n(n+1)} > \frac{1}{(n+1)^2}$, and therefore

$$\left(\frac{1}{n} - \frac{1}{(n+1)^2}, \frac{1}{n} + \frac{1}{(n+1)^2}\right) \cap F = \{1/n\}$$

for all $n \geq 2$. Hence $\left\{\left(\frac{1}{n} - \frac{1}{(n+1)^2}, \frac{1}{n} + \frac{1}{(n+1)^2}\right)\right\}_{n=2}^{\infty}$ is an open cover of F that does not contain a finite subcover, so F is not compact in X.

A metric space in which the converse of Theorem 3.2.1 is true is said to have the *Heine-Borel* property. The canonical example of such a space is $\mathbb{R}$.

Theorem (Heine-Borel). *A set $E \subset \mathbb{R}$ is compact if and only if it is closed and bounded.*

It follows from this theorem that every closed and bounded interval is compact in $\mathbb{R}$. These are the archetypical compact subsets of $\mathbb{R}$, and they play an important role in the proof of the theorem.

Proof. The first half of this statement, that compact sets are closed and bounded, has been proved in Theorem 3.2.1 for all metric spaces. We prove the converse, that closed and bounded sets in $\mathbb{R}$ are compact, by contradiction.

Suppose that E is closed and bounded in $\mathbb{R}$, and let $\mathcal{O} = \{O_\alpha\}_{\alpha \in A}$ be an open cover of E, that contains no finite subcover of E.

The boundedness of E implies that there is an interval $J_1 = [a, b]$, such that $E \subset J_1$. Let $J_1^1 = [a, (a+b)/2]$ be the left half of J_1 and let $J_1^2 = [(a+b)/2, b]$ be the right half. If both $E \cap J_1^1$ and $E \cap J_1^2$ admit finite subcovers from $\mathcal{O}$, then the union of these subcovers is a finite subcover of E (because $E = (E \cap J_1^1) \cup (E \cap J_1^2)$) so at least one of $E \cap J_1^1$ and $E \cap J_1^2$ fails to admit a finite subcover. We set $J_2 = J_1^1$ if $E \cap J_1^1$ does not admit a finite subcover from $\mathcal{O}$ and $J_2 = J_1^2$ otherwise.

Once again, we split J_2 into left and right closed half-intervals, J_2^1 and J_2^2. It follows as above that at least one of $E \cap J_2^1$ or $E \cap J_2^2$ must fail to admit a finite subcover from $\mathcal{O}$ and we set $J_3 = J_2^1$ if $E \cap J_2^1$ doesn't admit a finite subcover from $\mathcal{O}$ and $J_3 = I_2^2$ otherwise.

We continue in this way to recursively define a sequence of closed intervals $\{J_n\}$. Assume that $n \geq 1$ and that the intervals $J_1, J_2, \ldots, J_n$ have been chosen in such a way that (i) J_k is the closed left or right half of J_{k-1} for $2 \leq k \leq n$ and (ii) $E \cap J_k$ does not admit a finite subcover from $\mathcal{O}$ for $1 \leq k \leq n$. Letting J_n^1 and J_n^2 be the closed left and right half of J_n, we observe as before that either $E \cap J_n^1$ or $E \cap J_n^2$ must fail to admit a finite subcover from $\mathcal{O}$, and we set $J_{n+1} = J_n^1$ in the first case and $J_{n+1} = J_n^2$ otherwise.

This produces a sequence $\{J_n\}_{n\in\mathbb{N}}$, where each J_n is either the left or the right half of J_{n-1}, and for every n the set $E \cap J_n$ does not admit a finite subcover from $\mathcal{O}$. In particular, $E \cap J_n$ is not empty, and for each n, we choose a point $x_n \in E \cap J_n$.

For all n, we write $J_n = [a_n, b_n]$ (with $a_1 = a$ and $b_1 = b$), and note that by construction, $b_{n+1} - a_{n+1} = (b_n - a_n)/2$, and therefore $b_n - a_n = (b-a)/2^{n-1}$ for all $n \geq 1$, as follows by induction. Now, if $n, m \geq N$, then $x_n, x_m \in J_N$, because $J_n, J_m \subset J_N$ in this case, and therefore

$$|x_n - x_m| \leq b_N - a_N = \frac{b-a}{2^{N-1}}.$$

It follows that $\{x_n\}$ is a Cauchy sequence, and therefore converges to a limit $x^* \in \mathbb{R}$, because $\mathbb{R}$ is complete.

By Theorem 3.1.1, $x^* \in E$, because E is closed and $\{x_n\} \subset E$. Since $x^* \in E \subset \bigcup_{a\in A} O_a$, there exists $O_a \in \mathcal{O}$ such that $x^* \in O_a$, and so there is an $\varepsilon > 0$ such that $(x^* - \varepsilon, x^* + \varepsilon) \subset O_a$ (because O_a is open).

We now choose $N \in \mathbb{N}$ large enough so that both $|x_n - x^*| < \varepsilon/2$ for all $n \geq N$ and $(b-a)/2^{N-1} < \varepsilon/2$. With this choice, if $x \in J_N$, then $|x - x^*| \leq |x - x_N| + |x_N - x^*| < \varepsilon$, which implies that

$$E \cap J_N \subset J_N \subset (x^* - \varepsilon, x^* + \varepsilon) \subset O_a,$$

contradicting the assumption that $E \cap J_N$ does not admit a finite subcover from $\mathcal{O}$. This contradiction shows that E must be compact. □

From the Heine-Borel and Bolzano-Weierstrass theorems we have the following corollary.

Corollary. *A set $K \subset \mathbb{R}$ is compact if and only if every sequence $\{x_n\}$ in K has a subsequence that converges to a limit in K.*

I.e., compact subsets of $\mathbb{R}$ are *sequentially compact.*

Proof. Suppose that $K \subset \mathbb{R}$ is compact, and therefore closed and bounded. If $\{x_n\} \subset K$, then the sequence is bounded because K is bounded, and from Theorem 2.5.3 (Bolzano-Weierstrass) it follows that $\{x_n\}$ has a subsequence $\{x_{n_k}\}$ that converges to a limit $x^* \in \mathbb{R}$. Since K is also closed, it follows that $x^* \in K$ (by Theorem 3.1.1).

If K is not compact, then by the Heine-Borel theorem, K is not closed or K is not bounded. If K is not bounded, then for every $n \in \mathbb{N}$, there is an $x_n \in K$ such that $|x_n| > n$. The sequence $\{x_n\}$ does not have a convergent subsequence because for every $a \in \mathbb{R}$, the set $\{n \in \mathbb{N} : |x_n - a| < 1\}$ is finite.

If K is not closed, then there is a limit point x^* of K that does not belong to K. It follows that for every $n \in \mathbb{N}$, there is an $x_n \in K$ such that $|x_n - x^*| < 1/n$, and hence

$x_n \to x^*$, so $\lim_{k\to\infty} x_{n_k} = x^* \notin K$ for *every* subsequence $\{x_{n_k}\}$ of $\{x_n\}$. It follows that this sequence has no subsequence that converges to a point in K. □

*3.2.3. Sequential compactness and Lebesgue numbers

The property of compact subsets of $\mathbb{R}$ described in Corollary 3.2.2 has a name.

Definition. The subset E of the metric space (X, ρ) is *sequentially compact* if every sequence in E has a subsequence that converges to a point in E.

Using this language, Proposition 3.2.2 says that sets in $\mathbb{R}$ are compact if and only if they are sequentially compact. This statement generalizes to all metric spaces, but neither the Bolzano-Weierstrass theorem nor the Heine-Borel theorem are true in all metric spaces, so we need to use different tools to prove it. One of these tools is the *Lebesgue number lemma.*

Lemma. *If (X, ρ) is a metric space and $K \subset X$ is sequentially compact, then for every open cover $\mathcal{U} = \{U_a\}_{a\in A}$, there is a constant $\delta = \delta_{\mathcal{U},K} > 0$ such that for every $x \in K$, there is a $U \in \mathcal{U}$ that contains $B(x, \delta)$.*

The positive constant $\delta_{\mathcal{U},K}$ is called the *Lebesgue number* of the cover.

Proof. Let $K \subset X$ be sequentially compact and suppose to the contrary, that there is no $\delta > 0$ as claimed. Then for any $n \in \mathbb{N}$, there is an $x_n \in K$ such that $B(x_n, 1/n) \not\subset U_a$ for all $a \in A$.

By the sequential compactness of K, the sequence $\{x_n\}$ has a subsequence $\{x_{n_k}\}_{k=1}^{\infty}$ that converges to a limit $x^* \in K$. Since $\mathcal{U}$ is an open cover of K, there is a $U \in \mathcal{U}$ such that $x^* \in U$, and therefore there is an $\varepsilon > 0$ such that $B(x^*, \varepsilon) \subset U$.

Let $K \in \mathbb{N}$ be such that $\rho(x_{n_k}, x^*) < \varepsilon/2$ for all $k > K$. This and the triangle inequality together imply that $B(x_{n_k}, \varepsilon/2) \subset B(x^*, \varepsilon) \subset U$ for all $k > K$. Hence, if $k > \max(K, 2/\varepsilon)$, then $1/n_k \leq 1/k < \varepsilon/2$, and therefore $B(x_{n_k}, 1/n_k) \subset B(x_{n_k}, \varepsilon/2) \subset U$. This contradicts the assumption we made about the elements of the sequence $\{x_n\}$. □

We can now prove the generalization of Corollary 3.2.2.

Theorem. *If (X, ρ) is a metric space then for any $K \subset X$, K is compact if and only if K is sequentially compact.*

Proof. First suppose that $K \subset X$ is compact, and that the sequence $\{x_n\} \subset K$ has no convergent subsequences.

Let $A_m = \{x_k : k > m\}$ be the subsequence of $\{x_n\}$ obtained by removing the first m elements, then the assumption that $\{x_n\}$ has no convergent subsequences has two implications for the subsequences A_m. First,

$$\bigcap_{m=1}^{\infty} A_m = \varnothing, \tag{3.2.1}$$

because if $x \in \bigcap_{m=1}^{\infty} A_m$, then for every $m \in \mathbb{N}$, there must be an $n_m > m$ such that $x_{n_m} = x$, and the *constant* subsequence $\{x_{n_m}\}$ would converge to x.

Second, for every $m \in \mathbb{N}$, A_m has no convergent subsequence because such a subsequence would also be a convergent subsequence of $\{x_n\}_{n=1}^{\infty}$. Hence, A_m has no limit points (see **ex3.2.5**) and is therefore closed, so $O_m = A_m^c$ is open and it follows from (3.2.1) that

$$\bigcup_{m=1}^{\infty} O_m = \left(\bigcap_{n=1}^{\infty} A_m\right)^c = X.$$

In other words, $\{O_m\}_{m\in\mathbb{N}}$ is an open cover of X and hence, an open cover of $K \subset X$.

From the compactness of K it follows that $\{O_n : n \in \mathbb{N}\}$ contains a finite subcover of K, and since $O_1 \subset O_2 \subset \ldots \subset O_n \subset \ldots$, it follows that $K \subset O_N = A_N^c$ for some $N \in \mathbb{N}$. This implies that $x_j \notin K$ for all $j > N$, contradicting the assumption that $\{x_n\} \subset K$. This contradiction shows that K must be sequentially compact.

In the other direction, suppose that $K \subset X$ is sequentially compact and let $\mathcal{U} = \{U_a\}_{a\in A}$ be an open cover of K. We want to show that $\mathcal{U}$ contains a finite subcover of K, but the difficulty is that we don't know anything about the nature of the open sets in $\mathcal{U}$. This is where the Lebesgue number lemma comes in.

Let $\delta = \delta_{\mathcal{U},K} > 0$ be the Lebesgue number for $\mathcal{U}$. The family $\mathcal{O} = \{B(x,\delta)\}_{x\in K}$ is also an open cover of K, and it follows from the Lebesgue number theorem that if $\mathcal{O}$ admits a finite subcover of K, $\{B(x_1,\delta), \ldots, B(x_n,\delta)\}$, then so does $\mathcal{U}$. This is because for each i, there is an open set $U_i \in \mathcal{U}$ such that $B(x_i,\delta) \subset U_i$ for $1 \le i \le n$, so

$$K \subset \bigcup_{i=1}^{n} B(x_i,\delta) \subset \bigcup_{i=1}^{n} U_i.$$

Suppose that $\mathcal{O}$ does not admit a finite subcover of K, then in particular, K is infinite. Let $x_1 \in K$, then from our assumption, there must be an $x_2 \in K$ such that $x_2 \notin B(x_1,\delta)$. Likewise, there must be an $x_3 \in K$ such that $x_3 \notin B(x_1,\delta) \cup B(x_2,\delta)$, and more to the point, $\rho(x_1,x_2) > \delta$, $\rho(x_1,x_3) > \delta$ and $\rho(x_2,x_3) > \delta$. We continue in this way to recursively construct a sequence in K: having found $x_1, \ldots, x_n \in K$ satisfying $\rho(x_i,x_j) > \delta$ for $1 \le i < j \le n$, the assumption that $\mathcal{O}$ does not admit a finite subcover of K implies that there is an $x_{n+1} \in K$ such that $x_{n+1} \notin \bigcup_{i=1}^{n} B(x_i,\delta)$.

The sequence $\{x_i\}_{i=1}^{\infty} \subset K$, thus constructed, has the property that $\rho(x_i,x_j) > \delta$ for all $i \ne j$. It follows that for any $x \in X$, the ball $B(x,\delta/2)$ contains at most one element x_j from this sequence and therefore the sequence has no convergent subsequence, contradicting the assumption that K is sequentially compact. Hence, $\mathcal{O}$ admits a finite subcover of K and therefore, as explained above, so does $\mathcal{U}$. □

⋆3.2.4. The Heine-Borel theorem in $\mathbb{R}^n$

In any metric space, a compact set is closed and bounded, as we proved in Theorem 3.2.1. In $\mathbb{R}$, the converse is also true — this is the Heine-Borel theorem. This statement generalizes to closed and bounded sets in $\mathbb{R}^n$.

This can be proved by mimicking the proof of Theorem 3.2.2, but using n-dimensional boxes instead of (1-dimensional) intervals. We leave this approach as an exercise, and instead rely on the fact that compactness and sequential compactness are equivalent in any metric space and on the Bolzano-Weierstrass theorem (in Euclidean space).

Theorem. *If $K \subset \mathbb{R}^n$ is closed and bounded, then K is sequentially compact, and hence compact.*

Proof. Suppose that $K \subset \mathbb{R}^n$ is closed and bounded and let $\{\mathbf{x}_j\}_{j=1}^{\infty}$ be a sequence in K. The sequence is bounded because K is bounded, and hence by Theorem 2.5.6 (the Bolzano-Weierstrass theorem), it contains a subsequence $\{\mathbf{x}_{n_k}\}_{k=1}^{\infty}$ that converges to a limit $\mathbf{x}^*$. Since K is also closed, it follows that $\mathbf{x}^* \in K$. □

Exercises

ex3.2.1. Prove that finite subsets of $\mathbb{R}$ are compact.

ex3.2.2. Suppose that K_1 and K_2 are disjoint, compact sets in a metric space X. Show that there are disjoint, open sets $U, V \subseteq X$, such that $K_1 \subset U$ and $K_2 \subset V$.

ex3.2.3. Suppose that K_1 and K_2 are disjoint compact subsets of $\mathbb{R}$, with $d(K_1, K_2) = \delta > 0$. Show that if $(a, b) \cap (K_1 \cup K_2) \neq \emptyset$, then there is a closed subinterval $[c, d] \subset (a, b)$ such that $d - c > \delta/2$ and $[c, d] \cap (K_1 \cup K_2) = \emptyset$.

ex3.2.4. Let $\{K_n : n \in \mathbb{N}\}$ be a collection of compact subsets of $\mathbb{R}$ satisfying $\bigcap_{n=1}^{N} K_n \neq \emptyset$ for all $N \in \mathbb{N}$ (i.e., the intersection of any finite subcollection of these sets is nonempty).[4] Show that $\bigcap_{n=1}^{\infty} K_n \neq \emptyset$.

Hint: Prove this by contradiction: if $\bigcap_{n=1}^{\infty} K_n = \emptyset$, then $\mathbb{R} = \left(\bigcap_{n=1}^{\infty} K_n\right)^c$, so

$$K_1 \subset \mathbb{R} = \bigcup_{n=1}^{\infty} K_n^c,$$

which means that $\{K_n^c : n \in \mathbb{N}\}$ is an open cover of K_1 (why?). Now what?

ex3.2.5. Show that if K is a sequentially compact set in a metric space X, then K is closed and bounded. Do this directly from the definition of sequential compactness (not by quoting Theorems 3.2.3 and 3.2.1.)

ex3.2.6. Suppose that X is a complete metric space and that X is totally bounded. Show that X is a compact metric space.

3.3. Small sets in $\mathbb{R}$

We conclude this chapter with a brief discussion of different notions of "small" for subsets of $\mathbb{R}$.

From a purely set theoretic point of view, cardinality is the standard measure of size. The integers and rational numbers are both countable sets, while all intervals in $\mathbb{R}$ have the same uncountable cardinality.[5] We think of countable subsets of $\mathbb{R}$ as small and uncountable subsets as big, and as we will see, countable subsets of $\mathbb{R}$ are small in every other sense that we describe below.

[4] This property of a collection of sets is called the *finite intersection property*.

[5] See Appendix 2.2.

3.3.1. Sets of measure zero

There is another, more natural way to describe the size of an interval, namely its length. The length of an interval takes the order and field structure on $\mathbb{R}$ into account. This notion of size (and its generalization to *measure*, that we discuss in Chapter 8) is important to the development of the integral (that we discuss in Chapters 7 and 8).

For an interval $I \subset \mathbb{R}$, we denote its length by $\ell(I)$. If I is bounded with endpoints $a < b$, then $\ell(I) = b - a$, whether I is open, closed or half open. If I is unbounded, e.g., $I = (a, \infty), (-\infty, b]$ or $(-\infty, \infty)$, then $\ell(I) = \infty$. We think of shorter intervals as smaller sets and longer intervals as larger sets in this context.

More generally, if a set can be covered by (i.e., is contained in) a union of intervals, the sum of whose lengths is small, then the set is small. This intuition leads to the following definition.

Definition. A set $E \subset \mathbb{R}$ has *measure zero* if for every $\varepsilon > 0$, there is a countable set of intervals $\{I_n\}$ such that covers E (i.e., $E \subset \bigcup I_n$) and $\sum \ell(I_n) \leq \varepsilon$.

If $E \subset \mathbb{R}$ contains an interval (a, b), then the sum of the lengths of any (countable) collection of intervals covering E must also cover (a, b). It follows in this case that if $E \subset \bigcup I_n$, then $\sum \ell(I_n) \geq (b - a)$. In other words, sets of measure zero cannot contain any intervals.

Proposition. *If $E \subset \mathbb{R}$ is countable, then E has measure zero.*

Since intervals are uncountable, it is certainly true that a countable set cannot contain an interval, but this doesn't prove that a countable set must have measure zero.

Proof. Suppose first that $E = \{x_1, \dots, x_n\}$ is finite and let $\varepsilon > 0$. For $1 \leq j \leq n$, let $I_j = (x_j - \varepsilon/2n, x_j + \varepsilon/2n)$, then $x_j \in I_j$ for each $1 \leq j \leq n$, so $E \subset \bigcup_{j=1}^n I_n$. Moreover, $\sum_{j=1}^n \ell(I_n) = n \cdot 2\varepsilon/2n = \varepsilon$.

Now suppose that $E = \{x_n\}_{n\in\mathbb{N}}$ is countably infinite. Given $\varepsilon > 0$, let $I_n = (x_n - \varepsilon/3^n, x_n + \varepsilon/3^n)$ for $n \in \mathbb{N}$. Then $x_n \in I_n$ for each n, so $E \subset \bigcup_{n=1}^\infty I_n$, and

$$\sum_{n=1}^{\infty} \ell(I_n) = \sum_{n=1}^{\infty} \frac{2\varepsilon}{3^n} = 2\varepsilon \sum_{n=1}^{\infty} \frac{1}{3^n} = 2\varepsilon \cdot \frac{1}{2} = \varepsilon. \qquad \square$$

Thus for example, the set of rational numbers has measure zero, even though this set is dense in the set of real numbers.

A countable union of countable sets is countable.[6] The same is true for sets of measure zero.

Theorem. *If $E_n \subset \mathbb{R}$ has measure zero for every $n \in \mathbb{N}$, then $\bigcup_{n=1}^\infty E_n$ has measure zero.*

In other words, a countable union of sets of measure zero has measure zero.

[6] See Proposition A.2.1.

Proof. Let $\varepsilon > 0$ be given. For each $n \in \mathbb{N}$, E_n has measure zero, so there is a $J_n \subset \mathbb{N}$ and a collection of intervals $\{I_{n,j}\}_{j\in J_n}$ such that $E_n \subset \bigcup_{j\in J_n} I_{n,j}$ and

$$\sum_{j\in J_n} \ell(I_{n,j}) < \frac{\varepsilon}{2^n}.$$

The collection $\{I_{n,j} : n \in \mathbb{N} \text{ and } j \in J_n\}$ is a countable collection of intervals (as a countable union of countable sets) that covers $\bigcup_{n=1}^{\infty} E_n$ and

$$\sum_{n=1}^{\infty}\left(\sum_{j\in J_n} \ell(I_{n,j})\right) < \sum_{n=1}^{\infty} \frac{\varepsilon}{2^n} = \varepsilon.$$

It follows that $\bigcup_{n=1}^{\infty} E_n$ has measure zero, as claimed. □

*3.3.2. The Cantor set

There are uncountable sets of measure zero, the best known of which is the Cantor set. Starting with $C_0 = [0,1]$, let

$$C_1 = C_0 \setminus (1/3, 2/3) = [0,1/3] \cup [2/3,1],$$

i.e., C_1 is obtained by removing the *open middle third* of C_0. Next, C_2 is obtained by removing the open middle third of each of the two closed intervals comprising C_1:

$$C_2 = C_1 \setminus ((1/9,2/9) \cup (7/9,8/9)) = [0,1/9] \cup [2/9,1/3] \cup [2/3,7/9] \cup [8/9,1].$$

Likewise, we obtain C_3 from C_2 by removing the open middle thirds of each of the four (disjoint) closed intervals comprising C_2, so that C_3 is the union of eight disjoint closed intervals, etc.

Definition. Let $C_0 = [0,1]$ and for $n \in \mathbb{N}$, let C_n be obtained from by removing the open middle third of each of the closed intervals whose disjoint union comprises C_{n-1}. The *Cantor set* is the set

$$\mathscr{C} = \bigcap_{n=1}^{\infty} C_n.$$

Proposition. *For every $n \geq 0$, C_n is the union of 2^n disjoint, closed intervals, of length 3^{-n} each.*

Proof. The proof is by induction on n, beginning with the observation that statement is clearly true for $n = 0$. Now suppose that $n \geq 0$, the statement is true for n, and $I_1, \ldots, I_{2^n}$ are the disjoint closed intervals comprising C_n. If $I_j = [a_j, b_j]$ for $1 \leq j \leq 2^n$, then removing the open middle third from I_j leaves us with the two closed intervals $[a_j, c_j]$ and $[d_j, b_j]$, where $c_j = a_j + 3^{-(n+1)}$ and $d_j = a_j + 2\cdot 3^{-(n+1)}$, since $b_j - a_j = 3^{-n}$. Thus, C_{n+1} is the union of $2\cdot 2^n = 2^{n+1}$ closed intervals, each of length $3^{-(n+1)}$. Moreover, these intervals are disjoint because $I_1, \ldots, I_{2^n}$ are disjoint. □

It follows from this proposition that C_n is closed, for all n, as a finite union of closed sets, and hence $\mathscr{C}$ is closed as an intersection of closed sets.

In what follows, we say *the endpoints of* C_n, to mean the endpoints of the 2^n intervals comprising C_n. We observe that it follows from the definition (as detailed in the proof of the lemma) that if c is a left (right) endpoint of C_n, then c is also a left (right) endpoint of C_m for all $m \geq n$. Moreover, every endpoint of C_n is either an endpoint or an interior point of C_{n-1}, for $n \geq 1$. Hence, for every n, the set of endpoints of C_n is contained in $\mathscr{C}$.

The elements of the Cantor set are characterized by their ternary expansions (see §2.9.2). We show this first for the endpoints in $\mathscr{C}$.

Lemma. *The number x is a left endpoint of C_n if and only if*

$$x = \sum_{k=1}^{\infty} \frac{t_k(x)}{3^k},$$

with $t_k(x) = 0$ or 2 for all k and $t_k = 0$ for all $k > n$. Similarly, x is a right endpoint of C_n, if and only if

$$x = \sum_{k=1}^{\infty} \frac{t_k(x)}{3^k},$$

with $t_k(x) = 0$ or 2 for all k and $t_k = 2$ for all $k > n$.

Proof. We prove this by induction on n, starting with $n = 0$. The endpoints of C_0 are 0, which satisfies the claim with $t_k = 0$ for all k, and 1 which satisfies the claim because

$$1 = \sum_{k=1}^{\infty} \frac{2}{3^k}.$$

Now suppose that $n \geq 1$, that the claim is true for $n-1$ and that x is a right endpoint of C_n. The claim is true for right endpoints of C_{n-1}, so we may assume that x is not a right endpoint of C_{n-1}. In this case there is a left endpoint y of C_{n-1} such that

$$x = y + \frac{1}{3^n} = \sum_{k=1}^{n} \frac{t_k(y)}{3^k} + \frac{1}{3^n} = \sum_{k=1}^{n} \frac{t_k(y)}{3^k} + \sum_{k=n+1}^{\infty} \frac{2}{3^k}, \tag{3.3.1}$$

so x has the claimed ternary expansion. In the other direction, if x satisfies the identity (3.3.1), then $y = x - \frac{1}{3^n}$ must be a left endpoint of C_{n-1}, based on the induction hypothesis, so x is a right endpoint of C_n.

The inductive argument for left endpoints is analogous, and we leave it as an exercise. □

Corollary. *The number $x \in \mathscr{C}$ if and only if*

$$x = \sum_{n=1}^{\infty} \frac{t_n(x)}{3^n},$$

where $t_n(x) \in \{0, 2\}$ for all n.

Proof. Let $x \in [0,1]$ have the ternary expansion

$$x = \sum_{n=1}^{\infty} \frac{t_n(x)}{3^n},$$

where $t_n(x) \in \{0,1,2\}$ for all n. In the case that x has two ternary expansions, one that is finite and terminates with a $1/3^m$, and one that terminates with $\sum_{n=m+1}^{\infty} 2/3^n$, we choose the latter.

If $x \in \mathscr{C}$, then it follows from the construction that for every $n \in \mathbb{N}$, $x \in I_j$ for one of the closed intervals comprising C_n, so there is an endpoint $y_n \in C_n$ such that $|x - y_n| \le \frac{1}{2\cdot 3^n}$. It follows that $t_n(x) = t_n(y_n)$ so $t_n(x) = 0$ or 2, by the lemma, and hence x has the claimed expansion.

If $t_n(x) \in \{0,2\}$ for all n, then

$$y_n = \sum_{k=1}^{n} \frac{t_k(x)}{3^k}$$

is a left endpoint in $\mathscr{C}$, and $y_n \to x$. Hence $x \in \mathscr{C}$ because $\mathscr{C}$ is closed. □

Theorem. *The Cantor set is an uncountable, closed set of measure zero.*

Proof. We have already shown that $\mathscr{C}$ is closed.

Next, for each $n \in \mathbb{N}$, $\mathscr{C} \subset C_n$ and the sum of the lengths of the intervals comprising C_n is $(2/3)^n$, so $\mathscr{C}$ has measure zero because $(2/3)^n \to 0$.

Finally, to show that $\mathscr{C}$ is uncountable, we construct a bijection from $\mathscr{P}(\mathbb{N})$ to $\mathscr{C}$.[7] Specifically, let $\Lambda = \{n_k : k \in \mathbb{N}\}$ be a (finite or infinite) subset of $\mathbb{N}$, with the elements listed in ascending order. If $\Lambda = \varnothing$, we set $\varphi(\Lambda) = 0 \in \mathscr{C}$. If $\Lambda = \{n_1, \ldots, n_k, \ldots\}$ is nonempty, we set

$$\varphi(\Lambda) = \sum_{n_k \in \Lambda} \frac{2}{3^{n_k}},$$

so $\varphi(\Lambda) \in \mathscr{C}$ in this case too, by the preceding corollary. In particular, $\varphi(\Lambda)$ is a left endpoint if Λ is finite and $\varphi(\Lambda)$ is a right endpoint if the complement of Λ is finite.

For any $x \in \mathscr{C}$, we have $x = \varphi(\Lambda_x)$, where $\Lambda_x = \{n \in \mathbb{N} : t_n(x) = 2\}$, so the map φ is surjective. If $\Lambda_1 \ne \Lambda_2$, let $n_0 \in \mathbb{N}$ be the smallest integer that is not in $\Lambda_1 \cap \Lambda_2$. Then,

$$|\varphi(\Lambda_1) - \varphi(\Lambda_2)| \ge \frac{2}{3^{n_0}} - \sum_{n=n_0+1}^{\infty} \frac{2}{3^n} = \frac{1}{3^{n_0}} > 0,$$

since $\sum_{n=n_0+1}^{\infty} \frac{2}{3^n} = \frac{1}{3^{n_0}}$, so φ is injective as well. □

*3.3.3. Nowhere dense sets

A set E is dense in $\mathbb{R}$ if $E \cap O \ne \varnothing$ for every nonempty open set $O \subset \mathbb{R}$. The set of rational numbers is an important example of a dense subset of $\mathbb{R}$. Sets which have the opposite property, in the sense described below, are considered to be small, topologically speaking.

Definition. A subset F of a metric space X is *nowhere dense* if for every ball $B(x,\varepsilon) \subset X$, there is a ball $B(y,\delta) \subset B(x,\varepsilon)$ such that $F \cap B(x,\varepsilon) = \varnothing$. In other words a nowhere dense set misses an open piece of every open set.

[7] $\mathscr{P}(\mathbb{N})$ is the set of all subsets of $\mathbb{N}$, which is uncountable. See the Appendix.

A single point is clearly nowhere dense in $\mathbb{R}$, and more generally any finite set is nowhere dense. Some countable sets are nowhere dense in $\mathbb{R}$, like $\mathbb{Z}$, while others are not, like $\mathbb{Q}$. There are also uncountable sets that are nowhere dense.

Proposition. *The Cantor set is nowhere dense in* $\mathbb{R}$.

Proof. Let I be an arbitrary, nonempty open interval in $[0,1]$. We need to show that there is a nonempty open interval $J \subset I$ such that $J \subset \mathscr{C}^c$. If $I \cap \mathscr{C} = \varnothing$, then we can set $J = I$ and there is nothing more to prove. Otherwise, $I \cap \mathscr{C} \neq \varnothing$ and hence, $I \cap C_n \neq \varnothing$ for all $n \in \mathbb{N}$ (referring to the sets described in the previous section).

In this case, let $n \in \mathbb{N}$ be large enough that $\ell(I) > 1/3^n$, and denote by $\{I_k : 1 \leq k \leq 2^n\}$ the collection of intervals comprising C_n. Since $I \cap C_n \neq \varnothing$, it follows that $I \cap I_k \neq \varnothing$ for some k. The interval I cannot be properly contained in I_k because $\ell(I_k) = 1/3^n < \ell(I)$ and writing $I_k = [r_k, s_k]$, it follows that either $r_k \in I$ or $s_k \in I$.

Suppose that $r_k \in I$, then since I is open there is an $\varepsilon > 0$ such that $(r_k - \varepsilon, r_k + \varepsilon) \subset I$. Since the open interval $(r_k - 1/3^n, r_k) \subset \mathscr{C}^c$, it follows that $J = (r_k - \delta, r_k) \subset I \cap \mathscr{C}^c$, where $\delta = \min(\varepsilon, 1/3^n)$.[8] The same argument shows that if $s_k \in I$, then there is an $\eta > 0$ such that $J = (s_k, s_k + \eta) \subset I \cap \mathscr{C}^c$. □

Nowhere dense sets have certain properties in common with closed sets.

Theorem. *The closure of a nowhere dense set is nowhere dense and the union of finitely many nowhere dense sets is nowhere dense.*

Proof. Suppose (X, ρ) is a metric space and that $E \subset X$ is nowhere dense. Given any open ball $B_1 = B(x, \varepsilon) \subset X$, there is a second ball $B_2 = B(y, \delta) \subset B_1$, such that $B_2 \subset E^c$. If $z \in B_2$ then there is some $\eta > 0$, such that $B(z, \eta) \subset B_2 \subset E^c$, since B_2 is open. It follows that z is not a limit point of E for every $z \in B_2$, so $B_2 \cap \overline{E}$ is empty and hence $\overline{E}$ is nowhere dense.

We prove that a finite union of nowhere dense sets is nowhere dense by induction on the number n of sets in the union.

If $n = 1$ then there is nothing to prove, so we consider first the case that $n = 2$. Let E_1 and E_2 be nowhere dense sets and let $B = B(x, \varepsilon) \subset X$ be a nonempty open ball. Then there is a ball $B_1 = B(y_1, \delta) \subset B$ such that $B_1 \cap E_1 = \varnothing$. Likewise, there is a ball $B_2 = B(y_2, \eta) \subset B_1$ such that $B_2 \cap E_2 = \varnothing$. It follows that $B_2 \cap (E_1 \cup E_2) = \varnothing$, and since $B_2 \subset B$, it follows that $E_1 \cup E_2$ is nowhere dense.

Assume now that $n \geq 2$ and that the claim is true for n. If $E_1, \dots E_{n+1}$ are nowhere dense sets, then $\bigcup_{j=1}^n E_j$ is nowhere dense by the induction hypothesis, and therefore

$$\bigcup_{j=1}^{n=1} E_j = \left(\bigcup_{j=1}^{n} E_j \right) \cup E_{n+1}$$

is nowhere dense by the case $n = 2$. □

[8] See the paragraph preceding Theorem 3.3.2.

*3.3.4. Baire's category theorem

One of the indications that countable sets and sets of measure zero are small is that a countable union of such sets, is still small. I.e., A countable union of countable sets is still countable, and a countable union of sets of measure zero still has measure zero.

The union of finitely many nowhere dense sets is nowhere dense, but a countable union of nowhere dense sets need not be. For example $\mathbb{Q}$ is a countable union of singletons (each of which is nowhere dense), but $\mathbb{Q}$ is dense. Nonetheless, countable unions of nowhere dense sets are still topologically small, and they are important enough to have a name.

Definition. A subset E of a metric space X is of the *first category* if E is a countable union of nowhere dense sets. Otherwise, E is of the *second category*.

Since a countable union of countable sets is countable, it follows that a countable union of sets of first category is also a countable union of nowhere dense sets. This means that the collection of sets of first category is closed under countable unions like countable sets and sets of measure zero. Baire's category theorem shows that sets of first category are also small in the sense that no set of first category can cover $\mathbb{R}$. In fact this statement is true for all complete metric spaces.

Theorem (Baire). *If X is a complete metric space then X is of the second category.*

Proof. Let X be a complete metric space and let $\{F_n\}_{n\in\mathbb{N}}$ be a countable collection of nowhere dense sets. We want to show that $X \neq \bigcup F_n$. We will do this by constructing a Cauchy sequence $\{x_n\} \subset X$ whose limit x^* is not in $\bigcup F_n$. Throughout the proof, we rely on the fact that $\bigcup_{j=1}^n F_j$ is nowhere dense for each $n \in \mathbb{N}$

Let $x_0 \in X$ be an arbitrary point and let $B_0 = B(x_0, 1)$. Since F_1 is nowhere dense, there is an $x_1 \in B_0$ and an $\varepsilon_1 \in (0, 1/2)$, such that $B(x_1, \varepsilon_1) \subset B_0$ and $F_1 \cap B(x_1, \varepsilon_1) = \varnothing$.[9] We have $\rho(y, x_1) \geq \varepsilon_1$ for all $y \in F_1$, and setting $B_1 = B(x_1, \varepsilon_1/2)$, it follows from the triangle inequality that for all $y \in F_1$ and $x \in B_1$,

$$\rho(y, x) \geq \rho(y, x_1) - \rho(x, x_1) \geq \varepsilon_1 - \varepsilon_1/2 = \varepsilon_1/2.$$

Since $F_1 \cup F_2$ is nowhere dense, there is an $x_2 \in B_1$ and an $\varepsilon_2 \in (0, \varepsilon_1/2)$ such that $B(x_2, \varepsilon_2) \subset B_1$ and $(F_1 \cup F_2) \cap B(x_2, \varepsilon_2) = \varnothing$. Setting $B_2 = B(x_2, \varepsilon_2/2)$, the same reasoning we used above shows that $\rho(y, x) \geq \varepsilon_2/2$ for all $y \in F_1 \cup F_2$ and all $x \in B_2$.

Summarizing, after two steps we have open balls

$$B_2 = B(x_2, \varepsilon_2/2) \subset B(x_1, \varepsilon_1/2) = B_1,$$

with $\varepsilon_2/2 < (\varepsilon_1/2)/2 < 1/4$, such that $\rho(x, y) \geq \varepsilon_2/2$ for all $x \in B_2$ and $y \in F_1 \cup F_2$.

Suppose now that $n \geq 2$ and we have found points $x_1, x_2, \ldots, x_n \in X$ and balls $B_1 \supset B_2 \supset \ldots \supset B_n$, with $B_k = B(x_k, \varepsilon_k/2)$ for $1 \leq k \leq n$, and $\varepsilon_{k+1} < \varepsilon_k/2$ for $1 \leq k \leq n-1$, such that $\rho(x, y) \geq \varepsilon_n/2$ for all $x \in B_n$ and all $y \in \bigcup_{j=1}^n F_j$. Since $\bigcup_{j=1}^{n+1} F_j$ is nowhere dense, there is an $x_{n+1} \in B_n$ and an $\varepsilon_{n+1} \in (0, \varepsilon_n/2)$ such that $B(x_{n+1}, \varepsilon_{n+1}) \subset B_n$ and $\left(\bigcup_{j=1}^{n+1} F_j\right) \cap B(x_{n+1}, \varepsilon_{n+1}) = \varnothing$. Setting $B_{n+1} = B(x_{n+1}, \varepsilon_{n+1}/2)$, the same reasoning

[9] If $F_1 \cap B(x_1, \varepsilon) = \varnothing$ for some $\varepsilon \geq 1/2$, then it is certainly true that $F_1 \cap B(x_1, \varepsilon) = \varnothing$ for all $\varepsilon < 1/2$.

we have used twice before shows that $\rho(y,x) \geq \varepsilon_{n+1}/2$ for all $y \in \bigcup_{j=1}^{n+1} F_j$ and all $x \in B_{n+1}$.

In this way, we recursively construct a sequence of open balls, $\{B_n\}_{n=1}^{\infty} = \{B(x_n, \varepsilon_n/2)\}_{n=1}^{\infty}$, with $\varepsilon_n/2 > \varepsilon_{n+1}$ and $B_{n+1} \subset B_n$ for all n. For every $n \in \mathbb{N}$, these balls also have the property that for all $x \in B_n$ and for all $y \in \bigcup_{j=1}^{n} F_j$, we have $\rho(x,y) \geq \varepsilon_n/2$.

Since $\varepsilon_1 < 1/2$ and $\varepsilon_{n+1} < \varepsilon_n/2$ for all $n \in \mathbb{N}$, it follows by induction on n that $\varepsilon_n < 1/2^n$ for all n. Hence, if $n, m > N$, then $x_n, x_m \in B_N$ and therefore

$$\rho(x_n, x_m) \leq \rho(x_n, x_N) + \rho(x_N, x_m) < \varepsilon_N/2 + \varepsilon_N/2 = \varepsilon_N < 1/2^N.$$

It follows that $\{x_n\}$ is a Cauchy sequence in X and hence converges to a limit $x^* \in X$ because X is complete.

If $y \in \bigcup_{n=1}^{\infty} F_n$, then $y \in \bigcup_{n=1}^{n_0} F_n$ for some $n_0 \in \mathbb{N}$ and therefore $\rho(y,x) \geq \varepsilon_{n_0}/2$ for all $x \in B_{n_0}$. Let n be large enough that $n > n_0$ and $\rho(x_n, x^*) < \varepsilon_{n_0}/2$. With this choice of n, we have $x_n \in B_{n_0}$ and therefore

$$\rho(y, x^*) \geq \rho(y, x_n) - \rho(x_n, x^*) > \varepsilon_{n_0}/2 - \varepsilon_{n_0}/2 = 0,$$

so $y \neq x^*$. It follows that $x^* \notin \bigcup_{n=1}^{\infty} F_n$, so $\bigcup_{n=1}^{\infty} F_n \neq X$. □

⋆3.3.5. A partition of $\mathbb{R}$ into small sets

No countable union of sets of measure zero can cover $\mathbb{R}$ and no countable union of sets of first category can cover $\mathbb{R}$. One way of understanding this is that sets of measure zero are all small in the same way, so combining them in a countable union produces another set of the same type. The same can be said for sets of the first category.

On the other hand, sets of measure zero are small in a different way than sets of first category. In particular, a set of measure zero can be of second category, and a set of first category can have "full measure".[10] We illustrate this by finding a subset $A \subset \mathbb{R}$ of measure zero whose complement $B = \mathbb{R} \setminus A$ is of first category.

We begin with a simple observation.

Lemma. *A subset F of a metric space X is closed and nowhere dense in X if and only if F^c is open and dense in X.*

Note that the complement of an arbitrary dense set need not be nowhere dense. For example, the complement of $\mathbb{Q}$ in $\mathbb{R}$ is dense.

Proof. Suppose that F is closed and nowhere dense in X. Then $G = F^c$ is open and for any ball $B_0 = B(x, \varepsilon) \subset X$, there is a ball $B_1 = B(y, \delta) \subset B_0$ such that $F \cap B_1 = \varnothing$, so that $B_1 \subset G$, showing that G is dense.

In the other direction, if G is open and dense in X then $F = G^c$ is closed. If B_0 is an open ball in X, then there is a point $x_0 \in G \cap B_0$ since G is dense. Since $B_0 \cap G$ is open, there is a $\delta > 0$ such that $B_1 = B(x_0, \delta) \subset B_0 \cap G$, and hence $F \cap B_1 = \varnothing$. □

From this and Baire's theorem we have the following corollary.

[10] See Chapter 8.

Corollary. *If X is a complete metric space, and G_n is a dense open set in X for every $n \in \mathbb{N}$, then $\bigcap_{n\in\mathbb{N}} G_n$ is not empty.*

Proof. If G_n is a dense open set, then G_n^c is a nowhere dense and therefore, by Baire's category theorem,

$$\left(\bigcap_{n=1}^{\infty} G_n\right)^c = \bigcup_{n=1}^{\infty} G_n^c \neq X.$$

Hence $\bigcap_{n=1}^{\infty} G_n$ is not empty. □

Theorem. *There exist sets $A, B \subset \mathbb{R}$ such that A has measure zero, B is of first category and $\mathbb{R} = A \cup B$.*

Proof. The rational numbers form a countable set, so we can list them in a sequence: $\mathbb{Q} = \{r_k\}_{k=1}^{\infty}$. For each $n \in \mathbb{N}$, we use this sequence to form the set

$$G_n = \bigcup_{k=1}^{\infty} \left(r_k - \frac{1}{n^k}, r_k + \frac{1}{n^k}\right) = \bigcup_{k=1}^{\infty} I_{n,k}.$$

The set G_n is open for every n, as a union of open intervals, and it is dense in $\mathbb{R}$ because $\mathbb{Q} \subset G_n$. Writing $A = \bigcap_{n\in\mathbb{N}} G_n$, it follows from Lemma 3.3.5 that

$$B = A^c = \left(\bigcap_{n\in\mathbb{N}} G_n\right)^c = \bigcup_{n\in\mathbb{N}} G_n^c$$

is of first category. It remains to show that A has measure zero.

Let $\varepsilon > 0$ be given, and let $n_\varepsilon \in \mathbb{N}$ be such that $2/(n_\varepsilon - 1) < \varepsilon$. By construction, we have $A \subset G_{n_\varepsilon} = \bigcup_{k\in\mathbb{N}} I_{n_\varepsilon,k}$, and

$$\sum_{k=1}^{\infty} \ell(I_{n_\varepsilon,k}) = \sum_{k=1}^{\infty} \frac{2}{n_\varepsilon^k} = \frac{2}{n_\varepsilon - 1} < \varepsilon.$$

Hence, A has measure zero. □

Exercises

ex3.3.1. Complete the proof of Corollary 3.3.2.

ex3.3.2. Prove that $\mathscr{C}$ is uncountable by showing that

$$\psi(x) = \sum_{n=1}^{\infty} \frac{t_n(x)/2}{2^n}$$

is a *surjection* from $\mathscr{C}$ to $[0,1]$. It is understood here that if $x \in \mathscr{C}$ has two ternary expansions, we choose the one without 1s.

ex3.3.3. A closed subset F of a metric space X is a *perfect set* if every point in F is a limit point of F. Show that the Cantor set is perfect.

ex3.3.4. Generalize Theorem 3.3.3 to any metric space X.

ex3.3.5. Prove Corollary 3.3.5.

Chapter 4

Continuous Functions

The intuitive notion of a continuous function, $f : \mathbb{R} \to \mathbb{R}$, is that f is continuous if its graph, $\{(x, f(x)) : x \in \mathbb{R}\}$, is a continuous (i.e., unbroken) curve in $\mathbb{R}^2$. This description is important because it is easy to visualize, but it needs to be refined to obtain a definition of continuity that is useful. A better sense of what it means for a function $f : \mathbb{R} \to \mathbb{R}$ to be continuous is to say that f is continuous if, when x is close to y, $f(x)$ is close to $f(y)$. This idea is made precise in terms of the limit of a function at a point.

We will focus in this chapter on functions from $\mathbb{R}$ to $\mathbb{R}$, and more generally from subsets of $\mathbb{R}$ to $\mathbb{R}$. However much of what we discuss here generalizes directly to the study of functions from a metric space X to $\mathbb{R}$ and even more generally from one metric space to another. At the end of the chapter we will briefly survey these generalizations. In any case, except when specifically stated otherwise, we will assume throughout this chapter (and the remainder of the book) that the functions we discuss are *real-valued* — functions whose ranges are contained in $\mathbb{R}$.

4.1. Limits of functions

The limit of a function f at a point $a \in \mathbb{R}$ depends on the behavior of the function in a *neighborhood* of a.

Definition. The set $U \subset \mathbb{R}$ is a *neighborhood* of the point $x \in \mathbb{R}$ if there is an $\varepsilon > 0$ such that $B(x, \varepsilon) = (x - \varepsilon, x + \varepsilon) \subset U$. Equivalently, U is a neighborhood of x if x belongs to the interior of U. More generally, U is a neighborhood of a set $E \subset \mathbb{R}$ if E is a subset of the interior of U.

U is a *punctured neighborhood* of the point x if there is a neighborhood V of x such that $U = V \setminus \{x\}$.

Comment. Any open set $G \subset \mathbb{R}$ that contains x is a neighborhood of x, but not every neighborhood of x is an open set. For example, any closed set is a neighborhood of any point in its interior.

4.1.1. The limit of a function at a point

Definition. Suppose that the function f is defined in a punctured neighborhood of the point $a \in \mathbb{R}$. If there is an $L \in \mathbb{R}$ such that for every $\varepsilon > 0$, there is a $\delta > 0$ such that $0 < |x-a| < \delta$ implies $|f(x)-L| < \varepsilon$, then we say that the *limit* of $f(x)$ as *x approaches a* (or more briefly, the limit of f at a) exists and is equal to L, and we write

$$\lim_{x\to a} f(x) = L.$$

If there is no number L satisfying these conditions, then we say that $\lim\limits_{x\to a} f(x)$ *does not exist.*

Neither the existence nor the value of $\lim_{x\to a} f(x)$ depend on the value, or even the existence of $f(a)$. In fact, we make sure to exclude $f(a)$ from consideration in the definition of the limit. On the other hand, if $f(a)$ exists, then it is natural to guess that $f(a)$ might be the value of $\lim_{x\to a} f(x)$. This guess is not always correct, but for many of the most basic functions it is.

Proposition. *If $a \in \mathbb{R}$ and $n \in \mathbb{N}$, then $\lim\limits_{x\to a} x^n = a^n$. If $a > 0$, then $\lim\limits_{x\to a} x^{1/n} = a^{1/n}$.*

Proof. If $n \in \mathbb{N}$, then the function $f(x) = x^n$ is defined on all of $\mathbb{R}$ and the function $g(x) = x^{1/n}$ is defined for all $x > 0$. The statement now follows from Lemma 1.6.1 for x^n and Lemma 1.6.2 for $x^{1/n}$. □

The limit of a function at a point can also be described in terms of sequences converging to that point.

Theorem. *Suppose that f is defined in a punctured neighborhood U of $a \in \mathbb{R}$ then $\lim\limits_{x\to a} f(x) = L$ if and only if $\lim\limits_{n\to\infty} f(x_n) = L$ for every sequence $\{x_n\} \subset U$ that converges to a.*

Proof. Suppose that $\lim_{x\to a} f(x) = L$, let $\{x_n\} \subset U$ be a sequence converging to a and let $\varepsilon > 0$ be given. Since $\lim_{x\to a} f(x) = L$, it follows that there is a $\delta > 0$ such that if $0 < |x-a| < \delta$, then $|f(x) - L| < \varepsilon$, and since $x_n \to a$, it follows that there is an $N \in \mathbb{N}$ such that $0 < |x_n - a| < \delta$ if $n > N$. Hence, if $n > N$, then $|f(x_n) - L| < \varepsilon$, so $\lim_{n\to\infty} f(x_n) = L$.

Now suppose that $\lim_{x\to a} f(x) \neq L$. This means that either the limit doesn't exist or that the limit exists but is equal to $M \neq L$. In either case, it follows from the definition of a limit that there exists an $\varepsilon > 0$ such that for every $n \in \mathbb{N}$, there is an $x_m \in U$ such that $0 < |x_n - a| < 1/n$ and $|f(x_n)-L| \geq \varepsilon$. It follows that the sequence $\{x_n\}$ converges to a but $f(x_n)$ 6→ L. □

4.1.2. Limits and arithmetic

It follows from Theorem 4.1.1 that just as limits of sequences play nicely with arithmetic operations, so to do limits of functions.

Theorem. *Suppose that f and g are defined in a neighborhood of a. If $\lim_{x\to a} f(x) = L$ and $\lim_{x\to a} g(x) = M$, then*

$$\lim_{x\to a} f(x) \pm g(x) = L \pm M \quad \textit{and} \quad \lim_{x\to a} f(x)\cdot g(x) = L\cdot M.$$

Furthermore, if $M \neq 0$, then

$$\lim_{x\to a} f(x)/g(x) = L/M.$$

Proof. As we suggested above, this follows from Theorem 4.1.1 together with Theorem 2.3.2 for sums and products, and Proposition 2.3.2 (in combination with Theorem 2.3.2) for quotients. We leave the details to the exercises. □

The rules for sums and products may be extended to sums and products of any finite number of terms or factors.

Corollary. *If the functions $f_1, \ldots, f_n$ are all defined in a punctured neighborhood of $a \subset \mathbb{R}$, and $\lim_{x\to a} f_j(x) = L_j$, for $1 \le j \le n$, then*

$$\lim_{x\to a} (f_1(x) + \cdots + f_n(x)) = L_1 + \cdots + L_n$$

and

$$\lim_{x\to a} (f_1(x)\cdots f_n(x)) = L_1 \cdots L_n.$$

Proof. This is proved by induction on n, with the case $n = 2$ established by Theorem 4.1.2, above. We leave the details as an exercise. □

As a simple application, we have the following conclusion about polynomials and rational functions.

Proposition. *Suppose that $f(x)$ is a polynomial and $r(x) = p(x)/q(x)$ is a rational function, both with real coefficients. Then $\lim_{x\to a} f(x) = f(a)$ for all $a \in \mathbb{R}$ and $\lim_{x\to a} r(x) = r(a)$ at all $a \in \mathbb{R}$, such that $q(a) \neq 0$.*

Proof. To begin, we note that if $f(x)$ is a polynomial, then f is defined on all of $\mathbb{R}$, and therefore in a neighborhood of any $a \in \mathbb{R}$. If $f(x) = c$ then the claim is plainly true for f, since $f(a) - f(x) = 0$ for all $x \in \mathbb{R}$. It then follows from Proposition 4.1.1 and Theorem 4.1.2 (for products) that the claim is also true for $f(x) = cx^n$ for any $c \in \mathbb{R}$ and $n \in \mathbb{N}$. Finally, if $f(x) = c_n x^n + \cdots + c_1 x + c_0$, then the claim follows from Corollary 4.1.2 (for sums).

Suppose now that $r(x) = p(x)/q(x)$, where p and q are polynomials, and that $q(a) \neq 0$. By the truth of the statement for polynomials, there exists $\delta > 0$ such that $|x - a| < \delta$ implies $|q(x) - q(a)| < |q(a)|/2$, from which it follows that $q(x) \neq 0$ for $x \in (a - \delta, a + \delta)$. The claim for $r(x)$ now follows from Theorem 4.1.2 (for quotients). □

4.1.3. One-sided limits

Definition. Suppose that $f(x)$ is defined on the interval $(a-\eta, a)$, for some $\eta > 0$. If there is an $L \in \mathbb{R}$ such that for every $\varepsilon > 0$, there is a $\delta > 0$ such that $x \in (a-\delta, a)$ implies $|f(x) - L| < \varepsilon$, then *the limit as x approaches a from the left* is equal to L. We denote this by

$$\lim_{x\to a^-} f(x) = L,$$

and call L the *left-hand limit* of $f(x)$ as x approaches a.

Similarly, suppose that $f(x)$ is defined on the interval $(a, a+\eta)$, for some $\eta > 0$. If there is an $M \in \mathbb{R}$ such that for every $\varepsilon > 0$, there is a $\delta > 0$ such that $x \in (a, a+\delta)$ implies $|f(x) - M| < \varepsilon$, then *the limit as x approaches a from the right* is equal to M. We denote this by

$$\lim_{x\to a^+} f(x) = M,$$

and call M the *right-hand limit* of $f(x)$ as x approaches a.

As with (two-sided) limits, if there is no number L (or M) satisfying the condition above, then the left-hand (right-hand) limit of $f(x)$ at a does not exist.

Proposition. *If $f(x)$ is defined on a punctured neighborhood of $a \in \mathbb{R}$, then $\lim\limits_{x\to a} f(x) = L$ if and only if $\lim\limits_{x\to a^-} f(x) = \lim\limits_{x\to a^+} f(x) = L$.*

Proof. Exercise. □

4.1.4. Limits at infinity

Definition. Suppose that $f : \mathbb{R} \to \mathbb{R}$ is defined for all sufficiently large x. If for any $\varepsilon > 0$, there is an $M > 0$ such that $|f(x) - L| < \varepsilon$ for all $x > M$, then we say that *the limit of f as x tends to infinity is equal to L* and write $\lim_{x\to\infty} f(x) = L$.

Analogously, if f is defined for all sufficiently large $-x$ and for every $\varepsilon > 0$, there is an $N < 0$ such that $|f(x) - L| < \varepsilon$ for all $x < N$, we write $\lim_{x\to-\infty} f(x) = L$, and say that f tends to L as x tends to negative infinity.

Proposition. *If $f(x) = a_m x^m + \cdots + a_0$ and $g(x) = b_n x^n + \cdots + b_0$ are polynomials of degrees m and n, respectively. Then*

$$\lim_{x\to\infty} \frac{f(x)}{g(x)} = \begin{cases} a_n/b_n & : n = m \\ 0 & : n > m \\ \textit{does not exist} & : m > n. \end{cases}$$

Proof. Exercise. □

4.1.5. Bounded monotone functions

Definition. A function f defined on a set $E \subset \mathbb{R}$ is *monotone nondecreasing* on E if for all $x, y \in E$, $x < y$ implies $f(x) \le f(y)$. The function is *monotone increasing* on E if for all $x, y \in E$, $x < y$ implies $f(x) < f(y)$. A function f defined on a set $E \subset \mathbb{R}$ is *monotone nonincreasing* on E if for all $x, y \in E$, $x < y$ implies $f(x) \ge f(y)$. The function is *monotone decreasing* on E if for all $x, y \in E$, $x < y$ implies $f(x) > f(y)$.

If the function f is either monotone nonincreasing or monotone nondecreasing on a set E, f is said to be *monotone* on E.

The function f is *bounded* on E if the set $\{f(x) : x \in E\}$ is bounded.

Lemma. *If the function f is bounded and monotone in the interval $I = (a, b)$ then both $\lim_{x \to a^+} f(x)$ and $\lim_{x \to b^-} f(x)$ exist.*

Proof. We will prove the claim in the case that f is nondecreasing on I. The proof in the case that f is nonincreasing is analogous, and we leave it as an exercise.

From the assumption that f is bounded on I, it follows that the supremum $L = \sup\{f(x) : a < x < b\}$ exists. Hence given $\varepsilon > 0$, there exists $x_\varepsilon \in (a, b)$ such that $L - \varepsilon < f(x_\varepsilon) \leq L$, and we set $\delta_\varepsilon = b - x_\varepsilon$. If $b - \delta_\varepsilon < x < b$, then $x_\varepsilon < x < b$ and therefore $L - \varepsilon < f(x_\varepsilon) \leq f(x) \leq L$, since f is nondecreasing on I. This shows that $\lim_{x \to b^-} f(x) = L$. A similar argument (**ex4.1.5**) shows that $\lim_{x \to a^+} f(x) = \inf\{f(x) : a < x < b\}$. □

4.1.6. The lim sup and lim inf of a function

The fact that $\lim_{x \to a} f(x) = L$ gives information about of the behavior of the function f in the neighborhood of the point a, namely that f is approximately constant around a. To the extent possible, we would also like to understand the behavior of f near a in the case that the limit does not exist. One possibility is that f is unbounded as x approaches a.

Proposition. *If f is unbounded in every punctured neighborhood of a, then $\lim_{x \to a} f(x)$ does not exist.*

Proof. In this case, for any $L \in \mathbb{R}$, and any $\delta > 0$, there is an x such that $0 < |x - a| < \delta$ such that $|f(x)| > L + 1$. It follows that $\lim_{x \to a} f(x) \neq L$, and since this is true for all $L \in \mathbb{R}$, $\lim_{x \to a} f(x)$ cannot exist. □

Suppose on the other hand that f is bounded in some punctured neighborhood of a, i.e., there is a $\delta_0 > 0$ such that f is bounded in $(a - \delta_0, a) \cup (a, a + \delta_0)$. If $\lim_{x \to a} f(x)$ fails to exist in this case, then there is still some limiting behavior that we can describe.

If $0 < \delta < \delta_0$, then both

$$s_{a,\delta}(f) = \sup\{f(x) : 0 < |x - a| < \delta\}$$

and

$$i_{a,\delta}(f) = \inf\{f(x) : 0 < |x - a| < \delta\}$$

exist in $\mathbb{R}$ and if $0 < \delta_2 < \delta_1 < \delta_0$, then $s_{a,\delta_2}(f) \leq s_{a,\delta_1}(f)$ and $i_{a,\delta_2}(f) \geq i_{a,\delta_1}(f)$. It follows that $s_{a,\delta}(f)$ monotone nonincreasing in the interval $(0, \delta_0)$, and the function $i_{a,\delta}(f)$ is monotone nondecreasing there.

Definition. Suppose that f is bounded in $(a - \delta_0, a) \cup (a, a + \delta_0)$, for some $\delta_0 > 0$. The *limits superior* and *inferior* of $f(x)$ as x approaches a are given by

$$\limsup_{x \to a} f = \lim_{\delta \to 0^+} s_{a,\delta}(f) \quad \text{and} \quad \liminf_{x \to a} f = \lim_{\delta \to 0^+} i_{a,\delta}(f),$$

respectively. The limits exists by Lemma 4.1.5, since $s_{a,\delta}(f)$ and $i_{a,\delta}(f)$ are both monotone in $(0, \delta_0)$.

If f is unbounded in every punctured neighborhood of a, then either one or both of $\limsup_{x\to a} f$ and $\liminf_{x\to a} f$ will be infinite.

The following theorem is analogous to Lemma 2.5.4.

Theorem. *If f is defined and bounded in a punctured neighborhood of a, then $\lim_{x\to a} f(x)$ exists if and only if $\limsup_{x\to a} f(x) = \liminf_{x\to a} f(x) = L$, for some $L \in \mathbb{R}$.*

Proof. Suppose that $\lim_{x\to a} f(x) = L$, then for any $\varepsilon > 0$ there is a $\delta_\varepsilon > 0$ such that $|f(x) - f(a)| < \varepsilon/2$ in $(a - \delta_\varepsilon) \cup (a + \delta_\varepsilon)$. It follows that

$$f(a) + \varepsilon/2 > s_{a,\delta}(f) \geq i_{a,\delta}(f) > f(a) - \varepsilon/2,$$

so $s_{a,\delta}(f) - i_{a,\delta}(f) \leq \varepsilon$, for all $\delta < \delta_\varepsilon$. Hence

$$\limsup_{x\to a} f(x) - \liminf_{x\to a} f(x) \leq \varepsilon,$$

for any $\varepsilon > 0$, so $\limsup_{x\to a} f(x) = \liminf_{x\to a} f(x)$.

We leave the other direction as an exercise. □

4.1.7. Big O and little o

If $\lim_{x\to a} f(x) - g(x) = 0$, then $f(x) \approx g(x)$ for x sufficiently close to a, with the approximation becoming more accurate as x gets closer to a. To express this notion of approximation without writing "$\lim_{x\to a}$" explicitly, we can use "big O" or "little o" notation.

Definition. If $F(x)$ and $\phi(x)$ are defined in a neighborhood of 0, then we say that *F is big O of ϕ* as $h \to 0$, written $F(h) = O(\phi(h))$, if there exist constants $\delta, C > 0$ such that $|F(h)| \leq C|\phi(h)|$ for all $h \in (-\delta, +\delta)$.

We say that *F is little o of ϕ* as $h \to 0$, written $F(h) = o(\phi(h))$ if for every $\varepsilon > 0$, there is an $\eta > 0$ such that $|F(h)| < \varepsilon|\phi(h)|$ for all $h \in (-\eta, \eta)$.

If F and ϕ are both defined in (M, ∞) for some $M \in \mathbb{R}$, then $F = O(\phi)$ as $k \to \infty$, if there exist constants $N, C > 0$ such that $|F(k)| \leq C|\phi(k)|$ for all $k > N$, and $F = o(\phi)$ as $F \to \infty$, if for every $\varepsilon > 0$, there is an $N > 0$ such that $|F(k)| < \varepsilon|\phi(k)|$ for all $k > N$.

Comments. The letter "O" indicates "order of magnitude". In this notation, the function ϕ is typically a simple, well understood function that we are using to describe the magnitude of the function F near 0, or "near infinity". Also, we omit the expressions "$h \to 0$" or "$k \to \infty$" if they are easily understood from the context.

This notation is flexible and is frequently used to describe how well $f(x)$ is approximated by $g(x)$ as $x \to a$ or as $x \to \infty$. For example, if

$$\lim_{x\to a} f(x) = L = \lim_{x\to a} g(x),$$

we can say that $f(a+h) - g(a+h) = o(1)$. In cases like this, we typically substitute x for $a + h$ and write $f(x) = g(x) + o(1)$, (as $(x - a) \to 0$).

If we have more precise information about the convergence, we can include this as well. For example, writing $x^2 = 4x-4+o(x-2)$, means that the difference $|x^2-(4x-4)|$ goes to 0 "faster than" $|x-2|$ as $x \to 2$ (**ex4.1.8**).

Exercises

ex4.1.1. Show that if $f(x) = c$ or $f(x) = x$, then $\lim_{x\to a} f(x) = f(a)$ for any $a \in \mathbb{R}$.

ex4.1.2. Complete the inductive proof of Corollary 4.1.2.

ex4.1.3. Prove Proposition 4.1.3.

ex4.1.4. Prove Proposition 4.1.4. State and prove an analogous proposition about the limit at negative infinity of a rational function. What, if anything, changes?

ex4.1.5. Complete the proof of Lemma 4.1.5.

ex4.1.6. Complete the proof of Theorem 4.1.6.

ex4.1.7. Suppose that f and g are defined in a neighborhood U of a, $f(a) = 0$, and $g(x) \neq 0$ in $U\setminus\{a\}$. Show that $f(x) = o(g(x))$ (as $x \to a$) if and only if $\lim_{x\to a} f(x)/g(x) = 0$.

ex4.1.8. Show that $x^2 = 4x - 4 + o(x-2)$.

ex4.1.9. Let $R(x) = \dfrac{x^2+3x+2}{2x+5}$

(i) Show that $R(x) = O(x)$ as $x \to \infty$.

(ii) Find a linear function $L(x) = ax + b$ such that $R(x) = L(x) + O(1/x)$.

4.2. Continuity

A function f is continuous at the point $a \in \mathbb{R}$ if the values of $f(x)$ are close to $f(a)$ when x is close to a. The precise definition uses limits.

Definition. The function f is *continuous at* $a \in \mathbb{R}$ if f is defined in a neighborhood of a and and $\lim\limits_{x\to a} f(x) = f(a)$.

If $E \subset \mathbb{R}$ and f is defined in a neighborhood of E, then f is *continuous on* E if f is continuous *at* every point $a \in E$.

If $\lim_{x\to a} f(x)$ does not exist, or if $\lim_{x\to a} f(x) \neq f(a)$, then we say that $f(x)$ is *discontinuous* at a, or that a is a *point of discontinuity* of the function f.

Comment. We can leave the *language* of limits out of the definition of continuity by appealing directly to the idea of a limit. That is, we can say that *f is continuous at a if and only if f is defined in a neighborhood of a and for every $\varepsilon > 0$ there is a $\delta > 0$ such that $|x-a| < \delta$ implies $|f(x) - f(a)| < \varepsilon$.*

4.2.1. The arithmetic of continuous functions

We saw in Section 4.1.2 that the limits of sums, products and quotients are well-behaved and this implies that sums, products and quotients of continuous functions are also well-behaved.

Proposition. *If f and g are both continuous at a, then $f+g$ and fg are both continuous at a. If $g(a) \neq 0$, then f/g is also continuous at a.*

Proof. By assumption, we have $\lim_{x\to a} f(x) = f(a)$ and $\lim_{x\to a} g(x) = g(a)$, so it follows from Theorem 4.1.2 that

$$\lim_{x\to a}(f(x)+g(x)) = f(a)+g(a) \text{ and } \lim_{x\to a}(f(x)g(x)) = f(a)g(a).$$

It likewise follows from this theorem that if $g(a) \neq 0$, then

$$\lim_{x\to a}(f(x)/g(x)) = f(a)/g(a). \qquad \square$$

The composition of continuous functions is also well-behaved.

Theorem. *If f is continuous at a and g is continuous at $f(a)$, then $g \circ f$ is continuous at a.*

Proof. From the continuity of g at $f(a)$, it follows that given $\varepsilon > 0$, there is a $\eta > 0$ such that $|u - f(a)| < \eta$ implies $|g(u) - g(f(a))| < \varepsilon$. Likewise, from the continuity of f at a, there is an $\delta > 0$, such that $|x - a| < \delta$ implies $|f(x) - f(a)| < \eta$. Hence, if $|x - a| < \delta$, then $|g(f(x)) - g(f(a))| < \varepsilon$. $\square$

Since continuity on a set means continuous at every point in the set, we have the following immediate corollary.

Corollary. *If f is continuous on the set E and g is continuous on the set $f(E) = \{f(x) : x \in E\}$, then $g \circ f$ is continuous on E.*

4.2.2. Examples of continuous functions

Most commonly used functions are continuous where they are defined.

Theorem.

(a) *Polynomials are continuous in all of $\mathbb{R}$, and rational functions are continuous where they are defined.*

(b) *For any $u \in \mathbb{R}$, the function $g(x) = x^u$ is continuous in $(0, \infty)$.*

(c) *If $a > 0$, then $f(x) = a^x$ is continuous in $\mathbb{R}$.*

(d) *If $f(x)$ is continuous and injective on the set $E \subset \mathbb{R}$, then f^{-1} is continuous on $f(E)$.*

Proof.

(a) This follows from Proposition 4.1.2.

(b) The function $g(x)$ is continuous in $(1, \infty)$ by Theorem 1.7.3. If $0 < x < 1$, then $1/x > 1$ and $g(x) = (1/x)^{-u}$ is continuous in $(0, 1)$ by Corollary 4.2.1. Continuity at $a = 1$ follows from **ex1.7.3**.

(c) For $a > 1$, this follows from Theorem 1.7.3 and Corollary 4.2.1 extends this to the case that $0 < a < 1$. The function 1^x is constant.

(d) This follows from **ex4.2.4**. □

In Theorem 2.7.2 we proved that the series $\sum_{n=0}^{\infty} \frac{x^n}{n!}$ converges absolutely for all $x \in \mathbb{R}$, from which it follows that the exponential function

$$\exp(x) = \sum_{n=0}^{\infty} \frac{x^n}{n!}$$

is defined for all $x \in \mathbb{R}$. Moreover, if $x \geq 0$, then $\exp(x) \geq 1$.

Proposition. *The exponential function is continuous on all of* $\mathbb{R}$.

Proof. Suppose that $a, x \in \mathbb{R}$ and $|x - a| < 1$. Then $|x| < |a| + 1$ and hence,

$$\left|\sum_{j=1}^{n} x^{n-j} a^{j-1}\right| \leq \sum_{j=1}^{n} |x|^{n-j} |a|^{j-1} < n(|a|+1)^{n-1},$$

for any $n \in \mathbb{N}$. It follows from this and the identity (1.1.1) that for all $n \in \mathbb{N}$,

$$\left|\frac{x^n - a^n}{n!}\right| = \left|\frac{(x-a)\sum_{j=1}^{n} x^{n-j} a^{j-1}}{n!}\right| < |x-a| \frac{(|a|+1)^{n-1}}{(n-1)!}. \tag{4.2.1}$$

Now, by Theorem 2.6.4, we have

$$\exp(x) - \exp(a) = \sum_{n=0}^{\infty} \frac{x^n}{n!} - \sum_{n=0}^{\infty} \frac{a^n}{n!} = \sum_{n=1}^{\infty} \frac{x^n - a^n}{n!},$$

and therefore if $|x - a| < 1$, then it follows from (4.2.1) that

$$|\exp(x) - \exp(a)| < |x - a| \cdot \sum_{n=1}^{\infty} \frac{(|a|+1)^{n-1}}{(n-1)!} = |x - a| \exp(|a| + 1).$$

Hence, given $\varepsilon > 0$ if $\delta = \min(1, \varepsilon/\exp(|a|+1))$, then $|x - a| < \delta$ implies $|\exp(x) - \exp(a)| < \varepsilon$. □

Comment. The series defining the exponential function is an example of a *power series*, and the continuity of $\exp(x)$ is a special case of Theorem 5.4.1 about power series in general.

4.2.3. Continuity, more generally

The definition of continuity extends naturally to functions between any two metric spaces.

Definition. If (X, ρ_X) and (Y, ρ_Y) are metric spaces, $f : X \to Y$ is a function and $x_0 \in X$, then f is *continuous at* x_0 if for every $\varepsilon > 0$ there is a $\delta > 0$ such that if $\rho_X(x, x_0) < \delta$, then $\rho_Y(f(x), f(x_0)) < \varepsilon$.

If $E \subset X$, then f is *continuous on* E if f is continuous at every point $x \in E$.

Continuity may also be defined in terms of the metric topology.

Theorem. *If (X, ρ_X) and (Y, ρ_Y) are metric spaces, then the function $f : X \to Y$ is continuous on X if and only if $f^{-1}(U)$ is open in X for every open set $U \subset Y$.*

Proof. Suppose that $f : X \to Y$ is continuous in the metric sense, as given in the definition above, let $U \subset Y$ be open and let $x_0 \in f^{-1}(U)$. Then $f(x_0) \in U$ so there is an $\varepsilon > 0$ such that $B(f(x_0), \varepsilon) \subset U$, because U is open in Y. Since f is continuous in the metric sense, there is a $\delta > 0$ such that $\rho_X(x_0, x) < \delta$ implies $\rho_Y(f(x_0), f(x)) < \varepsilon$, and therefore $B(x_0, \delta) \subset f^{-1}(B(f(x_0), \varepsilon)) \subset f^{-1}(U)$, so $f^{-1}(U)$ is open in X.

Now suppose that $f^{-1}(U)$ is open in X for every open subset U of Y, and let $x_0 \in X$ and $\varepsilon > 0$ be given. The ball $B(f(x_0), \varepsilon)$ is open in Y, so by assumption, $f^{-1}(B(f(x_0), \varepsilon)$ is open in X. It follows that there is a $\delta > 0$ such that $B(x_0, \delta) \subset f^{-1}(B(f(x_0), \varepsilon)$, because $x_0 \in f^{-1}(B(f(x_0), \varepsilon)$. In other words, if $\rho_X(x, x_0) < \delta$, then $f(x) \in B(f(x_0), \varepsilon)$ so that $\rho_Y(f(x), f(x_0)) < \varepsilon$, and since x_0 and ε were arbitrary, it follows that f is continuous on X in the metric sense. □

Comment. The preceding theorem provides a definition of continuity in terms of open sets that is equivalent to the original definition on metric spaces, but with no reference to the metric. This allows us to generalize the concept of a continuous function to maps between arbitrary *topological* spaces. This definition is also quite useful, even in metric spaces as we shall see (e.g., **ex4.2.4** and Theorem 4.4.1).

Exercises

ex4.2.1. Show that any interval $I \subset \mathbb{R}$ may be expressed as a countable union of closed bounded intervals $\{I_N : N \in \mathbb{N}\}$.

Hint: If $I = (a, b)$ and $b - a = \delta > 0$, then for any $N \in \mathbb{N}$

$$[a + \delta/3N, b - \delta/3N] \subset (a, b).$$

Also, $\mathbb{R} = \bigcup_{N \in \mathbb{N}} [-N, N]$.

ex4.2.2. Show that if $f : \mathbb{R} \to \mathbb{R}$ is continuous and injective, then $f^{-1}(F)$ is closed for every closed set $F \subset \mathbb{R}$. Show by way of a counterexample that this statement is false if f is not injective.

ex4.2.3. Show that if $f : \mathbb{R} \to \mathbb{R}$ is continuous and injective, then $f(U)$ is open for every open set $U \subset \mathbb{R}$. Show by way of a counterexample that this statement is false if f is not injective.

ex4.2.4. Show that if f is continuous and injective on the set $E \subseteq \mathbb{R}$, then f^{-1} is continuous on $f(E)$.

ex4.2.5. Suppose that $f(x_1, \ldots, x_n)$ is a function from $\mathbb{R}^n$ to $\mathbb{R}$, and for a fixed point $(a_1, \ldots, a_n) \in \mathbb{R}^n$, let

$$f_k(x) = f(a_1, \ldots, a_{k-1}, x, a_{k+1}, \ldots, a_n),$$

for $1 \le k \le n$. Show that f is continuous at the point $(a_1, \ldots, a_n)$ if and only if f_k is continuous at a_k for each k.

4.3. Points of discontinuity

To understand continuity, it is useful to understand how it might fail.

Definition. If the function f is defined in a punctured neighborhood of $a \in \mathbb{R}$ and f is not continuous at a, then a is a *point of discontinuity* of f.

If a is a point of discontinuity of the function f, then there are three possibilities: (i) f is not defined at a, (ii) $\lim_{x\to a} f(x)$ does not exist or (iii) the limit exists but it is not equal to $f(a)$.

If $\lim_{x\to a} f(x) = L$, then in both cases (i) and (iii) we can define a function

$$f_1(x) = \begin{cases} f(x) & : x \neq a \\ L & : x = a \end{cases}$$

which is identical to f for all $x \neq a$ (in the domain of f), and which is continuous at a. In other words, we can 'remove the discontinuity' at a in these cases.

In the case that $\lim_{x\to a} f(x)$ does not exist, then there is no way to (re)define $f(a)$ that will make the function continuous at a. In this case, either f is unbounded in every neighborhood of a or f is bounded in some neighborhood of a, but $\limsup_{x\to a} f \neq \liminf_{x\to a}$ (see Proposition 4.1.6 and Theorem 4.1.6).

4.3.1. Monotone functions

Generally speaking, the behavior of a function near a point of discontinuity can be wild. However for monotone functions, discontinuity is relative tame.

Definition. If $\lim_{x\to a-} f(x) = L^-$ and $\lim_{x\to a+} f(x) = L^+$ both exist, but are not equal (so that $\lim_{x\to a} f(x)$ does not exist), then f is said to have a *jump* discontinuity at a. The difference $|L^- - L^+| > 0$ is the size of the jump.

Lemma. *If f is defined and monotone in (a, b), then every discontinuity of f in (a, b) is a jump discontinuity.*

Proof. Suppose that $x_0 \in (a, b)$, then there is a $\delta > 0$ such that f is defined and monotone in $(x_0-\delta, x_0+\delta)$. In particular, f is monotone in $(x_0-\delta, x_0)$ and in $(x_0, x_0+\delta)$, so both $\lim_{x\to x_0^-} f(x)$ and $\lim_{x\to x_0^+} f(x)$ exist by Lemma 4.1.5. Hence if x_0 is a point of discontinuity of f, then f must have a jump discontinuity at x_0. □

The fact that a monotone function f can only have jump discontinuities in an interval I, implies that f is almost continuous I, in the sense that the set of points of discontinuity of f is 'small'.

Theorem. *If f is monotone in the interval I, then f has at most countably many points of discontinuity in I.*

Proof. We may assume without loss of generality that f is monotone nondecreasing in I, because f and $-f$ have the same points of discontinuity and one of these is monotone nondecreasing.

We next observe that every interval $I \subset \mathbb{R}$ can be expressed as a countable union of closed and bounded intervals (see **exer4.2.1**). If I_N is closed and bounded for each $N \in \mathbb{N}$, $I = \bigcup I_N$, S is the set of points of discontinuity of f in I and S_N is the set of points of discontinuity of f in I_N, then $S = \bigcup S_N$. If S_N is countable for each N, then S is countable as a countable union of countable sets. Hence it is enough to prove the theorem in the case that $I = [a, b]$ is closed and bounded.

Denote by $S \subset I$ the set of points of discontinuity of f, then by the lemma above, f has a positive jump δ_x at each $x \in S$. We set $S_1 = \{x \in S : 1 < \delta_x\}$ and for $n \geq 2$, $S_n = \left\{x \in S : \frac{1}{n} < \delta_x \leq \frac{1}{n-1}\right\}$. If we show that S_n is finite for each $n \in \mathbb{N}$, then it follows that $S = \bigcup S_n$ is countable, as a countable union of finite sets.

For $n \in \mathbb{N}$, if a or b are the only points in S_n, then S_n is finite and there is nothing more to prove. Otherwise, suppose that $x \in S_n$, with $a < x < b$. If $a \leq u < x < v \leq b$, then because of the monotonicity of f and the jump at a,

$$f(u) \leq \lim_{t \to x^-} f(t) < \lim_{t \to x^-} f(t) + \delta_x = \lim_{t \to x^+} f(t) \leq f(v),$$

so that $f(v) - f(u) \geq \delta_x > \frac{1}{n}$. More generally, if $x_1, \ldots, x_k \in S$, with

$$a = u_0 < x_1 < u_1 < x_2 < u_2 \cdots < u_{k-1} < x_k < u_k = b,$$

then

$$f(b) - f(a) = \sum_{j=1}^{k} [f(u_k) - f(u_{k-1})] \geq n\delta_x > \frac{k}{n}.$$

Therefore $k < n(f(b) - f(a))$, so $\#(S_n) \leq n(f(b) - f(a)) + 2$ which is finite as claimed. □

4.3.2. A pure jump function

The following example shows that a monotone function may indeed have countably many points of discontinuity in a closed and bounded interval I. In fact the set of points of discontinuity can be *dense* in I.

Example. Let $I = [0, 1]$ and let $R = \mathbb{Q} \cap (0, 1)$, then R is countable (and dense in I) and we can enumerate its elements: $R = \{r_k\}_{k=1}^{\infty}$. For every $x \in (0, 1)$, we write $N_x = \{k \in \mathbb{N} : r_k < x\}$ and define the function $\varphi(x)$ by setting $\varphi(0) = 0$, $\varphi(1) = 1$ and for $0 < x < 1$,

$$\varphi(x) = \sum_{k \in N_x} \frac{1}{2^k}.$$

Since $\sum_{k \in \mathbb{N}} 2^{-k} = 1$ and $R \cap (0, x) \neq \varnothing$ for all $x \in (0, 1)$, we have $0 < \varphi(x) < 1$ for all $x \in (0, 1)$. Moreover, φ is monotone increasing in $[0, 1]$ because if $0 < x < y < 1$, then there exists $r_k \in (x, y) \cap R$ so

$$\varphi(0) = 0 < \varphi(x) < \varphi(x) + 2^{-k} \leq \varphi(y) < 1 = f(1).$$

We claim that for every $k \in \mathbb{N}$, there is a jump discontinuity at r_k of size at least 2^{-k}. Indeed, if $r_k \in R$, then $N_x \subseteq N_{r_k}$ for all $0 \leq x \leq r_k$, so $\varphi(x) \leq \varphi(r_k)$ for all $x \in [0, r_k]$. It follows that $\lim_{x \to r_k^-} \varphi(x) \leq \varphi(r_k)$. On the other hand, if $r_k < x \leq 1$, then R_{r_k} is a *proper* subset of R_x, since $r_k \in R_x$ but $r_k \notin R_{r_k}$, and therefore $\varphi(x) \geq \varphi(r_k) + 2^{-k}$ for all $x \in (r_k, 1]$. It follows that $\lim_{x \to r_k^+} \geq \varphi(r_k) + 2^{-k}$ so there is a jump at r_k of size

$$\delta_k = \lim_{x \to r_k^+} \varphi(x) - \lim_{x \to r_k^-} \varphi(x) \geq 2^{-k}.$$

*4.3.3. The Cantor function

The Cantor function is one of the most notorious examples in real analysis,[1] and its construction is based on the Cantor set that we discussed in §3.3.2.

The Cantor set, $\mathscr{C} \subset [0,1]$ is constructed recursively. First, we remove the open interval $I_{1,1} = (1/3, 2/3)$ from $[0,1]$, leaving the intervals $J_{1,1} = [0, 1/3]$ and $J_{1,2} = [2/3, 1]$. In the next step, we remove the open middle thirds, $I_{2,1} = (1/9, 2/9)$ and $I_{2,2} = (7/9, 8/9)$ from $J_{1,1}$ and $J_{1,2}$, respectively, leaving the intervals

$$J_{2,1} = [0, 1/9],\ J_{2,2} = [2/9, 1/3], J_{2,3} = [2/3, 7/9],\ J_{2,4} = [8/9, 1].$$

The construction continues recursively, where at each step, we remove the open middle thirds from the all the (closed) intervals that remained after the previous step. Denoting by $\{J_{n,k}\}_{k=1}^{2^n}$ the collection of pairwise disjoint closed intervals that remain after the nth stage of the construction of $\mathscr{C}$, the Cantor set is the closure of the set of endpoints of the collection of intervals. $\bigcup_{n,k} J_{n,k}$.

The Cantor set is uncountable, closed and has measure zero. It is also nowhere dense, which means that every open interval J in $[0,1]$ contains a subinterval J' such that $\mathscr{C} \cap J' = \varnothing$. In other words, the Cantor set misses an open piece of every open set. Finally, $\mathscr{C}$ is also a perfect set: if $x \in \mathscr{C}$, then there is a sequence $\{x_n\} \subset \mathscr{C}$ such that $x_n \to x$.

The Cantor set can also be described as the subset of $x \in [0,1]$ with ternary expansions that have no 1s among their ternary digits. I.e., $x \in \mathscr{C}$ if and only if x can be expressed in the form

$$x = \sum_{n=1}^{\infty} \frac{t_n(x)}{3^n},$$

with $t_n(x) = 0$ or 2 for all n. Some $x \in [0,1]$ have two ternary expansions, one that terminates (so $t_n(x) = 0$ from some point on), and one that ends in an infinite sequence of 2s.[2] It is this description of $\mathscr{C}$ that we use to construct the Cantor function.

Definition. The *Cantor function* is defined on $[0,1]$ by

$$\Phi(x) = \begin{cases} \sum_{n=1}^{\infty} t_n(x) \cdot 2^{-(n+1)} & : x \in \mathscr{C} \\ \sup\{\Phi(u) : u \in \mathscr{C} \cap (0, x]\} & : x \notin \mathscr{C}. \end{cases}$$

For $x \in \mathscr{C}$, if x has two ternary expansions, we choose the one with no 1's, e.g., for $x = 1/3$, we use the expansion $1/3 = \sum_{n=2}^{\infty} 2/3^n$, so $\Phi(1/3) = \sum_{n=2}^{\infty} 1/2^n = 1/2$.

Lemma. *$\Phi(x)$ is constant on $I_{n,k}$ for every $n \in \mathbb{N}$ and $1 \leq k \leq 2^n$ and monotone nondecreasing on $[0,1]$.*

[1] It has several names, including 'the devil's staircase'.

[2] For example, $1/3 = \sum_{n=2}^{\infty} 2/3^n$ and $4/9 = 1/3 + 1/9 = 1/3 + \sum_{n=3}^{\infty} 2/3^n$. Hence $1/3 \in \mathscr{C}$, but $4/9 \notin \mathscr{C}$, because both of the ternary expansions of 4/9 have a 1 in their ternary expansion.

Proof. If $x, y \in I_{n,k}$ for some n and k, then $\mathscr{C} \cap (0, x] = \mathscr{C} \cap (0, y]$, so $\Phi(x) = \Phi(y)$.

Now suppose that $x < y$, and consider three cases. If neither x nor y are in $\mathscr{C}$, then $\Phi(x) \le \Phi(y)$ because $\mathscr{C} \cap (0, x] \subseteq \mathscr{C} \cap (0, y]$. If $x \in \mathscr{C}$ and $y \notin \mathscr{C}$, then

$$\Phi(y) = \sup\{\Phi(u) : u \in \mathscr{C} \cap (0, y]\} \ge \sup\{\Phi(u) : u \in \mathscr{C} \cap (0, x]\} = \Phi(x).$$

If both $x, y \in \mathscr{C}$, and n_0 is the smallest natural number such that $t_n(x) \ne t_n(y)$, then $t_{n_0}(x) = 0$ and $t_{n_0}(y) = 2$, so

$$\Phi(y) - \Phi(x) = \frac{1}{2^{n_0}} + \sum_{n=n_0+1}^{\infty} \frac{t_n(y) - t_n(x)}{2^{n+1}} \ge \frac{1}{2^{n_0}} - \sum_{n=n_0+1}^{\infty} \frac{1}{2^n} \ge 0. \qquad \square$$

Corollary. *If x and y are the left and right endpoints of $I_{n,k}$ for some $n \in \mathbb{N}$ and $1 \le k \le 2^n$, then $\Phi(x) = \Phi(u) = \Phi(x)$ for all $u \in I_{n,k}$.*

Proof. The fact that that $\Phi(u) = \Phi(x)$ for all $u \in I_{n,k}$ follows from the definition and monotonicity of Φ. That $\Phi(x) = \Phi(y)$ follows from (3.3.1) in the proof Lemma 3.3.2, and the fact that $\sum_{k=n+1}^{\infty} \frac{1}{2^k} = \frac{1}{2^n}$. $\square$

Theorem. *The Cantor function is continuous on* $[0, 1]$.

Proof. To begin, if $x \notin \mathscr{C}$, then $x \in I_{n,k}$ for some n and k, so Φ is constant in a neighborhood of x, and hence continuous at x. To show that Φ is continuous at $x \in \mathscr{C}$, we need to work a little harder.

For any $x, y \in [0, 1]$, if $|x - y| < 3^{-m}$, then $t_k(x) = t_k(y)$ for all $k \le m$. It follows that if $x, u \in \mathscr{C}$ and $|x - u| < 3^{-m}$, then

$$|\Phi(x) - \Phi(u)| \le \sum_{k=m+1}^{\infty} \frac{1}{2^k} = \frac{1}{2^m}. \tag{4.3.1}$$

Next, if $x \in \mathscr{C}$, $y \notin \mathscr{C}$ and $x + 3^{-m} > y > x$, then by the monotonicity of Φ and (4.3.1), $\Phi(x) + 2^{-m} \ge \Phi(u) \ge \Phi(x)$ for all $u \in \mathscr{C} \cap [x, y]$, so

$$\Phi(x) + 2^{-m} \ge \Phi(y) = \sup\{\Phi(u) : u \in \mathscr{C} \cap [0, y]\} = \sup\{\Phi(u) : u \in \mathscr{C} \cap [x, y]\} \ge \Phi(x).$$

Finally, suppose that $x \in \mathscr{C}$, $y \notin \mathscr{C}$ and $y + 3^{-m} > x > y$. Let $u \in \mathscr{C}$ be the right endpoint of the interval $I_{n,k}$ such that $y \in I_{n,k}$. Then $\Phi(y) = \Phi(u)$, by the corollary, so

$$|\Phi(x) - \Phi(y)| = |\Phi(x) - \Phi(u)| \le \frac{1}{2^m}.$$

In other words, for all $x, y \in [0, 1]$, if $|x - y| < 3^{-m}$, then $|\Phi(x) - \Phi(y)| \le 2^{-}m$, so Φ is (uniformly) continuous. $\square$

We will meet the Cantor function again near the end of Chapter 8.

4.3.4. Oscillation

We can quantify the continuity of a function at a point (or lack thereof) in terms of its oscillation at that point.

Definition. If $f(x)$ is defined and bounded on a set $E \subset \mathbb{R}$, then *the oscillation of f on E* is given by

$$\omega_f(E) = \sup\{|f(x) - f(y)| : x, y \in E\}.$$

If f is unbounded on E, then $\omega_f(E) = \infty$.

If f is defined in a neighborhood U of the point a, then *the oscillation of f at a* is

$$\omega_f(a) = \lim_{\delta \to 0+} \omega_f(B_\delta(a)). \tag{4.3.2}$$

We observe that if $U \subset V$, then $\omega_f(U) \le \omega_f(V)$, so $\omega_f(B_\delta(a))$ is monotone nonincreasing as $\delta \to 0^+$, and the limit in (4.3.2) exists. This includes that possibility that f is unbounded in every neighborhood of a, in which case, $\omega_f(a) = \infty$.

If $\omega_f(a) > 0$, then f "bounces around" in every neighborhood of a, and the bigger the oscillation, the bigger the bounces. Not surprisingly, positive oscillation at a point is a bar to continuity at that point.

Theorem. *Suppose that f is defined in a neighborhood of the point $a \in \mathbb{R}$, then f is continuous at a if and only if $\omega_f(a) = 0$.*

Proof. If f is continuous at a, then for every $\varepsilon > 0$, there is a $\delta_\varepsilon > 0$ such that $|f(x) - f(a)| < \varepsilon/2$ in $B_{\delta_\varepsilon}(a)$. This implies that

$$|f(x) - f(y)| \le |f(x) - f(a)| + |f(a) - f(y)| < \varepsilon$$

for all $x, y \in B_{\delta_\varepsilon}(a)$, hence $\omega_f(B_\delta(a)) < \varepsilon$ for all $\delta < \delta_\varepsilon$, so $\omega_f(a) = 0$.

In the other direction, if $\omega_f(a) = 0$, then for every $\varepsilon > 0$, there is a $\delta_\varepsilon > 0$ such that $\omega_f(B_\delta(a)) < \varepsilon$. In particular, $|f(x) - f(a)| < \varepsilon$, for all $x \in B_{\delta_\varepsilon}(a)$ and it follows that f is continuous at a. □

The following observation is useful.[3]

Proposition. *If f is defined in the interval I, then $F_\eta = \{x \in I : \omega_f(x) \ge \eta\}$ is closed for any $\eta > 0$.*

Proof. For $a \in I$ with $\omega_f(a) < \eta$, let $\delta = (\eta - \omega_f(a))/2$. Then there is a neighborhood U of a such that $\omega_f(U) < \omega_f(a) + \delta$, so $\omega_f(x) < \omega_f(U) < \eta$ for every $x \in U$. It follows that $G_\eta = \{x \in I : \omega_f(x) < \eta\}$ is open, and therefore F_η is closed. □

Exercises

ex4.3.1. Show that $\lim_{x \to r_k^-} \varphi(x) = \varphi(r_k)$ and $\lim_{x \to r_k^+} \varphi(x) = \varphi(r_k) + 2^{-k}$ in Example 4.3.2, so the size of the jump at r_k is *equal* to 2^{-k}.

ex4.3.2. Show that the function $\varphi(x)$ in Example 4.3.2, is continuous at every *irrational* $x \in (0, 1)$.

[3]See the proof of the Vitali-Lebesgue theorem in Chapter 7.

Hint: Given $\varepsilon > 0$, let $k_0 \in \mathbb{N}$ be such that $2^{-k_0} < \varepsilon$. Show that if $x \notin \mathbb{Q}$, then there is a $\delta > 0$ such that if $r_k \in \mathbb{Q} \cap (x-\delta, x+\delta)$, then $k > k_0$. Use this to show that if $y \in (x-\delta, x+\delta)$, then $|\varphi(y) - \varphi(x)| < \varepsilon$.

ex4.3.3. If $J_{n,k} = [a,b]$ is one of the closed intervals left after the nth step in the construction of the Cantor set, show that $b - a = 3^{-n}$ and $\Phi(b) - \Phi(a) = 2^{-n}$.

ex4.3.4. Show that the size of the jump at a jump discontinuity of a function f at a point a is equal to the oscillation $\omega_a(f)$.

4.4. Continuous functions on bounded intervals

Functions that are continuous on bounded intervals have several important properties which follow from the fact that intervals are connected sets in $\mathbb{R}$ and that intervals that are both closed and bounded are compact.

In this section we will refer both implicitly and explicitly to the *relative* metric topology on subsets of $\mathbb{R}$. Briefly, if $E \subset \mathbb{R}$, then $U \subset E$ is open in E if and only if $U = E \cap O$ for some open set $O \subset \mathbb{R}$. See 3.1.6 for more details. To simplify exposition, for $x \in E$ we denote by $B_E(x,\varepsilon)$ the open ball of radius ε around x *in* E, i.e.,

$$B_E(x,\varepsilon) = \{y \in E : |x-y| < \varepsilon\} = B(x,\varepsilon) \cap E.$$

4.4.1. The continuous image of a compact set

We are mainly interested here in compact intervals in $\mathbb{R}$, but the following important result holds in complete generality. We state the theorem for maps between metric spaces, but the proof for maps between general topological spaces is identical.

Theorem. *If X and Y are metric spaces, $f : X \to Y$ is continuous, and $K \subset X$ is compact, then $f(K)$ is compact in Y.*

Proof. Let $\mathcal{O}_Y = \{U_a\}_{a\in A}$ be an open cover of $f(K)$. If $x \in K$, then $f(x) \in \bigcup_{a\in A} U_a$, so $f(x) \in U_a$ for some $a \in A$, and therefore $x \in f^{-1}(U_a)$ for this a. In other words, $K \subset \bigcup_{a\in A} f^{-1}(U_a)$, and since $f^{-1}(U_a)$ is open in X for it follows from this and Theorem 4.2.3 that $\mathcal{O}_X = \{f^{-1}(U_a)\}_{a\in A}$ is an open cover of K.

Since K is compact, $\mathcal{O}_X$ admits a finite subcover $\{f^{-1}(U_{a_i})\}_{i=1}^n$, of K. This implies that $f(x) \in \bigcup_{i=1}^n U_{a_i}$ for all $x \in K$, so that $f(K) \subset \bigcup_{i=1}^n U_{a_i}$. In other words, the open cover $\mathcal{O}_Y$ of $f(K)$ admits the finite subcover $\{U_{a_i}\}_{i=1}^n$, and since $\mathcal{O}_Y$ was arbitrary, $f(K)$ is compact. □

Corollary. *If $f(x)$ is continuous on a compact set $K \subset \mathbb{R}$, then f attains both a maximum and a minimum value in K.*

Proof. Since $f(K)$ is compact, both $\sup f(K)$ and $\inf f(K)$ exist since $f(K)$ is bounded and belong to K since $f(K)$ is closed. □

4.4.2. Uniform continuity

Definition. A function $f : \mathbb{R} \to \mathbb{R}$ is *uniformly continuous* on a set $E \subset \mathbb{R}$ if for every $\varepsilon > 0$, there is a $\delta > 0$ such that if $x, y \in E$ and $|x - y| < \delta$, then $|f(x) - f(y)| < \varepsilon$.

The important feature here is that the choice of δ depends only on ε and not on the particular point $x \in E$.

Theorem. *If $f : \mathbb{R} \to \mathbb{R}$ is continuous on the compact set $K \subset \mathbb{R}$, then f is uniformly continuous on K.*

Proof. Let $\varepsilon > 0$ be given. For every $x \in K$, there is a $\delta_x > 0$ such that $y \in B_K(x, \delta_x)$ implies $|f(x) - f(y)| < \varepsilon/2$. The collection $\{B_K(x, \delta_x/2) : x \in K\}$ is an open cover of K and so admits a finite subcover, $\{B_K(x_1, \delta_{x_1}/2), \ldots, B_K(x_n, \delta_{x_n}/2)\}$. Let $\delta_0 = \min_{1\le j\le n} \delta_{x_j}/2$.

Suppose that $x, y \in K$ with $|x - y| < \delta_0$. We have $x \in B_K(x_j, \delta_{x_j}/2)$ for some j, and hence $|f(x) - f(x_j)| < \varepsilon/2$. We also have

$$|y - x_j| \le |y - x| + |x - x_j| < \delta_0 + \delta_{x_j}/2 \le \delta_{x_j},$$

so $|f(y) - f(x_j)| < \varepsilon/2$ and hence

$$|f(x) - f(y)| \le |f(x) - f(x_j)| + |f(y) - f(x_j)| < \varepsilon.$$

□

4.4.3. The continuous image of connected sets

Theorem. *If $I \subset \mathbb{R}$ is connected and f is continuous on I, then $f(I)$ is connected.*

As we have already seen, connected subsets of $\mathbb{R}$ are intervals. On the other hand the statement is true in greater generality, so we give a proof that refers only to the connectedness of the sets involved.

Proof. Suppose that $I \subset \mathbb{R}$ is connected and f is continuous on I. We set $f(I) = E \subset \mathbb{R}$ and suppose that $E \subset U \cup V$ for some disjoint pair of open sets U and V. We want to show that either $E \cap U = \varnothing$ or $E \cap V = \varnothing$.

Let $O_U = \{x \in I : f(x) \in U\}$ and $O_V = \{x \in I : f(x) \in V\}$ be the pre-images of $E \cap U$ and $E \cap V$, respectively. Since $E \subset U \cup V$, it follows that $I \subset O_U \cup O_V$. Furthermore, since $U \cap V = \varnothing$ by assumption, it follows that $O_U \cap O_V = \varnothing$. If we can show that O_U and O_V are both open, then one of them must be empty because I is connected, and this will imply that either $E \cap U$ or $E \cap V$ is empty.

We will show that O_U is open; the proof for O_V is identical. If $x \in I$ and $f(x) \in U$, then there exists an $\varepsilon > 0$ such that $B(f(x), \varepsilon) \subset U$ because U is open. Since f is continuous at x, it follows that there is a $\delta > 0$, such that $y \in B_I(x, \delta)$ implies that $f(y) \in B(f(x), \varepsilon) \subset U$. Hence the set $B_I(x, \delta) \subset O_U$, so O_U is open. □

Corollary. *Suppose that f is continuous in a neighborhood of the interval $[a, b]$. If $f(a) < y < f(b)$ or $f(b) < y < f(a)$, then there exists $x \in (a, b)$ such that $f(x) = y$.*

This is called the *intermediate value theorem*.

Proof. It follows from Theorem 4.4.3, above, that $I = f([a, b])$ is connected and is therefore an interval in $\mathbb{R}$. Since $f(a), f(b) \in I$, it follows that if $f(a) < y < f(b)$ or $f(b) < y < f(a)$, then $y \in I$ (by Proposition 1.5.1) so $y = f(x)$ for some $x \in (a, b)$. □

Exercises

ex4.4.1. Prove Theorem 4.4.1 for the case that $X = Y = \mathbb{R}$, by showing directly that if $K \subset \mathbb{R}$ is compact, then $f(K)$ is closed and bounded in $\mathbb{R}$.

*4.5. The space $C[0, 1]$

If f and g are continuous, real-valued functions on a compact interval $I \subset \mathbb{R}$, then $\alpha f + \beta g$ is also continuous on I for any constants $\alpha, \beta \in \mathbb{R}$. It follows that the set $C(I)$ of all such functions on I is a *vector space* over the real numbers. More is true: $C(I)$ is a *normed* vector space (in a sense to be made precise below) which allows us to define a metric on this space. In this section, we introduce the basic ideas, and in future sections we show that with the metric we define below, $C(I)$ is both complete and separable.[4]

4.5.1. The supremum norm and metric on $C(I)$

Let $I = [a, b] \subset \mathbb{R}$, and recall that a real-valued function f is continuous on I if it is continuous with respect to the metric on I inherited from the metric on $\mathbb{R}$ (or equivalently, if f is continuous with respect to the relative metric topology on I).

If f is continuous on I, then $|f|$ is also continuous on I, because

$$||f(x)| - |f(y)|| \leq |f(x) - f(y)|$$

for all $x, y \in I$. Hence, if $f \in C(I)$, then $|f| \in C(I)$ and so attains a maximum value in I.

Definition. If $f \in C(I)$, then the *supremum norm* (or sup norm) of f is denoted by $||f||_\infty$, $||f||$ or simply $||f||$ if there is no chance of confusion, and defined by $||f|| = \max_{x \in I} |f(x)|$.

We observe that it follows from this definition that $|f(x)| \leq ||f||$ for all $x \in I$.

Theorem. *For all $f, g \in C(I)$ and $\alpha \in \mathbb{R}$, (i) $||f|| \geq 0$ and $||f|| = 0$ if and only if $f = 0$ (on I), (ii) $||\alpha f|| = |\alpha| \cdot ||f||$, and (iii) $||f + g|| \leq ||f|| + ||g||$.*

Proof. We leave the proofs of (i) and (ii) as an exercise and prove (iii).[5] Suppose that $x_0 \in I$ is a point where $|f + g|$ attains its maximum value, then it follows from the observation above that

$$||f + g|| = |f(x_0) + g(x_0)| \leq |f(x_0)| + |g(x_0)| \leq ||f|| + ||g||. \qquad \square$$

[4] The notation $C(X)$ is commonly used to denote the space of all continuous *complex-valued* functions on the set X. The space of continuous *real-valued* functions on X, that we are discussing here, is often denoted by $C_{\mathbb{R}}(X)$ in this broader context.

[5] The triangle inequality.

A vector space with a norm satisfying these properties is called a *normed vector space* or a normed space.

The supremum norm plays the same role in $C(I)$ that the absolute value plays in $\mathbb{R}$, and we use it to define a metric on $C(I)$ analogously to the way we defined the standard metric on $\mathbb{R}$.

Corollary. *The function $\rho(f,g) = \|f-g\|$ defines a metric on $C(I)$.*

Proof. We need to show that for all $f,g,h \in C(I)$, (i) $\|f-g\| = \|g-f\|$, (ii) $\|f-g\| \geq 0$ and $\|f-g\| = 0$ if and only if $f = g$, and (iii) $\|f-h\| \leq \|f-g\| + \|g-h\|$.

Property (i) follows from the fact that $|f(x)-g(x)| = |g(x)-f(x)|$ for all x, (ii) follows from part (ii) of Theorem 4.5.1, and (iii) follows from part (iii) of the theorem, with $f-g$ in place of f and $g-h$ in place of g. □

4.5.2. The space $C[0,1]$

If $I = [0,1]$, we denote $C(I)$ by $C[0,1]$. The theorem below shows that $C(I)$ is essentially a copy of $C[0,1]$, for any compact interval $I \subset \mathbb{R}$. Hence to understand the structure and properties of $C(I)$, it is enough to study $C[0,1]$.

If $I = [a,b]$, then the linear function $\phi_I(x) = (x-a)/(b-a)$ maps I bijectively onto the interval $[0,1]$ with inverse $\phi_I^{-1}(x) = a + (b-a)x$. We next define a map $\varphi_I : C[0,1] \to C(I)$ using composition: $\varphi_I(f) = f \circ \phi_I$. I.e., $\varphi_I(f)(x) = f(\phi(x))$ for any $f \in C[0,1]$ and $x \in I$. Since the composition of continuous functions is continuous (and linear functions are continuous), it follows that $\varphi_I(f) \in C(I)$ for every $f \in C[0,1]$.

Lemma. *The map $\varphi_I : C[0,1] \to C(I)$ is a bijection.*

Proof. For the sake of concreteness, let $I = [a,b]$. Suppose that $f,g \in C[0,1]$ and $f \neq g$, then there is an $x_0 \in [0,1]$ such that $f(x_0) \neq g(x_0)$. Let $\phi_I^{-1}(x_0) = u_0$, then $\phi(u_0) = x_0$, and therefore $\varphi_I(f)(u_0) = f(x_0) \neq g(x_0) = \varphi_I(g)(u_0)$, so $\varphi_I(f) \neq \varphi_I(g)$, i.e., φ_I is injective.

If $g \in C(I)$ and $f(x) = g(\phi_I^{-1}(x))$, then $f \in C[0,1]$ and $\varphi_I(f)(x) = g(x)$, so φ_I is surjective. □

The next theorem shows that the bijection $\varphi_I : C[0,1] \to C(I)$ is a linear, distance-preserving map. Such a map between normed vector spaces is called an *isometry*.

Theorem. *Let $f,g \in C[0,1]$, then $\|\varphi_I(f) - \varphi_I(g)\| = \|f-g\|$, and for any $u,v \in \mathbb{R}$, $\varphi_I(uf+vg) = u\varphi_I(f) + v\varphi_I(g)$.*

It is understood that $\|\varphi_I(f) - \varphi_I(g)\|$ refers to the norm on $C(I)$ and $\|f-g\|$ refers to the norm on $C[0,1]$.

Proof. For $x \in I$, we use x_ϕ to denote $\phi_I(x) \in [0,1]$, so that for any $f \in C[0,1]$, $\varphi_I(f)(x) = f(x_\phi)$. It follows that for any $u,v \in \mathbb{R}$, $f,g \in C[0,1]$, and any $x \in I$, we have

$$\varphi_I(uf+vg)(x) = (uf+vg)(x_\phi) = uf(x_\phi) + vg(x_\phi) = u\varphi_I(f)(x) + v\varphi_I(g)(x).$$

Moreover, as x ranges over all of I, x_ϕ ranges over all of $[0,1]$ and therefore
$$\|\varphi_I(f)-\varphi_I(g)\| = \max_{x\in I}|\varphi_I(f)-\varphi_I(g)| = \max_{x_\phi\in[0,1]}|f(x_\phi)-g(x_\phi)| = \|f-g\|. \qquad \square$$

Exercises

ex4.5.1. Prove parts (i) and (ii) of Theorem 4.5.1.

ex4.5.2. Prove that the functions f_n in the example above are continuous.

ex4.5.3. For $\mathbf{a} = (a_0, a_1, \ldots, a_n) \in \mathbb{R}^{n+1}$, let $\psi(\mathbf{a}) = a_0 + a_1x + \cdots + a_nx^n \in C[0,1]$. Show that for any $\varepsilon > 0$, there is a $\delta > 0$ such that if $|\mathbf{a}-\mathbf{b}| < \delta$, then $\|\psi(\mathbf{a})-\psi(\mathbf{b})\| < \varepsilon$. Conclude that the map $\psi : \mathbb{R}^{n+1} \to C[0,1]$ is uniformly continuous.

ex4.5.4. Suppose that $f(x)$ is a polynomial of degree n. Show that for any $\varepsilon > 0$, there exists a polynomial $g(x)$ of degree n with rational coefficients such that $\|f-g\| < \varepsilon$, (where $\|\cdot\|$ refers to the sup norm on $[0,1]$).

Chapter 5

Sequences and Series of Functions

In this chapter we study sequences and series of functions and consider the question of when and in what sense does a given sequence of functions converge to another function. We focus here on real-valued functions defined on $\mathbb{R}$ or intervals $I \subset \mathbb{R}$, and many of the technical details we need here were developed in Chapter 2.

5.1. Sequences of functions

When we say that a sequence of functions $\{f_n\}$ converges to another function f, we mean that for n sufficiently large f_n is *close* to f in some sense. To make the claim of convergence precise, we need to make clear the sense in which we mean that f_n and f are close.

5.1.1. Pointwise convergence

The most basic sense of convergence for a sequence of functions is *pointwise convergence.*

Definition. Let $I \subset \mathbb{R}$ be an interval and suppose that f_n is a real-valued function on I for each $n \in \mathbb{N}$. The sequence $\{f_n\}$ *converges pointwise* to the function f if for every $u \in I$, the numerical sequence $\{f_n(u)\}$ converges to the limit $f(u)$. In other words, $\{f_n\}$ converges pointwise to f in I if and only if for every $x \in I$ and every $\varepsilon > 0$, there is an $n_{x,\varepsilon} \in \mathbb{N}$ such that $n > n_{x,\varepsilon}$ implies $|f_n(x) - f(x)| < \varepsilon$.

The function f is called the *pointwise-limit* of the sequence $\{f_n\}$.

Examples.

(i) Let $f_n(x) = x^n$ for $n \in \mathbb{N}$. If $0 \le x < 1$, then $x^n \to 0$, while $1^n = 1$ for all n, so the sequence $\{x^n\}$ converges pointwise in $[0, 1]$ to the function

$$w(x) = \begin{cases} 0 & : 0 \le x < 1 \\ 1 & : x = 1 \end{cases}$$

(ii) Suppose that $g_n(x) = 1 + x + \cdots + x^{n-1}$ for all $n \in \mathbb{N}$. If $x \neq 1$, then $g_n(x) = (1 - x^n)/(1 - x)$ and for fixed $x \in (-1, 1)$,

$$\left| \frac{1 - x^n}{1 - x} - \frac{1}{1 - x} \right| = \frac{|x^n|}{1 - x} \to 0,$$

so the sequence $\{g_n(x)\}$ converges pointwise to $1/(1 - x)$ in the interval $(-1, 1)$. Note that $g_n(1) = n \to \infty$ and that $g_n(-1)$ alternates between 1 and 0, so the sequence $\{g_n(x)\}$ does not converge pointwise in $[-1, 1]$.

(iii) If $e_n(x) = \sum_{k=0}^{n} \frac{x^n}{n!}$, then Theorem 2.7.2 can be interpreted to say that the sequence $\{e_n(x)\}$ converges pointwise to the function $\exp(x)$ in all of $\mathbb{R}$.

The next proposition is a direct consequence of the analogous result for numerical sequences and we leave the proof as an exercise.

Proposition. *If the sequences $\{f_n(x)\}$ and $\{g_n(x)\}$ converge pointwise in the interval I to limits $f(x)$ and $g(x)$ respectively, then*

(i) *$\{af_n(x) + bg_n(x)\}$ converges pointwise to $af(x) + bg(x)$ in I.*

(ii) *$\{f_n(x)g_n(x)\}$ converges pointwise to $f(x)g(x)$ in I.*

Proof. Exercise. □

5.1.2. Uniform convergence

When a sequence $\{f_n\}$ converges pointwise to f in the interval I the point in the sequence beyond which $|f_n(x) - f(x)| < \varepsilon$ depends on *both* x and ε. The sequence converges *uniformly* in I if the same n_ε "works" for all $x \in I$.

Definition. Let $I \subset \mathbb{R}$ be an interval and suppose that $\{f_n\}$ is a sequence of real-valued functions on I. The sequence $\{f_n\}$ *converges uniformly* in I to the function f if and only if for every $\varepsilon > 0$, there is an $n_\varepsilon \in \mathbb{N}$ such that $n > n_\varepsilon$ implies $|f_n(x) - f(x)| < \varepsilon$ *for all* $x \in I$. The function f is called the *uniform limit* of the sequence $\{f_n\}$.

The sequence $\{f_n\}$ is *uniformly bounded* in I if there is an $M > 0$ such that $|f_n(x)| < M$ for all $n \in \mathbb{N}$ and all $x \in I$.

Examples.

(i) The sequence $\{x^n\}$ converges pointwise in $[0, 1]$ to the function $w(x)$ in Example 5.1.1, but *not* uniformly so. To see this, note that $0 < \sqrt[n]{1/2} < 1$ for all $n \in \mathbb{N}$, and if $\sqrt[n]{1/2} < x < 1$, then $1/2 < x^n$, so $|x^n - w(x)| > 1/2$. I.e., there is no $n \in \mathbb{N}$ such that $|x^n - w(x)| < 1/2$ for all $x \in [0, 1]$.

(ii) We saw in Example 5.1.1 (ii) that $g_n \to 1/(1-x)$ pointwise in $(-1,1)$, but this convergences is also not uniform (see **ex5.1.2**). On the other hand, if $|x| \le a < 1$, then

$$\left|g_n(x) - \frac{1}{1-x}\right| = \frac{|x^n|}{1-x} < \frac{a^n}{1-a} \to 0.$$

Therefore, given $\varepsilon > 0$, there is therefore an $n_{a,\varepsilon} \in \mathbb{N}$ such that $\left|g_n(x) - \frac{1}{1-x}\right| \le \varepsilon$ for all $x \in [-a,a]$, so $\{g_n\}$ converges uniformly to $1/(1-x)$ in $[-a,a]$.

Proposition 5.1.1 extends to uniformly convergent sequences with the added condition of uniform boundedness for part (ii). The assumptions of uniform convergence and uniform boundedness allow us to draw another conclusion as well.

Proposition. *If the sequences $\{f_n(x)\}$ and $\{g_n(x)\}$ converge uniformly in the interval I to limits $f(x)$ and $g(x)$ respectively, then*

(i) *$\{af_n(x) + bg_n(x)\}$ converges uniformly to $af(x) + bg(x)$ in I.*

(ii) *If $\{f_n\}$ and $\{g_n\}$ are both uniformly bounded in I, then $\{f_n(x)g_n(x)\}$ converges uniformly to $f(x)g(x)$ in I.*

(iii) *If $\{f_n\}$ is uniformly bounded in I and $|g(x)| > \eta > 0$ in I, then for some $N_0 \in \mathbb{N}$, the sequence $\{f_n(x)/g_n(x)\}_{n=N_0}^{\infty}$ converges uniformly to $f(x)/g(x)$ in I.*

Proof. We leave the proof of part (i) as an exercise.

To prove part (ii), the assumption of uniform boundedness implies that there exists an $M > 0$ such that $|f_n(x)| < M$ and $|g_n(x)| < M$ for all $x \in I$ and all $n \in \mathbb{N}$. Furthermore for some $N \in \mathbb{N}$, $|f_n(x) - f(x)| < 1$ and $|g_n(x) - g(x)| < 1$ for all $x \in I$ and all $n > N$, because of the assumption of uniform convergence. This implies that $|f(x)| < M + 1$ and $|g(x)| < M + 1$ for all $x \in I$. Now, given $\varepsilon > 0$, there exists $N_\varepsilon \in \mathbb{N}$ such that for all $n > N_\varepsilon$ and all $x \in I$,

$$|f_n(x) - f(x)| < \frac{\varepsilon}{2(M+1)} \quad \text{and} \quad |g_n(x) - g(x)| < \frac{\varepsilon}{2(M+1)}$$

and hence

$$\begin{aligned}|f_n(x)g_n(x) - f(x)g(x)| &\le |f_n(x)g_n(x) - f(x)g_n(x)| + |f(x)g_n(x) - f(x)g(x)| \\ &\le |g_n(x)||f_n(x) - f(x)| + |f(x)||g_n(x) - g(x)| \\ &< (M+1)\,(|f_n(x) - f(x)| + |g_n(x) - g(x)|) < \varepsilon,\end{aligned}$$

for all $n > N_\varepsilon$ and all $x \in I$.

To prove (iii), we note that there is an $N_0 \in \mathbb{N}$ such that $|g_n(x) - g(x)| < \eta/2$ for $n > N_0$ and all x in I because the convergence is uniform, and it follows from this that $|g_n(x)| > \eta/2 > 0$ in I for all $n > N_0$. Given $\varepsilon > 0$, there is an $N_\varepsilon > N_0$ such that $|g_n(x) - g(x)| < 2\eta^2\varepsilon$ for all $n > N_\varepsilon$ and all $x \in I$, and hence

$$\left|\frac{1}{g_n(x)} - \frac{1}{g(x)}\right| = \left|\frac{g(x) - g_n(x)}{g_n(x)g(x)}\right| < \varepsilon.$$

I.e., $\{1/g_n(x)\}_{n=N_0}^{\infty}$ converges uniformly to $1/g(x)$ in I. Part (iii) now follows from part (ii) applied to the sequences $\{f_n(x)\}_{n=N_0}^{\infty}$ and $\{1/g_n(x)\}_{n=N_0}^{\infty}$, and the observation that the latter sequence is uniformly bounded by $2/\eta$ in I. □

For a sequence of *continuous* functions $\{f_n\}$ that are converging to a function f in an interval $I \subset \mathbb{R}$, there is an important benefit to the convergence being uniform.

Theorem. *If $f_n(x)$ is continuous in the interval I for each $n \in \mathbb{N}$, and $f_n \to f$ uniformly in I, then f is continuous in I.*

Proof. Let $x_0 \in I$ and $\varepsilon > 0$ be given. By assumption there exists an n_0 such that if $n \geq n_0$, then $|f_n(x) - f(x)| < \varepsilon/3$ for all $x \in I$. From the continuity of f_{n_0} in I, it follows that there exists a $\delta > 0$ such that if $x \in I$ and $|x - x_0| < \delta$, then $|f_{n_0}(x) - f_{n_0}(x_0)| < \varepsilon/3$. Hence, if $x \in I$ and $|x - x_0| < \delta$, then

$$|f(x) - f(x_0)| \leq |f(x) - f_{n_0}(x)| + |f_{n_0}(x) - f_{n_0}(x_0)| + |f_{n_0}(x_0) - f(x_0)| < \varepsilon.$$

This shows that f is continuous at x_0, so f is continuous in I, because x_0 was an arbitrary point in I. □

Comment. Example 5.1.1 (i) illustrates the fact that if the convergence of a sequence of continuous functions is not uniform, then the (pointwise) limit of the sequence need not be continuous. Moreover, if the limit of a sequence of continuous functions is not continuous, then the convergence cannot be uniform.

5.1.3. Cauchy criteria

A sequence $\{u_n\}$ of real numbers converges to a limit if and only if it is a Cauchy sequence. This gives rise to the following definitions.

Definition. A sequence of functions $\{f_n\}$ is *pointwise Cauchy* in the interval $I \subset \mathbb{R}$ if for each $x \in I$, the numerical sequence $\{f_n(x)\}$ is a Cauchy sequence in $\mathbb{R}$. I.e., for any $x \in I$ and any $\varepsilon > 0$, there is an $N_{x,\varepsilon} \in \mathbb{N}$ such that if $n, m \geq N_{x,\varepsilon}$, then $|f_n(x) - f_m(x)| < \varepsilon$.

A sequence of functions $\{f_n\}$ is *uniformly Cauchy* in the interval $I \subset \mathbb{R}$ if for any $\varepsilon > 0$, there is an $N_\varepsilon \in \mathbb{N}$ such that if $n, m \geq N_\varepsilon$, then $|f_n(x) - f_m(x)| < \varepsilon$ for all $x \in I$. I.e., if N_ε does not depend on x.

Theorem. *The sequence of functions $\{f_n\}$ on an interval I converges pointwise to the function f if and only if $\{f_n\}$ is pointwise Cauchy in I. The sequence converges to f uniformly in I if and only if it is uniformly Cauchy in I.*

Proof. The first claim of the theorem follows from the completeness of $\mathbb{R}$ and the definitions of pointwise convergent sequences and pointwise Cauchy sequences.

If $f_n \to f$ uniformly in I, then for any $\varepsilon > 0$, there is an N_ε such that $|f_n(x) - f(x)| < \varepsilon/2$ for all $n \geq N_\varepsilon$ and all $x \in I$. It follows that if $n, m \geq N_\varepsilon$, then

$$|f_n(x) - f_m(x)| \leq |f_n(x) - f(x)| + |fx) - f_m(x)| < \varepsilon$$

for all $x \in I$, so $\{f_n\}$ is uniformly Cauchy in I.

Conversely, suppose that $\{f_n\}$ is uniformly Cauchy in I. then the sequence is pointwise Cauchy and hence converges pointwise to a function f on I. We need to show that the convergence is in fact uniform. Given $\varepsilon > 0$, there exists $N_\varepsilon \in \mathbb{N}$ such that $|f_n(x) - f_{N_\varepsilon}(x)| < \varepsilon/3$ for all $n \geq N_\varepsilon$ and all $x \in I$, because $\{f_n\}$ is uniformly Cauchy. Additionally, for any $x \in I$, there exists an N_x such that $|f_n(x) - f(x)| < \varepsilon/3$ for all

$n \geq N_x$, because $f_n \to f$ pointwise. Hence, for any $x \in I$, if we set $N = \max(N_\varepsilon, N_x)$, then

$$|f_{N_\varepsilon}(x) - f(x)| \leq |f_{N_\varepsilon}(x) - f_N(x)| + |f_N(x) - f(x)| < 2\varepsilon/3,$$

and therefore

$$|f_n(x) - f(x)| \leq |f_n(x) - f_{N_\varepsilon}(x)| + |f_{N_\varepsilon}(x) - f(x)| < \varepsilon$$

for all $x \in I$ and $n > N_\varepsilon$, so $f_n \to f$ uniformly in I. □

The criterion for uniform convergence given here is useful because it is sometimes easier to show that a sequence of functions is uniformly Cauchy on a given interval than to show directly that it converges uniformly there.

5.1.4. Equicontinuity

The Cauchy criteria above describe a sequence of functions without reference to a possible limit function. *Equicontinuity* is another characteristic of a sequence of functions that makes no reference to a limit function.

Definition. The sequence $\{f_n\}$ of real-valued functions on $I \subset \mathbb{R}$ is *equicontinuous* at a point $x \in I$ if for every $\varepsilon > 0$ the exists $\delta_{x,\varepsilon} > 0$ such that if $|x - y| < \delta_{x,\varepsilon}$ then $|f_n(x) - f_n(y)| < \varepsilon$ for all n.

The sequence $\{f_n\}$ is *pointwise equicontinuous* in I, if it is equicontinuous at every point in I.

The sequence is *uniformly equicontinuous* in I if for every $\varepsilon > 0$ there exists $\delta_\varepsilon > 0$ such that for all $x, y \in I$, if $|x - y| < \delta_\varepsilon$ then $|f_n(x) - f_n(y)| < \varepsilon$ for all n.

Recall that if f is continuous on the closed interval $[a, b]$ then f is uniformly continuous there (Theorem 4.4.2). There is an analogous result for equicontinuous sequences.

Theorem. *The sequence $\{f_n\}$ is pointwise equicontinuous in the interval $[a, b]$, if and only if it is uniformly equicontinuous there.*

Proof. That uniform equicontinuity of the sequence implies pointwise equicontinuity follows directly from the definition.

To prove the converse, suppose that $\{f_n\}$ is pointwise equicontinuous in $[a, b]$ and let $\varepsilon > 0$ be given. From the pointwise equicontinuity of $\{f_n\}$, it follows that for each $x \in [a, b]$, there is a $\delta_x > 0$ such that if $|y - x| < \delta_x$, then $|f_n(y) - f_n(x)| < \varepsilon/2$, for all $n \in \mathbb{N}$. The intervals $\{(x - \delta_x/2, x + \delta_x/2) : x \in [a, b]\}$ form an open cover of $[a, b]$, so it follows from the compactness of $[a, b]$ that there exist $x_1, x_2, \ldots, x_k \in [a, b]$ such that $[a, b] \subset \bigcup_{i=1}^k (x_i - \delta_{x_i}/2, x_i + \delta_{x_i}/2)$. Let $\delta_0 = \min_{1 \leq i \leq k} \delta_{x_i}/2$.

Thus, given $x \in [a, b]$, there is a $j \in \{1, \ldots, k\}$, such that $|x - x_j| < \delta_{x_j}/2$, and therefore $|f_n(x) - f_n(x_j)| < \varepsilon/2$ for all n. If $y \in [a, b]$ and $|x - y| < \delta_0$, then

$$|y - x_j| \leq |y - x| + |x - x_j| < \delta_0 + \delta_{x_j}/2 \leq \delta_{x_j},$$

so $|f_n(y) - f_n(x_j)| < \varepsilon/2$ for all n. Hence if $|x - y| < \delta_0$, then

$$|f_n(x) - f_n(y)| \leq |f_n(x) - f_n(x_j)| + |f_n(y) - f_n(x_j)| < \varepsilon,$$

so the sequence is uniformly equicontinuous. □

5.1.5. The Arzelà-Ascoli theorem

The Bolzano–Weierstrass theorem says that a bounded sequence in $\mathbb{R}^n$ has a convergent subsequence. The Arzelà-Ascoli theorem, in its most basic formulation, is an analog of this result for bounded sequences of equicontinuous functions on a interval $[a, b]$.[1] A bounded set in $\mathbb{R}^n$ is one such that the coordinates of all its points lie in a bounded set in $\mathbb{R}$. The analogous condition on a set of functions is called *uniform boundedness.* A sequence $\{f_n(x)\}$ of functions is *uniformly bounded* in a set U if there is an $M > 0$ such that $|f_n(x)| < M$ for all $x \in U$ and all $n \in \mathbb{N}$.

The proofs are also analogous in that in both, we show that a convergent subsequence exists by a "filtering" process on the "coordinates" of the elements of the bounded sequence.[2] An element in $\mathbb{R}^n$ has n coordinates, so this "filtering" process requires only finitely many steps. A function on an interval $[a, b]$ has uncountably many "coordinates" — one for each point in the interval. Uncountable sets are too big to manage in this context, so we first establish the existence of a subsequence that converges at every *rational* point in $[a, b]$. The rational points form a *countable, dense* subset of $[a, b]$. The countability of the set of rational points makes the filtering process feasible and its density, together with the equicontinuity of the original sequence and the compactness of $[a, b]$, allows us to conclude that the subsequence we have found converges uniformly on all of $[a, b]$.

Theorem (Arzelà-Ascoli). *If $\{f_n\}$ is uniformly bounded and equicontinuous on the interval $[a, b]$ then it has a subsequence that converges uniformly to a continuous function on $[a, b]$.*

Proof. The uniform limit of continuous functions is continuous, so it is enough to show that a uniformly convergent subsequence exists, and by Theorem 5.1.3, it is therefore enough to find a subsequence that it is uniformly Cauchy in $[a, b]$. This we do as outlined above.

Let $\{q_j\}_{j \in \mathbb{N}}$ be an enumeration of the set $Q = \mathbb{Q} \cap [a, b]$. The sequence $\{f_n(q_1)\}$ is bounded, because $\{f_n\}$ is uniformly bounded in $[a, b]$, and therefore contains a convergent subsequence $\{f_n(q_1)\}_{n \in \Lambda_1}$, where Λ_1 is a strictly increasing subsequence of $\mathbb{N}$. Likewise, the sequence $\{f_n(q_2)\}_{n \in \Lambda_1}$ is bounded, so it contains a convergent subsequence $\{f_n(q_2)\}_{n \in \Lambda_2}$, where $\Lambda_2 \subset \Lambda_1$ is also strictly increasing. Furthermore, since $\Lambda_2 \subset \Lambda_1$, it follows that $\{f_n(q_1)\}_{n_2 \in \Lambda_2}$ is also a convergent sequence.

We continue in this way to recursively establish the existence of subsequences that converge at more and more of the points $q_j \in Q$. For $k \geq 1$ we assume that there exists a strictly increasing $\Lambda_k \subset \mathbb{N}$ with the property that $\{f_n(q_j)\}_{n \in \Lambda_k}$ converges for $1 \leq j \leq k$. By the Bolzano-Weierstrass theorem in $\mathbb{R}$, the bounded sequence

[1] By Theorem 5.1.4, above, pointwise equicontinuity and uniform equicontinuity are equivalent on a compact interval, so we simply say equicontinuity in this case.

[2] By filtering, we mean recursively finding subsequences of subsequences that satisfy useful conditions.

$\{f_n(q_{k+1})\}_{n\in\Lambda_k}$ has a convergent subsequence. I.e., there is a strictly increasing subsequence $\Lambda_{k+1} \subset \Lambda_k$, such that $\{f_n(q_{k+1})\}_{n\in\Lambda_{k+1}}$ converges. And since $\Lambda_{k+1} \subset \Lambda_k$, it follows that $\{f_n(q_j)\}_{n\in\Lambda_{k+1}}$ also converges for $1 \le j \le k$. Hence, there exists a nested set of strictly increasing sequences

$$\mathbb{N} \supset \Lambda_1 \supset \Lambda_2 \supset \cdots \supset \Lambda_k \supset \cdots$$

such that for each $k \in \mathbb{N}$, $\{f_n(q_j)\}_{n\in\Lambda_k}$ converges for $1 \le j \le k$.

We now construct a subsequence $\{f_{n_m}\} \subset \{f_n\}$ that converges pointwise at *all* $q_j \in Q$ as follows. For each $m \in \mathbb{N}$, let n_m be the mth element in Λ_m. The subsequence $\{n_m : m = 1, 2, 3, \ldots\} \subset \mathbb{N}$ is strictly increasing because $\Lambda_{m+1} \subset \Lambda_m$, so the $(m+1)^{st}$ element of Λ_{m+1} is at least as big as the $(m+1)^{st}$ element of Λ_m and therefore bigger than the mth element of Λ_m. For any $q_j \in Q$, if $m \ge j$, then $n_m \in \Lambda_m \subset \Lambda_j$, so $\{n_m : m \ge j\}$ is a subsequence of Λ_j. It follows that $\{f_{n_m}(q_j)\}_{m=j}^{\infty}$ converges and therefore $\{f_{n_m}(q_j)\}_{m=1}^{\infty}$ converges. It remains to show that $\{f_{n_m}\}$ is uniformly Cauchy in $[a, b]$.

Let $\varepsilon > 0$ be given. From the equicontinuity of the original sequence, and hence of the subsequence $\{f_{n_m}\}$, it follows that there is a $\delta > 0$ such that if $x, y \in [a, b]$ and $|x - y| < \delta$, then $|f_{n_m}(x) - f_{n_m}(y)| < \varepsilon/3$, for all $m \in \mathbb{N}$. The collection $\{(x - \delta/2, x + \delta/2) : x \in [a, b]\}$ forms an open cover of $[a, b]$ and therefore admits a finite subcover, $\{J_i\}_{i=1}^{N} = \{(x_i - \delta/2, x_i + \delta/2) : 1 \le i \le N\}$.

Since Q is dense in $[a, b]$, each interval J_i in this finite subcover contains a rational point q_i. By construction, the sequence $\{f_{n_m}(q_i)\}$ converges for each $1 \le i \le N$, so there is an M_i, such that $|f_{n_m}(q_l) - f_{n_l}(q_i)| < \varepsilon/3$ for all $m, l > M_i$. Setting $M_\varepsilon = \max\limits_{1\le i\le N} M_i$, it follows that $|f_{n_m}(q_i) - f_{n_l}(q_i)| < \varepsilon/3$ for $1 \le i \le N$ and all $m, l > M_\varepsilon$.

Finally, if $x \in [a, b]$, then $x \in J_i$ for some i, so $|x - q_i| < \delta$ (since the diameter of J_i is δ). It follows that $|f_{n_m}(x) - f_{n_m}(q_i)| < \varepsilon/3$ for all m, by equicontinuity, and therefore, for all $m, l \ge M_\varepsilon$,

$$|f_{n_m}(x) - f_{n_l}(x)| \le |f_{n_m}(x) - f_{n_m}(q_i)| + |f_{n_m}(q_i) - f_{n_l}(q_i)| + |f_{n_l}(q_i) - f_{n_l}(x)| < \varepsilon,$$

showing that the subsequence $\{f_{n_m}\}$ is uniformly Cauchy in $[a, b]$. □

Exercises

ex5.1.1. Prove Proposition 5.1.1.

ex5.1.2. Show that the sequence $\{g_n\}$ from Example 5.1.1 (ii) does not converge uniformly to $1/(1-x)$ in $(-1, 1)$.

Hint: Show that if $0 < a < x < 1$, then $|g_n(x) - 1/(1-x)| > a^n/(1-x)$. Now what?

ex5.1.3. Show that $x^n \to 0$ uniformly in $[0, a]$ for any $0 < a < 1$.

ex5.1.4. Show that $e_n(x) = \sum_{k=0}^{n} \frac{x^n}{n!}$ converges uniformly in $(-a, a)$ for any $a > 0$ and conclude that $\exp(x) = \lim_{n\to\infty} e_n(x)$ is continuous on $\mathbb{R}$. (This reproduces the conclusion of Proposition 4.2.2.)

ex5.1.5. Prove Proposition 5.1.1.

ex5.1.6. Prove Proposition 5.1.2, parts (i) and (ii). Show by example that part (iii) might fail if the pointwise convergence is *not* uniform.

ex5.1.7. Show that the sequence $\{x^n\}$ is *not* equicontinuous in $[0, 1]$. Do this in two ways: (i) directly and (ii) by relying on the Arzèla-Ascoli theorem.

***ex5.1.8.** Show that $\{f_n\}$ is uniformly bounded in $[a, b]$ if and only if $\{f_n\}$ is bounded in $C[a, b]$ with respect to the sup norm.

*5.2. Completeness and separability of $C[0, 1]$

In Chapter 4, we introduced the metric space $C[0, 1]$ of continuous functions on $[0, 1]$ with the supremum norm $\|f\| = \max_{[0,1]} |f(x)|$, for $f \in C[0, 1]$, and the corresponding metric, $d(f, g) = \|f - g\|$, for all $f, g \in C[0, 1]$.[3]

A sequence $\{f_n\}$ of continuous functions on $[0, 1]$ converging uniformly to a continuous limit f (by Theorem 5.1.2), is the same thing as $\|f_n - f\| \to 0$. The theorems of the previous sections can therefore be used to understand more about the space $C[0, 1]$.

5.2.1. Completeness

Theorem. *The space $C[0, 1]$ is complete.*

Proof. Let $\{f_n\}$ be a Cauchy sequence in $C[0, 1]$ with respect to the supremum norm. This means that for any $\varepsilon > 0$, there is an $N \in \mathbb{N}$ such that $\|f_n - f_m\| < \varepsilon$ for all $n, m > N$. This implies that for all $n, m > N$, $\max_{x\in[0,1]} |f_n(x) - f_m(x)| < \varepsilon$, so that $|f_n(x) - f_m(x)| < \varepsilon$ for all $x \in [0, 1]$ and all $n, m > N$. In other words, $\{f_n\}$ is uniformly Cauchy in $[0, 1]$ and hence converges uniformly there to a continuous function f on $[0, 1]$ by Theorems 5.1.3 and 5.1.2.

Since $f_n \to f$ uniformly, it follows that given $\varepsilon > 0$, $|f_n(x) - f(x)| < \varepsilon$ for all $x \in [0, 1]$ if n is sufficiently large. I.e.,

$$\|f_n - f\| = \max_{x\in[0,1]} |f_n(x) - f(x)| < \varepsilon$$

for n sufficiently large, which means that $\{f_n\}$ converges to $f \in C[0, 1]$ with respect to the supremum norm. I.e., every Cauchy sequence in $C[0, 1]$ converges to a limit in $C[0, 1]$. □

In addition to completeness, $C[0, 1]$ satisfies a weak form of the Bolzano-Weierstrass theorem, as follows from the Arzèla-Ascoli theorem. A bounded sequence $\{f_n\} \subset C[0, 1]$ that is also equicontinuous in $[0, 1]$ has a convergent subsequence. Absent equicontinuity though, bounded sequences need not have convergent subsequences, as the next example shows.

[3]The norm is also denoted $\|f\|_\infty$ or $\|f\|_{\text{sup}}$, to avoid confusion when other norms are also being considered.

For $n \in \mathbb{N}$, let

$$L_n(x) = \begin{cases} 1 & : 0 \le x \le 1/2^n \\ 2 - 2^n x & : 1/2^n \le x \le 1/2^{n-1} \\ 0 & : 1/2^{n-1} \le x \le 1. \end{cases}$$

Each of these functions is piecewise linear and so continuous, and $\|L_n\| = 1$ for all $n \in \mathbb{N}$, so the sequence $\{L_n\}$ is *bounded* in $C[0,1]$. On the other hand, if $n > m$, then $L_n(1/2^m) = 0$ and $L_m(1/2^m) = 1$, so $\|L_n - L_m\| = 1$. This means that the sequence $\{L_n\}$ has no convergent subsequence because it has no subsequence that is a Cauchy sequence.

The space $C[0,1]$ also fails to have the Heine-Borel property. The sequence $\{L_n\}$, above, is closed in $C[0,1]$ because it has no limit points in $C[0,1]$ and bounded as we have shown, but it is not compact. The open cover $\{B(L_n, 1/2) : n \in \mathbb{N}\}$ admits no finite subcover since $L_m \notin B(L_n, 1/2)$ when $m \ne n$.

5.2.2. Separability

Theorem. *The space $C[0,1]$ is separable.*

Proof. We need to find a set of functions $P = \{p_\lambda : \lambda \in \Lambda\} \subset C[0,1]$ such that Λ is countable and for every $f \in C[0,1]$ and every $\varepsilon > 0$, there is a $p_\lambda \in P$ such that $\|f - p_\lambda\| < \varepsilon$.

For $n \in \mathbb{N}$ let $\lambda = \{r_0, \dots, r_n\}$ be an $n+1$-tuple of rational numbers. We define the function p_λ on $[0,1]$ by setting $p_\lambda(k/n) = r_k$ for $0 \le k \le n$ and

$$p_\lambda(x) = r_k + (r_{k+1} - r_k)(nx - k)$$

for each $0 \le k \le n-1$ and every $x \in [k/n, (k+1)/n]$. These functions are continuous on $[0,1]$ because they are piecewise linear. Moreover, the collection of all such functions is countable because for each $n \in \mathbb{N}$, the set Λ_n of rational $n+1$-tuples is countable, so the union $\Lambda = \bigcup_{n\in\mathbb{N}} \Lambda_n$ is countable, being a countable union of countable sets.

Given $f \in C[0,1]$ and $\varepsilon > 0$, there is a $\delta > 0$ such that $|f(x) - f(y)| < \varepsilon/5$ if $|x - y| < \delta$, because f is uniformly continuous on $[0,1]$. Let $n \in \mathbb{N}$ be such that $1/n < \delta$. For each $0 \le k \le n$, let r_k be a rational number satisfying $|r_k - f(k/n)| < \varepsilon/5$, set $\lambda = \{r_0, \dots, r_k\}$ and let $p_\lambda(x)$ be the piecewise linear function constructed as above.

If $0 \le k \le n-1$ and $x \in [k/n, (k+1)/n]$, we have $|f(x) - f(k/n)| < \varepsilon/5$ because $|x - k/n| \le 1/n < \delta$. In particular, $|f((k+1)/n) - f(k/n)| < \varepsilon/5$, from which it follows that $|r_k - r_{k+1}| < 3\varepsilon/5$, because of the way we chose $r_0, r_1, \dots, r_n$, and the triangle inequality:

$$|r_k - r_{k+1}| \le |r_k - f(k/n)| + |f(k/n) - f((k+1)/n)| + |f((k+1)/n) - r_{k+1}|.$$

This implies that $|p_\lambda(x) - r_k| < 3\varepsilon/5$ because $p_\lambda(x)$ falls between r_k and r_{k+1}, and therefore

$$|f(x) - p_\lambda(x)| \le |f(x) - f(k/n)| + |f(k/n) - r_k| + |r_k - p_\lambda(x)| < \varepsilon,$$

for all $x \in [k/n, (k+1)/n]$. Since $[0,1] = \bigcup_{k=0}^{n-1} [k/n, (k+1)/n]$, it follows that $\|f - p_\lambda\| < \varepsilon$, so $\{p_\lambda : \lambda \in \Lambda\}$ is a dense countable subset of $C[0,1]$. □

5.2.3. The Weierstrass approximation theorem

The countable, dense subset $\{p_\lambda : \lambda \in \Lambda\}$ of $C[0,1]$ that we constructed in the previous section consists of piecewise linear functions. In this section, we show that the (countable) set $\mathbb{Q}[x]$ of polynomials with rational coefficients is likewise dense in $C[0,1]$.

This is a conclusion of the Weierstrass approximation theorem.

Theorem (Weierstrass). *For any $f \in C[0,1]$ and any $\varepsilon > 0$, there is a polynomial p_ε such that $\|f - p_\varepsilon\| < \varepsilon$.*

There are a number of proofs of this important theorem, including a broad generalization due to Marshall Stone,[4] which in this general form is called the Stone-Weierstrass theorem. The proof we give below is due to Sergei Bernstein, and uses the eponymous *Bernstein polynomials.*

Definition. For $f \in C[0,1]$ and $n \in \mathbb{N}$, the nth *Bernstein polynomial* for f is defined by

$$B_{f,n}(x) = \sum_{k=0}^{n} f\left(\frac{k}{n}\right) \cdot \binom{n}{k} \cdot x^k(1-x)^{n-k},$$

for $0 < x < 1$, with $B_{f,n}(0) = f(0)$ and $B_{f,n}(1) = f(1)$.

To prove the theorem, we show that if $f \in C[0,1]$ and $\varepsilon > 0$, then $\|f - B_{f,n}\| < \varepsilon$ for n sufficiently large.

Bernstein's original proof relies on probabilistic arguments. The elementary proof below does not refer directly to these arguments, but they underlie it. We begin with some simple properties of Bernstein polynomials

Proposition. *For $f, g \in C[0,1]$ and $a, b \in \mathbb{R}$:*

(i) $B_{f,0}(x) = f(0)$.

(ii) $B_{f,1}(x) = (1-x)f(0) + xf(1)$.

(iii) $B_{af+bg,n}(x) = aB_{f,n}(x) + bB_{g,n}(x)$, *for all $n \geq 0$.*

(iv) *If $f \geq 0$ in $[0,1]$, then $B_{f,n}(x) \geq 0$ in $[0,1]$, for all $n \geq 0$.*

Proof. Exercise. □

The next lemma is somewhat technical.

[4] Stone proved the analog of Weierstrass's theorem for the space of continuous functions, $C(X)$, on a compact *Hausdorff space* X. A Hausdorff space is a topological space with the property that for any two distinct points in the space, there exist disjoint neighborhoods of each. In Stone's theorem, the set of polynomials on $[0,1]$ is replaced by an algebra $A \subseteq C(X)$ that contains the constant functions, and *separates points*: if $x \neq y$, there is an $f \in A$ such that $f(x) \neq f(y)$.

Lemma. *For $n \geq 0$*

$$B_{1,n}(x) = 1, \tag{5.2.1}$$

$$B_{x,n}(x) = x, \tag{5.2.2}$$

$$B_{x^2,n}(x) = \frac{n-1}{n}x^2 + \frac{1}{n}x. \tag{5.2.3}$$

$$B_{(x-u)^2,n}(x) = x^2 + \frac{1}{n}(x - x^2) - 2ux + u^2, \tag{5.2.4}$$

for any $u \in [0,1]$.

Proof. For $n = 0$ or 1, the lemma follows from Proposition 5.2.3(i) and (ii), so we may assume that $n \geq 2$.

For (5.2.1), we have

$$B_{1,n} = \sum_{k=0}^{n} \binom{n}{k} \cdot x^k(1-x)^{n-k} = \big(x + (1-x)\big)^n = 1,$$

by the binomial theorem.

Next, we observe that $\frac{k}{n}\binom{n}{k} = \binom{n-1}{k-1}$ for $n \geq 2$ and $0 \leq k \leq n$, and therefore

$$B_{x,n}(x) = \sum_{k=0}^{n} \frac{k}{n} \cdot \binom{n}{k} \cdot x^k(1-x)^{n-k} = \sum_{k=1}^{n} \binom{n-1}{k-1} \cdot x^k(1-x)^{n-k},$$

since the term in $B_{x,n}(x)$ corresponding to $k = 0$ is 0. Pulling out a factor of x and setting $l = k-1$ in the sum on the right shows that

$$B_{x,n}(x) = x \sum_{l=0}^{n-1} \binom{n-1}{l} \cdot x^l(1-x)^{n-1-l} = x \cdot B_{1,n-1}(x) = x,$$

proving (5.2.2).

Next, we replace k^2/n^2 by

$$\frac{k}{n} + \frac{k^2}{n^2} - \frac{k}{n} = \frac{k}{n} + \frac{k(k-n)}{n^2}$$

in the sum for $B_{x^2,n}(x)$ to see that

$$B_{x^2,n}(x) = \sum_{k=0}^{n} \frac{k^2}{n^2} \cdot \binom{n}{k} \cdot x^k(1-x)^{n-k} = x + \sum_{k=0}^{n} \frac{k(k-n)}{n^2} \cdot \binom{n}{k} \cdot x^k(1-x)^{n-k},$$

as follows from Proposition 5.2.3(iii) and the identity (5.2.2). The terms in the sum on the right above for $k = 0$ and $k = n$ are both zero, and for $1 \leq k \leq n-1$,

$$\begin{aligned}
\frac{k(k-n)}{n^2} \cdot \binom{n}{k} &= \frac{k(k-n)}{n^2} \cdot \frac{n!}{k!\,(n-k)!} = -\frac{1}{n} \cdot \frac{(n-1)!}{(k-1)!\,(n-1-k)!} \\
&= -\frac{n-1}{n} \cdot \frac{(n-2)!}{(k-1)!\,((n-2)-(k-1))!} = -\frac{n-1}{n}\binom{n-2}{k-1}.
\end{aligned}$$

From this, and the expression for $B_{x^2,n}(x)$ above that, it follows that for $n \geq 2$,

$$\begin{aligned} B_{x^2,n} &= x - \frac{n-1}{n}\sum_{k=1}^{n-1}\binom{n-2}{k-1}x^k(1-x)^{n-k} \\ &= x - \frac{n-1}{n}x(1-x)\sum_{k=1}^{n-1}\binom{n-2}{k-1}x^{k-1}(1-x)^{(n-2)-(k-1)} \\ &= x - \frac{n-1}{n}x(1-x)\sum_{l=0}^{n-2}\binom{n-2}{l}x^l(1-x)^{(n-2)-l} = \frac{n-1}{n}x^2 + \frac{1}{n}x, \end{aligned}$$

substituting $l = k-1$ in the sum on the second line and observing that the sum on the third line is $B_{1,n-2}(x)$.

Finally, (5.2.4) follows from Proposition 5.2.3(iii), the previous three identities and the fact that $(x-u)^2 = x^2 - 2ux + u^2$. □

We can now complete Bernstein's proof of Weierstrass's approximation theorem.

Proof. Let $f(x) \in C[0,1]$, then $|f|$ is bounded in $[0,1]$ by $\|f\|$, so

$$|f(x) - f(y)| \leq 2\|f\|$$

for all $x, y \in [0,1]$. Next, for fixed $\varepsilon > 0$ and arbitrary $u \in [0,1]$, there is a $\delta > 0$ such that $|f(x) - f(u)| < \varepsilon/2$ for all $x \in [0,1]$ with $|x-u| < \delta^{1/2}$. On the other hand, if $|x-u| \geq \delta^{1/2}$, then

$$|f(x) - f(u)| \leq 2\|f\| \leq \frac{(x-u)^2}{\delta}\cdot 2\|f\|,$$

and therefore

$$|f(x) - f(u)| < \frac{(x-u)^2}{\delta}\cdot 2\|f\| + \frac{\varepsilon}{2} = C\cdot(x-u)^2 + \frac{\varepsilon}{2} \tag{5.2.5}$$

for all $x \in [0,1]$ (where $C = 2\|f\|/\delta$).

From Proposition 5.2.3(iii) and the identity (5.2.1), we have

$$B_{f-f(u),n}(x) = B_{f,n}(x) - B_{f(u),n}(x) = B_{f,n}(x) - f(u),$$

since $f(u)$ is constant as a function of x. Hence, for all $x \in [0,1]$,

$$|B_{f,n}(x) - f(u)| = |B_{f-f(u),n}(x)| \leq B_{|f-f(u)|,n}(x) < B_{C(x-u)^2+\frac{\varepsilon}{2},n}(x),$$

where the first inequality follows from the triangle inequality and the second one follows from (5.2.5) and Proposition 5.2.3(iv). It next follows from Proposition 5.2.3(iii), (5.2.1) and (5.2.4) that

$$B_{C(x-u)^2+\frac{\varepsilon}{2},n}(x) = C\cdot B_{(x-u)^2,n}(x) + \frac{\varepsilon}{2} = Cx^2 + \frac{C}{n}(x-x^2) - 2Cux + Cu^2 + \frac{\varepsilon}{2},$$

and hence

$$|B_{f,n}(x) - f(u)| < Cx^2 + \frac{C}{n}(x-x^2) - 2Cux + Cu^2 + \frac{\varepsilon}{2}.$$

In particular, setting $x = u$, shows that

$$|B_{f,n}(u) - f(u)| < \frac{C}{n}(u-u^2) + \frac{\varepsilon}{2} \leq \frac{C}{4n} + \frac{\varepsilon}{2},$$

because the quadratic function $v = u - u^2$ has its maximum at $u = 1/2$. The choice of $u \in [0, 1]$ was arbitrary, and it follows that

$$\|B_{f,n} - f\| < \frac{C}{4n} + \frac{\varepsilon}{2} < \varepsilon$$

for all $n > \frac{C}{2\varepsilon} = \frac{\|f\|}{\varepsilon \cdot \delta}$. □

Corollary. *If $f(x) \in C[0, 1]$ then for any $\varepsilon > 0$, there is a polynomial $q(x)$ with rational coefficients such that $\|q - f\| < \varepsilon$.*

Proof. Exercise. □

Exercises

ex5.2.1. Show that $C[0, 1]$ is not *locally* compact.

ex5.2.2. Prove Proposition 5.2.3.

ex5.2.3. Show that Weierstrass's theorem is true in $C[a, b]$, for any interval $[a, b] \subset \mathbb{R}$.

Hint: The map $f(x) \mapsto \varphi(x) = f(a + (b - a)x)$ is an *isometry* (with respect to the sup norm) from $C[a, b]$ to $C[0, 1]$.

ex5.2.4. Prove Corollary 5.2.3.

Hint: Show that if $p(x) = \alpha_n x^n + \cdots + \alpha_1 x + \alpha_0$ is a polynomial with real coefficients and $\varepsilon > 0$, you can find rational numbers $a_0, a_1, \ldots, a_n$, with a_j sufficiently close to α_j for all j, such that the polynomial $q(x) = a_n x^n + \cdots + a_1 x + a_0$ satisfies $|q(x) - p(x)| < \varepsilon/2$ for all $x \in [0, 1]$.

5.3. Series of functions

Just as in the case of numerical series, series of function are defined as the limits of the sequences of their partial sums.

Definition. Let $\{f_n\}_{n=1}^{\infty}$ be a sequence of real-valued functions defined in an interval $I \subset \mathbb{R}$, and let $F_n(x) = \sum_{j=1}^{n} f_n(x)$. The series $\sum_{n=1}^{\infty} f_n(x)$ converges pointwise in I, if the sequence $\{F_n(x)\}$ converges pointwise in I, and $\sum_{n=1}^{\infty} f_n(x)$ converges uniformly in I, if the sequence $\{F_n(x)\}$ converges uniformly in I.

Example. It follows from Example 5.1.2 (ii) that the series $\sum_{n=1}^{\infty} x^n = \frac{1}{1-x}$ converges pointwise in $(-1, 1)$, but not uniformly so. On the other hand, it also follows from the same example that this series does converge uniformly in $[-a, a]$ for any $0 < a < 1$.

5.3.1. The criteria of Cauchy and Dirichlet

The Cauchy and Dirichlet criteria for the convergence of numerical sequences can be reformulated to give criteria for uniform convergence of series of functions.

Theorem. *The series $\sum f_n(x)$ converges uniformly in the interval I, if and only if for every $\varepsilon > 0$, there is an $N_\varepsilon \in \mathbb{N}$ such that $\left|\sum_{j=m}^{n} f_n(x)\right| < \varepsilon$ for all $n \geq m > N_\varepsilon$ and all $x \in I$.*

Proof. The series $\sum f_n(x)$ converges uniformly in I if and only if the sequence of its partial sums converges uniformly in I, and by Theorem 5.1.3, the sequence of partial sums converges uniformly in I if and only it is uniformly Cauchy in I. The claim follows from this. □

Corollary. *If the series $\sum f_n(x)$ converges uniformly in the interval I, then the sequence $\{f_n\}$ converges uniformly to 0 in I.*

Proof. Exercise. □

The proof of Dirichlet's criterion, below, also relies on Cauchy's criterion.

Proposition. *Suppose that the sequence $\{f_n(x)\}$ converges uniformly to 0 in the interval I, with $f_n(x) \geq f_{n+1}(x) \geq 0$ for all $x \in I$ and $n \in \mathbb{N}$, and that the partial sums $\sum_{j=1}^{n} g_j(x)$ are uniformly bounded in I. Then the series $\sum_{n=1}^{\infty} f_n(x)g_n(x)$ converges uniformly in I.*

Proof. Since $f_n \to 0$ uniformly in I and $f_n(x) \geq 0$ for all $x \in I$, there exists an $N \in \mathbb{N}$ such that $0 \leq f_n(x) < 1$ for all $n > N$ and all $x \in I$. It follows that for $n > N$, $a_n = \sup\{f_n(x) : x \in I\}$ exists $0 \leq a_n \leq 1$. Additionally, $a_n \to 0$ and $a_{n+1} \leq a_n$ for all $n > N$, because $0 \leq f_{n+1}(x) \leq f_n(x)$ for all n and all $x \in I$.

Let $\left|\sum_{j=1}^{n} g_j(x)\right| < E$ for all n and all $x \in I$, then given $\varepsilon > 0$, there is an $m_0 > N$ such that $a_m < \varepsilon/3E$ for all $m > m_0$, and it follows that for any $x \in I$, and $n \geq m > m_0$, we have

$$\left|\sum_{j=m}^{n} f_j(x)g_j(x)\right| \leq \left|\sum_{j=m}^{n} a_j g_j(x)\right| \leq 3Ea_m < \varepsilon,$$

as can be shown by mimicking the proof of Theorem 2.7.5. Hence, $\sum f_n(x)g_n(x)$ converges uniformly in I by the Cauchy criterion. □

5.3.2. Absolute convergence

Definition. The series $\sum_{n=1}^{\infty} f_n(x)$ converges (uniformly and) *absolutely* in the interval I if the series $\sum_{n=1}^{\infty} |f_n(x)|$ converges (uniformly) in the interval I.

Neither uniform convergence nor absolute convergence of a series implies the other. For example, the series $\sum_{n=0}^{\infty} x^n = \frac{1}{1-x}$ converges absolutely in $(-1, 1)$ but not uniformly, as follows from Example 5.1.2.

On the other hand, the series $\sum_{n=1}^{\infty} \frac{(-|x|)^n}{n}$ converges uniformly in $[-1, 1]$ by Dirichlet's criterion with $f_n(x) = |x|^n/n$ and $g_n(x) = (-1)^n$, but does not converge absolutely there, since $\sum 1/n$ diverges.

On the other hand, absolute convergence does imply convergence.

Proposition. *If the series $\sum_{n=1}^{\infty} f_n(x)$ converges absolutely in the interval I, then it converges there.*

Proof. Exercise. □

Exercises

ex5.3.1. Prove Corollary 5.3.1.

ex5.3.2. Complete the details in the proof of Proposition 5.3.1.

ex5.3.3. Give an example of a sequence of continuous functions on $[0, 1]$ which converges everywhere to a function f which is not continuous at $x = 1/2$. (The convergence clearly can't be *uniform*). Generalize your example to give a sequence of continuous functions that converges everywhere in $[0, 1]$ to a function that is discontinuous at the points $\{j/k : 1 \leq j \leq k-1\}$.

Can you find an example of a sequence of continuous function that converges everywhere in $[0, 1]$ to a function that has countably many points of discontinuity?

ex5.3.4. Prove Proposition 5.3.2.

ex5.3.5. Let $\{f_n\}$ be a sequence of positive continuous functions on $[0, 1]$. Show that if the series $\sum_{n=1}^{\infty} f_n(x)$ converges everywhere in $[0, 1]$ to a continuous function F, then the convergence is uniform.

Hint: For fixed $\varepsilon > 0$ and $n \in \mathbb{N}$, consider the sets

$$O_n(\varepsilon) = \{x \in [0, 1] : F(x) - \sum_{k=1}^{n} f_k(x) < \varepsilon\}.$$

Show that O_n is open and $O_n \subset O_{n+1}$ for every n, and that $\bigcup O_n(\varepsilon) = [0, 1]$. Now use the Heine-Borel theorem to show... What?

5.4. Power series

The series $\sum_{n=0}^{\infty} x^n$, $\sum_{n=1}^{\infty} x^n/n$ and $\sum_{n=0}^{\infty} x^n/n!$ are all examples of *power series*.

Definition. A *power series* is an expression of the form

$$\sum_{n=0}^{\infty} a_n(x - x_0)^n,$$

where $x = x_0$ is the *center* of the series, and $\{a_n\}$ are the coefficients of the series. Power series with complex coefficients (and complex variables) are very useful, but we restrict our attention here to power series with real coefficients.

5.4.1. Radius of convergence

Suppose that the series $\sum a_n(x-x_0)^n$ converges absolutely at $x = x_1$, then by the comparison test, the series converges absolutely at x if $|x-x_0| \le |x_1 - x_0|$. This observation leads to the following definition.

Definition. The *radius of convergence* of the series $\sum_{n=0}^{\infty} a_n(x-x_0)^n$ is

$$R = \sup\Big\{|x-x_0| : \sum a_n(x-x_0)^n \text{ converges absolutely}\Big\}.$$

If the series converges absolutely for all x, then $R = \infty$, and if it diverges for all $x \neq x_0$, then $R = 0$.

If $R > 0$, the interval $(x_0 - R, x_0 + R)$ is called the *interval of convergence* of the series.

From the definition, it follows that $\sum a_n(x-x_0)^n$ converges *absolutely* for all x in its interval of convergence. It follows from this that if the radius of convergence of the series, R, is strictly positive, then the function $f(x) = \sum a_n(x-x_0)^n$ is well defined throughout the interval of convergence, $(x_0 - R, x_0 + R)$.

Theorem. *If the power series $\sum a_n(x-x_0)^n$ has radius of convergence $R > 0$, then its sum $f(x)$ is continuous in $(x_0 - R, x_0 + R)$.*

Proof. The function $f(x) = \sum a_n(x-x_0)^n$ is defined for all $x \in (x_0 - R, x_0 + R)$, so we need only show that $f(x)$ is continuous in this interval.

If $0 < r < R$, then the series converges absolutely at $x_0 + r$. Hence for any $\varepsilon > 0$, there exists by the Cauchy criterion for a numerical sequence (Theorem 2.6.2), an $N_{r,\varepsilon} \in \mathbb{N}$ such that $\sum_{k=m}^{n} |a_n r^n| < \varepsilon$ for all $n \ge m \ge N_{r,\varepsilon}$. It follows that if $|x - x_0| \le r$, then

$$\left|\sum_{k=m}^{n} a_n(x-x_0)^n\right| \le \sum_{k=m}^{n} |a_n(x-x_0)^n| \le \sum_{k=m}^{n} |a_n r^n| < \varepsilon$$

for all $n \ge m \ge N_{r,\varepsilon}$. By the Cauchy criterion for a series of functions, the power series $\sum a_n(x-x_0)^n$ converges uniformly in $[x_0 - r_2, x_0 + r_2]$ to $f(x)$. Hence, $f(x)$ is a uniform limit of continuous functions (because the partial sums of a power series are polynomials) and is therefore continuous in $[x_0 - r, x_0 + r]$ by Theorem 5.1.2. Finally, as the previous statement is true for all $0 < r < R$, $f(x)$ is continuous in all of $(x_0 - R, x_0 + R)$. □

The definition of the radius of convergence of a power series is not useful for determining its value for any given series. Nor does it say anything about the behavior of the series outside its interval of (absolute) convergence. The following theorem addresses both these issues.

Proposition (Cauchy-Hadamard). *Let $R = \left(\limsup_{n\to\infty} \sqrt[n]{|a_n|}\right)^{-1}$, where we set $R = 0$ if $\limsup_{n\to\infty} \sqrt[n]{|a_n|} = \infty$ and $R = \infty$ if $\limsup_{n\to\infty} \sqrt[n]{|a_n|} = 0$. Then the series $\sum a_n(x-x_0)^n$ converges absolutely if $|x-x_0| < R$ and diverges if $|x - x_0| > R$.*

In other words R, as given in the proposition, is the radius of convergence of the series $\sum a_n(x-x_0)^n$.

Proof. By the root test for convergence (Theorem 2.7.3) the series converges absolutely at x if

$$\limsup \sqrt[n]{|a_n(x-x_0)^n|} = |x-x_0|\left(\limsup \sqrt[n]{|a_n|}\right) < 1,$$

and diverges if $|x-x_0|\left(\limsup \sqrt[n]{|a_n|}\right) > 1$.

If $\limsup \sqrt[n]{|a_n|} = \infty$, then $|x-x_0|\left(\limsup \sqrt[n]{|a_n|}\right) > 1$ for all $x \neq x_0$, so $R = 0$; If $\limsup \sqrt[n]{|a_n|} = 0$, then the $|x-x_0|\left(\limsup \sqrt[n]{|a_n|}\right) < 1$ for all x, so $R = \infty$.

Finally, if $0 < \limsup \sqrt[n]{|a_n|} < \infty$, then $|x-x_0|\left(\limsup \sqrt[n]{|a_n|}\right) < 1$ if and only if $|x-x_0| < 1/\limsup \sqrt[n]{|a_n|}$. □

The behavior of a series at the endpoints of its interval of convergence varies. The series $\sum x^n$, $\sum x^n/n$ and $\sum x^n/n^2$ all have radius of convergence $R = 1$, as you can verify. The series $\sum x^n$ diverges at both endpoints, the series $\sum x^n/n$ diverges at $x = 1$ but converges *conditionally* at $x = -1$, and the series $\sum x^n/n^2$ converges absolutely at both endpoints by Proposition 2.7.4.

5.4.2. Combining power series

Simple combinations of continuous functions are continuous, and since power series converge to continuous functions in their intervals of convergence, it is natural to assume that the same combinations of convergent power series are also convergent power series.

Proposition. *Suppose that the power series $\sum a_n(x-x_0)^n$ and $\sum b_n(x-x_0)^n$ have radii of convergence $R_a > 0$ and $R_b > 0$, and converge to $f_a(x)$ and $f_b(x)$ in their respective intervals of convergence. Then for any $u, v \in \mathbb{R}$, $\sum(ua_n + vb_n)(x-x_0)^n$ converges absolutely to $uf_a(x) + vf_b(x)$ in the interval $(x_0 - R, x_0 + R)$, where $R = \min(R_a, R_b)$.*

Proof. Exercise □

If $f(x) = a_0 + a_1x + \cdots + a_mx^m$ and $g(x) = b_0 + b_1x + \cdots b_mx^m$, their product is the polynomial

$$f(x)g(x) = a_0b_0 + (a_1b_0 + a_0b_1)x + \cdots + (a_mb_0 + \cdots + a_0b_m)x^{2m} = \sum_{n=0}^{2m} c_nx^n.$$

The coefficients of the product of two power series are defined analogously.

Definition. Given the power series $\sum a_n(x-x_0)^n$ and $\sum b_n(x-x_0)^n$, their *Cauchy product* is the series $\sum c_n(x-x_0)^n$, where for all $n \geq 0$.

$$c_n = \sum_{k=0}^{n} a_kb_{n-k}.$$

This is nothing more than the definition of the Cauchy product given in §2.6.6 applied to the given power series.

Theorem. *Suppose that the series $A(x) = \sum a_n(x-x_0)^n$ and $B(x) = \sum b_n(x-x_0)^n$ have radii of convergence $R_a, R_b > 0$ respectively, and let $R_1 = \min(R_a, R_b)$. Then the Cauchy product of these series, $C(x) = \sum c_n(x-x_0)^n$, converges to $A(x)B(x)$ in the interval $(x_0 - R_1, x_0 + R_1)$.*

Comment. It follows from Theorem 5.4.1 that the radius of convergence of $C(x)$ is $\geq R_1$.

Proof. If $|x - x_0| < R_1$, then *both* $\sum a_n(x-x_0)^n$ and $\sum b_n(x-x_0)^n$ converge absolutely, and the theorem is therefore a direct consequence of Theorem 2.6.6 (Mertens's theorem). □

5.4.3. The exponential function

We saw in Theorem 2.7.2 that the series $\exp(x) = \sum \frac{x^n}{n!}$ converges for all $x \in \mathbb{R}$. The following proposition is another consequence of Theorem 2.6.6.

Proposition. *For all $x_1, \ldots, x_m \in \mathbb{R}$,*

$$\exp(x_1)\exp(x_2)\cdots\exp(x_m) = \exp(x_1 + x_2 + \cdots + x_m).$$

Proof. For the case $m = 2$, we have $\exp(x_1)\exp(x_2) = \sum_{n=0}^{\infty} c_n(x_1, x_2)$, where

$$c_n(x_1, x_2) = \sum_{k=0}^{n} \frac{x_1^k}{k!} \cdot \frac{x_2^{n-k}}{(n-k)!} = \frac{1}{n!}\sum_{k=0}^{n} \frac{n!}{k!\,(n-k)!} x_1^k x_2^{n-k} = \frac{(x_1 + x_2)^n}{n!},$$

by Mertens's theorem and the binomial theorem. The general case follows by induction. We leave the details as an exercise. □

Theorem. *For all $x \in \mathbb{R}$, $\exp(x) = e^x$ where $e = \exp(1)$.*

Proof. We first observe that $\exp(0) = 1 = e^0$. Next, It follows from the proposition above that for $n \in \mathbb{N}$,

$$\exp(n) = \exp(\overbrace{1 + \cdots + 1}^{n}) = \exp(1)^n = e^n,$$

and $\exp(1/n) = e^{1/n}$ because $\exp(1/n) > 0$ and

$$e = \exp(1) = \exp(\overbrace{1/n + \cdots + 1/n}^{n}) = \exp(1/n)^n.$$

Continuing in the same vein, if $m, n \in \mathbb{N}$, then

$$\exp(m/n) = \exp(\overbrace{1/n + \cdots + 1/n}^{m}) = \left(e^{1/n}\right)^m = e^{m/n}.$$

It follows that if $r > 0$ is rational, then $\exp(r) = e^r$ and

$$1 = \exp(0) = \exp(r - r) = \exp(-r)e^r,$$

so $\exp(-r) = 1/e^r = e^{-r}$. Thus $\exp(r) = e^r$ for all $r \in \mathbb{Q}$.

Finally, if u is any real number, let $\{r_n\}$ be a sequence of rational numbers approaching u. Then $\exp(r_n) = e^{r_n}$ for all n, and since the function e^x is continuous on $\mathbb{R}$ (by Theorem 1.7.3(iv)) and $\exp(x)$ is continuous on $\mathbb{R}$ (as a convergent power series), it follows that $\exp(u) = \lim_{n\to\infty} \exp(r_n) = \lim_{n\to\infty} e^{r_n} = e^u$. □

Corollary. *The function* $\exp(x)$ *is strictly monotone increasing on all of* $\mathbb{R}$ *with range* $(0, \infty)$.

Proof. The strict monotonicity follows from Theorem 1.7.3 and the fact that $e = \exp(1) > 1$. Since $\mathbb{R}$ is an interval and exp is continuous, the range of exp is also an interval, by Theorem 4.4.3. Finally, since $\exp(x)$ is unbounded and positive, and $\exp(-x) = 1/\exp(x)$, it follows that the range is $(0, \infty)$. $\square$

Exercises

ex5.4.1. Show that the radius of convergence of the series $\sum a_n(x - x_0)^n$ is the same as $\sup\{r : \text{ the sequence } \{|a_n r^n|\} \text{ is bounded}\}$.

ex5.4.2. Prove Proposition 5.4.2.

ex5.4.3. Complete the proof of Proposition 5.4.3.

Chapter 6

Differentiation

6.1. The derivative

In beginning calculus classes, the derivative of a function $f(x)$ at a point a is typically motivated by the problem of (defining and) finding the slope of the curve $y = f(a)$ at the point $(a, f(a))$. This visually intuitive introduction to the derivative is followed by the observation that $f(x)$ is well-approximated by the linear function

$$L_a(x) = f(a) + f'(a)(x - a)$$

if x is sufficiently close to a.

In practice, the fact that $f(x)$ has a very good linear approximation in the vicinity of the point $x = a$ is what makes the derivative so useful. So here, we approach the definition of the derivative through the problem of finding the best linear approximation to a function at a point.

6.1.1. Best linear approximation

Suppose that $f(x)$ is a real-valued function defined in a neighborhood of the point a. We say that the function $g(x)$ is a better approximation to $f(x)$ at the point a than the function $h(x)$, if there is a neighborhood of U of a such that $|f(x)-g(x)| \leq |f(x)-h(x)|$ for all $x \in U$. To go from *better* approximation to *the best* approximation, we need to be flexible about the neighborhoods.

Definition. The linear function $L_{f,a}(x)$ is the *best linear approximation to $f(x)$ at $x = a$* if for every linear function $L(x)$, there is a $\delta > 0$ such that

$$|f(x) - L(x)| \geq |f(x) - L_{f,a}(x)|,$$

for all $x \in (a - \delta, a + \delta)$.

Not every function has a best linear approximation at every point. For example, the function $f(x) = |x|$ has no best linear approximation at $x = 0$ (**ex6.1.1**). On the other hand, if there is a best linear approximation as defined above, then it must be unique (**ex6.1.2**), justifying the use of the definite article in the definition.

Proposition. *If $L_1(x)$ and $L_2(x)$ are linear functions, with $L_1(a) = f(a)$ and $L_2(a) \neq f(a)$, then L_1 is a better approximation than L_2 to f at a.*

Proof. Exercise. □

It follows that when looking for the best linear approximation to f at a, we need only consider linear functions L satisfying $L(a) = f(a)$. Any such function has the form

$$L_\mu(x) = f(a) + \mu(x - a),$$

for some $\mu \in \mathbb{R}$.

Lemma. *If $\frac{f(x)-f(a)}{x-a}$ is unbounded in every punctured neighborhood of a, then there is no best linear approximation to f at a.*

Proof. If $\mu \geq 0$, then from the unboundedness assumption, it follows that there is a sequence $U = \{x_n\}$ such that $0 < |x_n - a| < 1/n$ and

$$|f(x_n) - f(a)| > (\mu + 1)|x_n - a|,$$

for every $n \in \mathbb{N}$. Consider the subsequences

$$A_1 = \{x_n \in U : x_n > a \text{ and } f(x_n) - f(a) > (\mu + 1)(x_n - a)\}.$$
$$A_2 = \{x_n \in U : x_n > a \text{ and } f(x_n) - f(a) < -(\mu + 1)(x_n - a)\}.$$
$$A_3 = \{x_n \in U : x_n < a \text{ and } f(x_n) - f(a) > (\mu + 1)(x_n - a)\}.$$
$$A_4 = \{x_n \in U : x_n < a \text{ and } f(x_n) - f(a) < -(\mu + 1)(x_n - a)\}.$$

If $x_n \in A_1$, we have

$$f(x_n) > f(a) + (\mu + 1)(x_n - a) > f(a) + \mu + (x_n - a),$$

so $L_{\mu+1}(x_n)$ is closer to $f(x_n)$ than $L_\mu(x_n)$. Likewise, if $x_n \in A_4$, then $L_{\mu+1}(x_n)$ is closer to $f(x_n)$ than $L_\mu(x_n)$. Similarly, if $x_n \in A_2$ or A_3, then $L_{-(\mu+1)}(x_n)$ is closer to $f(x_n)$ than $L_\mu(x_n)$. Since) one of these four subsequences converges to a, it follows that L_μ fails to be better linear approximation to f at a than one of $L_{\mu+1}$ or $L_{-(\mu+1)}$. The same argument works when $\mu < 0$ (**ex6.1.4**). □

The "little o" notation from §4.1.7. is useful here. Recall that $f(x) = L_{\mu_0}(x)+o(x-a)$ means that for any $\varepsilon > 0$ there is a $\delta > 0$ such that $|f(x) - L_{\mu_0}(x)| < \varepsilon|x - a|$ for all $x \in (a - \delta, a + \delta)$.

Theorem. *$L_{\mu_0}(x)$ is the best linear approximation to $f(x)$ at $x = a$ if and only if*

$$f(x) = L_{\mu_0}(x) + o(x - a).$$

Proof. Suppose first that $f(x) = L_{\mu_0}(x) + o(x - a)$ and let $\mu \neq \mu_0$. It follows that $\varepsilon = |\mu - \mu_0| > 0$, and $|L_{\mu_0}(x) - L_\mu(x)| = \varepsilon|x - a|$ for all x. On the other hand, there is a δ_μ such that if $|x - a| < \delta_\mu$, then

$$|f(x) - L_{\mu_0}(x)| < \frac{\varepsilon}{2}|x - a|.$$

Hence for all $x \in (a - \delta_\mu, a + \delta_\mu)$,

$$\begin{aligned}|f(x) - L_\mu(x)| &\geq |L_{\mu_0}(x) - L_\mu(x)| - |f(x) - L_{\mu_0}(x)| \\ &> \frac{\varepsilon}{2}|x - a| > |f(x) - L_{\mu_0}(x)|,\end{aligned}$$

by the triangle inequality. Since μ was arbitrary, $L_{\mu_0}(x)$ is the best linear approximation to $f(x)$ at a.

If $f(x) - L_\mu(x) \neq o(x - a)$, then there is an $\varepsilon > 0$ such that for every $n \in \mathbb{N}$, there is an x_n such that $0 < |x_n - a| < 1/n$ and

$$|f(x_n) - L_\mu(x_n)| > \varepsilon|x_n - a|. \tag{6.1.1}$$

From the preceding lemma, we only need consider the case that $\frac{f(x)-f(a)}{x-a}$ is bounded in a punctured neighborhood of a, and it follows from this assumption that

$$\limsup_{n\to\infty} \frac{f(x_n) - f(a)}{x_n - a} = \beta$$

is finite. Thus, there is a subsequence $\{x_{n_k}\}$ such that $\lim_{k\to\infty} \frac{f(x_{n_k})-f(a)}{x_{n_k}-a} = \beta$. In particular, there is a K_0, such that if $k > K_0$, then

$$\left|\frac{f(x_{n_k}) - f(a)}{x_{n_k} - a} - \beta\right| < \varepsilon/2,$$

so that

$$|f(x_{n_k}) - L_\beta(x_{n_k})| < \frac{\varepsilon}{2}|x_{n_k} - a|,$$

for all such k. On the other hand, it follows from (6.1.1) that

$$|L_\mu(x_{n_k}) - f(x_{n_k})| \geq |L_\mu(x_{n_k}) - L_\beta(x_{n_k})| - |L_\beta(x_{n_k}) - f(x_{n_k})| > \frac{\varepsilon_0}{2}|x_{n_k} - a|.$$

Therefore L_μ is not a better approximation to f at a than L_β (since $x_{n_k} \to a$), so L_μ is not the best linear approximation to f at a. □

Corollary. *If f has a best linear approximation at a, then f is continuous at a.*

Proof. By assumption, $f(x) = L_\mu(x) + o(x - a)$ for some $\mu \in \mathbb{R}$, so given $\varepsilon > 0$, there is a $\delta > 0$ such that $|f(x) - L_\mu(x)| < \varepsilon|x - a|$, if $|x - a| < \delta$. Hence,

$$|f(x) - f(a)| \leq (\varepsilon + |\mu|)|x - a| < \varepsilon$$

if $|x - a| < \min(\delta, \varepsilon/(\varepsilon + |\mu|))$. □

6.1.2. The derivative

The derivative is defined in terms of the *difference quotient* $\frac{f(x)-f(a)}{x-a}$ that played a central role in the discussion of best linear approximations.

Definition. The *derivative* of f at a is defined as the limit

$$\lim_{x\to a} \frac{f(x) - f(a)}{x - a} = f'(a) \tag{6.1.2}$$

if the limit exists. If $f'(a)$ exists, then f is said to be *differentiable* at a. If the limit (6.1.2) does not exist, then f is *not differentiable* at a.

A function f is differentiable in the set E, if f is differentiable at every point $x \in E$. The derivative (function) of $f(x)$ is defined by

$$f'(x) = \lim_{\xi \to x} \frac{f(\xi) - f(x)}{\xi - x} = \lim_{h \to 0} \frac{f(x+h) - f(x)}{h},$$

setting $h = \xi - x$ for convenience, defined at all points where the limit exists.

Theorem. *If $f(x)$ is differentiable at a, then $T(x) = f(a) + f'(a)(x - a)$ is the best linear approximation to f at a. If f is not differentiable at a, then f has no best linear approximation at a.*

Proof. The function f is differentiable at a if and only if the limit (6.1.2) exists. If this limit exists, then for every $\varepsilon > 0$, there is a $\delta > 0$ such that

$$\left| \frac{f(x) - f(a)}{x - a} - f'(a) \right| < \varepsilon$$

for $0 < |x - a| < \delta$. Hence, $|f(x) - f(a) - f'(a)(x - a)| < \varepsilon |x - a|$ for all such x, so $f(x) = T(x) + o(x - a)$, and $T(x)$ is the best linear approximation to $f(x)$ at a by Theorem 6.1.1.

On the other hand, if $f(x) = L_\mu(x) + o(x - a)$, then for any $\varepsilon > 0$, there is a $\delta > 0$ such that $|f(x) - L_\mu(x)| < \varepsilon |x - a|$ if $|x - a| < \delta$. It follows that if $0 < |x - a| < \delta$, then

$$\left| \frac{f(x) - f(a)}{x - a} - \mu \right| < \varepsilon,$$

so the limit (6.1.2) exists and $f'(a) = \mu$. □

Corollary. *If f is differentiable at a, then f is continuous at a.*

Proof. This follows from Corollary 6.1.1. □

Examples. Since $\lim_{h \to 0} \frac{c-c}{h} = 0$ for any constant c, it follows that if $f(x) = c$ in some interval (a, b), then $f'(x) = 0$ for all $x \in (a, b)$. More generally, a function that is linear in an open interval is differentiable there, a fact that can be shown without reference to the limit (6.1.2) (**ex6.1.4**). In most cases however, we rely on the limit to establish differentiability, as in the following proposition.

Proposition. *If $f(x) = x^n$ with $n \in \mathbb{N}$, then $f'(x) = nx^{n-1}$.*

Proof. For $n \in \mathbb{N}$ and $x, \xi \in \mathbb{R}$ with $x \neq \xi$, we have

$$\frac{\xi^n - x^n}{\xi - x} = \sum_{k=1}^{n} \xi^{n-k} x^{k-1},$$

by the identity (1.1.1). Therefore

$$\lim_{\xi \to x} \frac{\xi^n - x^n}{\xi - x} = \sum_{k=1}^{n} \left(\lim_{\xi \to x} \xi^{n-k} x^{k-1} \right) = \sum_{k=1}^{n} \left(x^{k-1} \lim_{\xi \to x} \xi^{n-k} \right) = nx^{n-1}$$

for all $x \in \mathbb{R}$ and $n \in \mathbb{N}$. □

Notation. The difference quotient $\frac{f(\xi)-f(x)}{\xi-x}$ is often expressed as $\frac{\Delta f}{\Delta x}$, where $\Delta f = f(\xi)-f(x)$ and $\Delta x = \xi - x$. This leads to the other familiar notation for the derivative

$$f'(x) = \lim_{\Delta x \to 0} \frac{\Delta f}{\Delta x} = \frac{df}{dx}.$$

We also use d/dx to denote the *differentiation operator*. E.g., for $n \in \mathbb{N}$ we write

$$\frac{d}{dx}(x^n) = nx^{n-1}.$$

6.1.3. Points of increase, points of decrease and relative extrema

The derivative of a function at a point indicates the "direction" of the function at that point, i.e., whether it is increasing or decreasing around that point. We use this information to find the relative extrema of the function.

Definition. Let f be defined in a neighborhood of x_0. Then x_0 is a *point of increase* (*decrease*) of f if there is a $\delta > 0$ such that $f(d) > f(x_0) > f(c)$ ($f(d) < f(x_0) < f(c)$) for all $x_0 - \delta < c < x_0 < d < x_0 + \delta$. If x_0 is a point of increase (decrease), we also say that f is *increasing* (*decreasing*) *around* x_0.

If there is a $\delta > 0$ such that $f(x_0) \geq f(x)$ ($f(x) \leq f(x_0)$) for all $x \in (x_0 - \delta, x_0 + \delta) \subseteq (a, b)$, then $f(x_0)$ is a *relative maximum* (*minimum*) *value* of f. If $f(x_0)$ is relative maximum or minimum value, then we say that $f(x_0)$ is a *relative extreme value* or *relative extremum* of the function.

If $f'(x_0) = 0$, then x_0 is a *stationary point* of f.

Theorem. *If f is defined in a neighborhood of x_0, f is differentiable at x_0 and $f'(x_0) > 0$, then x_0 is a point of increase of f. If $f'(x_0) < 0$, then x_0 is a point of decrease.*

Proof. Suppose that $f'(x_0) > 0$. Then, since $f(x_0+h) = f(x_0)+f'(x_0)h+o(h)$, there is a $\delta > 0$ such that

$$|f(x_0 + h) - (f(x_0) + f'(x_0)h)| < \frac{|f'(x_0)|}{2} \cdot |h| = \frac{f'(x_0)}{2} \cdot |h|$$

if $|h| < \delta$. Hence, for $h \in (0, \delta)$

$$f(x_0 + h) - f(x_0) > \frac{f'(x_0)}{2}h > 0 \quad \text{and} \quad f(x_0 - h) - f(x_0) < \frac{f'(x_0)}{2}(-h) < 0,$$

from which it follows that x_0 is a point of increase. An analogous argument shows that if $f'(x_0) < 0$, then x_0 is a point of decrease. □

We note that a point x_0 can be a point of increase/decrease of a function f even though $f'(x_0) = 0$. For example, if $f(x) = x^3$, then $f'(0) = 0$, but $f(x)$ is an increasing function on all of $\mathbb{R}$. On the other hand ...

Corollary. *If f is defined in a neighborhood of x_0, differentiable at x_0, and $f(x_0)$ is a relative extreme value, then x_0 is a stationary point of f.*

Proof. If $f(x_0)$ is a relative extreme value, then it is neither a point of increase nor a point of decrease, so $f'(x_0) \not> 0$ and $f'(x_0) \not< 0$. □

Exercises

ex6.1.1. Referring only to Definition 6.1.1, show that the function $f(x) = |x|$ has no best linear approximation at $a = 0$.

ex6.1.2. Referring only to Definition 6.1.1, show that the best linear approximation to $f(x)$ at $x = a$, if it exists, is unique.

Hint: What happens if there are two linear functions satisfying the definition?

ex6.1.3. Prove Proposition 6.1.1.

ex6.1.4. Complete the proof of Lemma 6.1.1 by filling in the details for the case that $\mu < 0$.

ex6.1.5. Referring only to Definition 6.1.1 (i.e., without computing any limits) show every linear function is its own best linear approximation and conclude that all linear functions are differentiable.

ex6.1.6. Prove Proposition 6.1.2 (again) using the binomial theorem, and the definition $f'(x) = \lim_{h\to 0} \frac{f(x+h)-f(x)}{h}$.

6.2. The arithmetic of differentiable functions

Simple combinations (sums, products, etc.) of differentiable functions are differentiable and the derivatives of the combinations are combinations of the derivatives of the original functions.

6.2.1. Linear combinations

Lemma. *If f_1 and f_2 are differentiable at x_0 and $c_1, c_2 \in \mathbb{R}$, then $c_1 f_1(x) + c_2 f_2(x)$ is differentiable at x_0 and $(c_1 f_1 + c_2 f_2)'(x_0) = c_1 f_1'(x_0) + c_2 f_2'(x_0)$.*

Proof. Exercise. □

Theorem. *If $f_1(x), \ldots, f_n(x)$ are differentiable at x_0, then for any real numbers $c_1, \ldots, c_n$, $g(x) = \sum_{k=1}^n c_k f_k(x)$ is differentiable at x_0, and $g'(x) = \sum_{k=1}^n c_k f_k'(x)$.*

Proof. The statement follows from the preceding lemma and induction on n. □

It follows in particular that if $P(x) = \sum_{k=0}^n c_k x^k$ is a polynomial, then for all $x \in \mathbb{R}$,

$$P'(x) = \sum_{k=1}^{n} k c_k x^{k-1},$$

by Proposition 6.1.2 and the fact that $\frac{d}{dx}(c_0) = 0$ for any constant c_0.

6.2.2. Products and quotients

Theorem (product rule). *If f and g are differentiable at x_0, then fg is differentiable at x_0 and*

$$(fg)'(x_0) = f'(x_0)g(x_0) + f(x_0)g'(x_0).$$

Proof. We have $g(x) \to g(x_0)$, because differentiability at x_0 implies continuity there, so

$$\begin{aligned}
(fg)'(x_0) &= \lim_{x\to x_0} \frac{f(x)g(x) - f(x_0)g(x_0)}{x - x_0} \\
&= \lim_{x\to x_0} \frac{f(x)g(x) - f(x_0)g(x) + f(x_0)g(x) - f(x_0)g(x_0)}{x - x_0} \\
&= \lim_{x\to x_0} \frac{f(x) - f(x_0)}{x - x_0} \cdot \lim_{x\to x_0} g(x) + f(x_0) \cdot \lim_{x\to x_0} \frac{g(x) - g(x_0)}{x - x_0} \\
&= f'(x_0)g(x_0) + f(x_0)g'(x_0).
\end{aligned}$$

□

Lemma. *If g is differentiable at x_0 and $g(x_0) \neq 0$, then $1/g$ is differentiable at x_0 and $(1/g)'(x_0) = -g'(x_0)/g(x_0)^2$.*

Proof. We note that $g(x) \neq 0$ in some neighborhood x_0 because $g(x_0) \neq 0$, and g is continuous at x_0. It follows that $1/g(x)$ is defined in a neighborhood of x_0, and

$$\begin{aligned}
\lim_{x\to x_0} \frac{\dfrac{1}{g(x)} - \dfrac{1}{g(x_0)}}{x - x_0} &= \lim_{x\to x_0} \frac{\dfrac{g(x_0) - g(x)}{g(x)g(x_0)}}{x - x_0} \\
&= -\lim_{x\to x_0} \frac{g(x) - g(x_0)}{x - x_0} \cdot \lim_{x\to x_0} \frac{1}{g(x)g(x_0)} = -\frac{g'(x_0)}{g(x_0)^2}.
\end{aligned}$$

□

Corollary (quotient rule). *If f and g are both differentiable at x_0 and $g(x_0) \neq 0$, then f/g is differentiable at x_0 and*

$$(f/g)'(x_0) = \frac{f'(x_0)g(x_0) - f(x_0)g'(x_0)}{g(x_0)^2}.$$

Proof. This follows from the product rule, the lemma above and the fact that $f/g = f \cdot 1/g$. □

6.2.3. Composition

To show that the composition of differentiable functions is differentiable, and find the 'rule' for the derivative of $f \circ g$, using the "arithmetic" of limits as we did for the preceding rules will not work. Instead, we must rely on the concept of "best linear approximation".

Theorem (chain rule). *If g is differentiable at x_0 and f is differentiable at $g(x_0)$, then $f \circ g$ is differentiable at x_0 and*

$$\frac{d}{dx}(f \circ g)\bigg|_{x=x_0} = f'(g(x_0)g'(x_0).$$

Proof. It follows from the comment following Theorem 6.1.2 that for h sufficiently small, $g(x_0 + h) = g(x_0) + (g'(x_0) + o(1))h$ and for ϵ sufficiently small, $f(g(x_0) + \epsilon) = f(g(x_0)) + (f'(g(x_0)) + o(1))\epsilon$. Setting $\epsilon = (g'(x_0) + o(1))h$, it follows that for $h \neq 0$,

$$f(g(x_0 + h)) = f(g(x_0) + \epsilon) = f(g(x_0)) + (f'(g(x_0)) + o(1))(g'(x_0) + o(1))h,$$

and therefore

$$\frac{f(g(x_0 + h)) - f(g(x_0))}{h} = f'(g(x_0))g'(x_0) + o(1),$$

since $o(1)(g'(x_0) + o(1)) = o(1)$. □

For example, if $n \in \mathbb{Z}$, then

$$\frac{d}{dx}(f(x)^n) = nf(x)^{n-1} \cdot f'(x),$$

at all points where f is differentiable (and $f(x) \neq 0$, in case $n < 0$).

6.2.4. Derivatives of inverse functions

Recall that if f is strictly monotone on an interval I, then there is a function f^{-1} defined on $J = f(I)$ such that $f^{-1} \circ f$ is the identity on I and $f \circ f^{-1}$ is the identity on J. Furthermore, if f is continuous, then J is also an interval and f^{-1} is continuous on J.

If f is strictly monotone and differentiable on the interval I, it is tempting to use the chain rule to differentiate f^{-1}. If $x_0 \in I$, $f'(x_0) \neq 0$ and $y_0 = f(x_0)$, then by the chain rule

$$1 = \frac{d}{dx}(x)\Big|_{x=x_0} = \frac{d}{dx}(f^{-1} \circ f)(x)\Big|_{x=x_0} = (f^{-1})'(y_0) \cdot f'(x_0),$$

implying that $(f^{-1})'(y_0) = 1/f'(x_0)$. This is true but the argument is flawed, because it assumes that f^{-1} is differentiable, an assumption that is not yet proved.

Theorem. *Suppose that f is continuous and strictly monotone on the open interval I, and differentiable at the interior point $x_0 \in I$, with $f'(x_0) \neq 0$. Then f^{-1} is differentiable at $y_0 = f(x_0)$ and $(f^{-1})'(y_0) = 1/f'(x_0)$.*

Proof. If $f'(x_0) \neq 0$, then

$$\lim_{h\to 0} \frac{h}{f(x_0 + h) - f(x_0)} = \frac{1}{\lim_{h\to 0} \frac{f(x_0+h)-f(x_0)}{h}} = \frac{1}{f'(x_0)}. \tag{6.2.1}$$

If $|\epsilon| > 0$ is sufficiently small, then $y_0 + \epsilon \in J$, so $y_0 + \epsilon = f(x_0 + h_\eta)$ where $|h_\eta| > 0$ because of the monotonicity of f. Moreover, $\epsilon \to 0$ if and only if $h_\eta \to 0$ because f is continuous at x_0 and f^{-1} is continuous at y_0. Therefore

$$\lim_{\eta\to 0} \frac{f^{-1}(y_0 + \epsilon) - f^{-1}(y_0)}{\epsilon} = \lim_{\eta\to 0} \frac{(x_0 + h_\eta) - x_0}{f(x_0 + h_\eta) - f(x_0)} = \lim_{h_\eta\to 0} \frac{h_\eta}{f(x_0 + h_\eta) - f(x_0)},$$

so $(f^{-1})'(y_0)$ exists and is equal to $1/f'(x_0)$ by (6.2.1). □

6.2.5. Differentiating rational powers

Lemma. *If $n \in \mathbb{N}$ and $x > 0$, then*

$$\frac{d}{dx}\left(x^{1/n}\right) = \frac{1}{n}x^{(1-n)/n}. \tag{6.2.2}$$

Proof. For $n \in \mathbb{N}$, the function $f(x) = x^n$ is monotone increasing and differentiable on $(0, \infty)$, with $f'(x) = nx^{n-1} > 0$ for all x there. Moreover, since $x^n > 0$ for all $x > 0$ and every positive real number has an nth root, we have $f((0, \infty)) = (0, \infty)$. It therefore follows from Theorem 6.2.4 that $f^{-1}(x) = x^{1/n}$ is also differentiable on $(0, \infty)$, and writing $x = \xi^n$, we have

$$\frac{d}{dx}\left(x^{1/n}\right) = \left(f^{-1}\right)'(x) = \frac{1}{f'(\xi)} = \frac{1}{n\xi^{n-1}} = \frac{1}{n}\left(x^{1/n}\right)^{1-n}. \qquad \square$$

Theorem. *If $r \in \mathbb{Q}$ then x^r is differentiable for all $x > 0$ and*

$$\frac{d}{dx}\left(x^r\right) = rx^{r-1},$$

for such x.

Proof. If $r \in \mathbb{Q}$, then $r = m/n$ with $m \in \mathbb{Z}$ and $n \in \mathbb{N}$, and $x^r = \left(x^{1/n}\right)^m$. Hence the theorem follows from Propositions 6.1.2 and 6.2.2, the chain rule and the lemma above. $\square$

Exercises

ex6.2.1. Prove Lemma 6.2.1.

ex6.2.2. Show that if $n \in \mathbb{N}$, and $f(x) = x^{-n}$, then $f'(x) = -nx^{-(n+1)}$ for all $x \neq 0$.

ex6.2.3. Prove Lemma 6.2.5 directly from the definition of the derivative.

Hint: Note that $\xi - x = \left(\xi^{1/n}\right)^n - \left(x^{1/n}\right)^n$ and use the identity (1.1.1).

6.3. Functions that are differentiable in an interval

We saw in §4.4 that functions that are continuous in a bounded interval are well-behaved in certain senses. Functions that are differentiable in an interval, have additional useful properties.

6.3.1. The mean value theorem

Proposition (Rolle's Theorem). *If f is continuous on $[a, b]$, f is differentiable in (a, b) and $f(a) = f(b)$, then $f'(c) = 0$ for some $c \in (a, b)$.*

Proof. If $f(x) = f(a)$ for all $x \in [a, b]$, then $f'(x) = 0$ for all $x \in (a, b)$. Otherwise, since f is continuous and $[a, b]$ is compact, f attains its maximum or minimum value at a point $c \in (a, b)$. Hence $f'(c) = 0$ by Corollary 6.1.3, since $f(c)$ is a relative extreme value in this case. $\square$

Rolle's theorem is equivalent to the following, more general statement.

Theorem (Mean Value Theorem). *If f is continuous on $[a, b]$ and differentiable in (a, b), then there is a point $c \in (a, b)$ such that*

$$\frac{f(b) - f(a)}{b - a} = f'(c).$$

Proof. If $g(x) = f(x)(b - a) - x(f(b) - f(a))$, then g is continuous in $[a, b]$, differentiable in (a, b) and $g(a) = bf(a) - af(b) = g(b)$. Hence there is a point $c \in (a, b)$ such that

$$0 = g'(c) = f'(c)(b - a) - (f(b) - g(a)),$$

by Rolle's theorem, and the statement follows. □

Corollary. *If f is differentiable in the open interval I, and $f'(x) = 0$ for all $x \in I$, then f is constant in I.*

Proof. Let $x_0 \in I$, then for any $x \neq x_0$ in I, there is a c between x and x_0 such that

$$\frac{f(x) - f(x_0)}{x - x_0} = \frac{f(x_0) - f(x)}{x_0 - x} = f'(c) = 0,$$

by the mean value theorem, so $f(x) = f(x_0)$ for all $x \in I$. □

6.3.2. Cauchy's mean value theorem

The following statement, known as Cauchy's mean value theorem is a further generalization of Rolle's theorem.

Theorem. *Suppose that $f(x)$ and $g(x)$ are both continuous in the closed interval $[a, b]$ and differentiable in (a, b), then there exists $c \in (a, b)$ such that*

$$f'(c)(g(b) - g(a)) = g'(c)(f(b) - f(a)). \tag{6.3.1}$$

Cauchy's mean value theorem reduces to the basic version by setting $g(x) = x$.

Proof. If $g(b) = g(a)$, then $g'(c) = 0$ for some $c \in (a, b)$, by Rolle's theorem, and the claim follows immediately. If $g(b) - g(a) \neq 0$, let $\varphi(x) = f(x) - \alpha g(x)$, where $\alpha = (f(b) - f(a))/(g(b) - g(a))$. It then follows that $\varphi(x)$ is continuous in $[a, b]$, differentiable in (a, b) and

$$\varphi(b) = f(b) - \frac{f(b) - f(a)}{g(b) - g(a)} g(b) = f(a) - \frac{f(b) - f(a)}{g(b) - g(a)} g(a) = \varphi(a).$$

Once again by Rolle's theorem, there exists $c \in (a, b)$ such that $\varphi'(c) = 0$, and hence

$$f'(c) - \frac{f(b) - f(a)}{g(b) - g(a)} g'(c) = 0,$$

from which the statement follows. □

Corollary. *Suppose that f and g are both differentiable in a neighborhood I of the point a, $f(a) = g(a) = 0$, and $g'(x) \neq 0$ for $x \in I \setminus \{a\}$. If $\lim_{x \to a} \frac{f'(x)}{g'(x)}$ exists, then*

$$\lim_{x \to a} \frac{f(x)}{g(x)} = \lim_{x \to a} \frac{f'(x)}{g'(x)}.$$

Proof. We first observe that the condition that $g(a) = 0$ and $g'(x) \neq 0$ for $x \neq a$ in I, implies that $g(x) \neq 0$ for $x \neq a$. Next, by Cauchy's mean value theorem, for every $x \neq a$ in I, there is a u_x between a and x such that $f'(u_x)g(x) = g'(u_x)f(x)$, because $f(a) = g(a) = 0$, and hence

$$\frac{f'(u_x)}{g'(u_x)} = \frac{f(x)}{g(x)}.$$

As x approaches a, so does the point u_x, and the result follows. □

Comment. This corollary is a special case of L'Hôpital's rule, which in its general form weakens the requirements on f and g, and includes the case that $\lim_{x\to a} f(x) = \lim_{x\to a} g(x) = \pm\infty$.

6.3.3. Lipschitz continuity

Differentiability can confer a stronger form of continuity in certain cases.

Definition. A function f satisfies a *Lipschitz condition* or is *Lipschitz continuous* in the interval I if there is a constant $\lambda > 0$ such that $|f(x) - f(y)| < \lambda|x - y|$ for all $x, y \in I$.

Lipschitz continuity implies uniform continuity.

Proposition. *If f is Lipschitz continuous in the interval I, then f is uniformly continuous in I.*

Proof. Suppose that f is Lipschitz continuous in I with Lipschitz constant λ. Given $\varepsilon > 0$, if $|x - y| < \varepsilon/\lambda$, then $|f(x) - f(y)| < \lambda|x - y| < \varepsilon$. □

Theorem. *If f is differentiable in the interval I and f' is bounded on I, then f is Lipschitz continuous in I.*

In particular, if I is compact and f' is continuous in I, then f is Lipschitz continuous in I.

Proof. Let $\lambda = \max_I |f'(\xi)|$. For any $x, y \in I$, it follows that there is a ξ between x and y such that $|f(x) - f(y)| = |f'(\xi)||x - y|$, by the mean value theorem, and hence $|f(x) - f(y)| \leq \lambda|x - y|$. □

6.3.4. An intermediate value theorem for derivatives

The derivative of a function f may be defined in an interval (a, b) without being continuous there.

Example. Let $f(x) = x^2 \sin(1/x)$ for $x \neq 0$ and $f(0) = 0$. If $x \neq 0$, then

$$f'(x) = 2x\sin(1/x) - \cos(1/x),$$

assuming for the moment that we know that $\sin' u = \cos u$. Hence, as x approaches 0, $f'(x) \approx \cos(1/x)$ and is therefore not continuous at $x = 0$. On the other hand,

$$f'(0) = \lim_{h\to 0} \frac{h^2\sin(1/h) - 0}{h} = \lim_{h\to 0} h\sin(1/h) = 0,$$

so $f'(0)$ exists.

More generally, using a series of appropriately scaled and translated copies of the function f, above, it is possible to construct functions that are differentiable everywhere in $[0, 1]$, whose derivatives are discontinuous on the (uncountable) Cantor set.

On the other hand, derivatives cannot have jump discontinuities, as follows from the next theorem.

Theorem. *If f is differentiable in a neighborhood of the interval $[a, b]$ and ϵ lies between $f'(a)$ and $f'(b)$, then $f'(c) = \epsilon$ for some $c \in (a, b)$.*

Proof. Suppose that $f'(a) < \epsilon < f'(b)$, and let $g(x) = f(x) - \epsilon x$. It follows that $g'(a) < 0$ and $g'(b) > 0$ so that g is decreasing around a and increasing around b. Hence, the minimum value of g in $[a, b]$ occurs at some point c in (a, b), so $g'(c) = f'(c) - \epsilon = 0$. The same argument works in the case that $f'(a) > \epsilon > f'(b)$. □

Corollary. *The derivative of a differentiable function cannot have 'jump' discontinuities.*

Proof. Exercise. □

6.3.5. Sequences of differentiable functions

If $\{f_n\}$ is a sequence of continuous functions on the interval $[a, b]$ that is converging uniformly in $[a, b]$ to a function f, then f is also continuous in $[a, b]$.

The analogous claim is not generally true for a sequence of differentiable functions. I.e., a uniform limit of differentiable functions need not be differentiable. This follows, for example, from Weierstrasse's approximation theorem (see §5.2.3), which says that *every* continuous function in $[a, b]$ is a uniform limit of polynomials. All polynomials are differentiable, but many continuous functions are not.

The problem is that uniform convergence of a sequence of differentiable functions doesn't imply the convergence of their derivatives. Moreover, the uniform convergence of $\{f_n'\}$ in $[a, b]$ doesn't guarantee the convergence (uniform or otherwise) of $\{f_n\}$. For example, if f is differentiable in $[a, b]$ and $f_n(x) = f(x) + n$ for $n \in \mathbb{N}$, then $\{f_n'\}$ converges uniformly to f' (why?), but $\{f_n\}$ diverges for all $x \in [a, b]$.

On the other hand, if the sequence of derivatives converges uniformly and the original functions converge at one point, then the original functions converge uniformly too.

Lemma. *Suppose that $\{f_n\}$ is a sequence of differentiable functions on the interval $[a, b]$ and that $f_n' \to \varphi$ uniformly in $[a, b]$. If $\{f_n(x_0)\}$ converges for some $x_0 \in [a, b]$, then $\{f_n\}$ converges uniformly in $[a, b]$.*

Proof. To show that $\{f_n\}$ is converging uniformly in $[a, b]$, it is enough to show that $\{f_n(x)\}$ is uniformly Cauchy in $[a, b]$, by Theorem 5.1.3. For any $n, m \in \mathbb{N}$, the function $f_n - f_m$ is differentiable, and fixing $x \in [a, b]$, we have

$$(f_n(x) - f_m(x)) - (f_n(x_0) - f_m(x_0)) = (f_n'(u) - f_m'(u))(x - x_0),$$

for some u between x and x_0, by the mean value theorem, and hence

$$|f_n(x) - f_m(x)| \leq |f_n(x_0) - f_m(x_0)| + |f_n'(u) - f_m'(u)||x - x_0| \tag{6.3.2}$$

Given $\varepsilon > 0$, there is an $N_1 \in \mathbb{N}$ such that $|f_n(x_0) - f_m(x_0)| < \varepsilon/2$ for $n, m > N_1$, since $\{f_n(x_0)\}$ is a Cauchy sequence. Likewise, there is an $N_2 \in \mathbb{N}$ such that $|f_n'(u) - f_m'(u)| < \varepsilon/2(b-a)$ for $n, m > N_2$ and all $u \in [a,b]$, since $\{f_n'\}$ is converging uniformly in $[a,b]$. It follows from these bounds and (6.3.2) that

$$|f_n(x) - f_m(x)| \le |f_n(x_0) - f_m(x_0)| + |f_n'(u) - f_m'(u)|(b-a) < \varepsilon$$

for $n, m > N_\varepsilon = \max(N_1, N_2)$. Since N_ε does not depend on x, the sequence $\{f_n(x)\}$ is uniformly Cauchy in $[a,b]$. □

More is true, under the conditions stated in the lemma, the limit of the sequence $\{f_n\}$ is differentiable and equal to the limit of the sequence of the derivatives.

Theorem. *Suppose that $\{f_n\}$ is a sequence of differentiable functions on the interval $[a,b]$ and that $f_n' \to \varphi$ uniformly in $[a,b]$. If $\{f_n(x_0)\}$ converges for some $x_0 \in [a,b]$, then $\{f_n\}$ is converging uniformly in $[a,b]$ to a differentiable function f and $f' = \varphi$ in $[a,b]$.*

Proof. It follows from the lemma above that $\{f_n\}$ is converging uniformly in $[a,b]$ to some function f, and it remains to prove that f is differentiable and that $f' = \varphi$.

To prove that f is differentiable at x_0 and that $f'(x_0) = \varphi(x_0)$, for $x_0 \in [a,b]$, we show that for any $\varepsilon > 0$, there is a $\delta_\varepsilon > 0$

$$\left|\frac{f(x) - f(x_0)}{x - x_0} - \varphi(x_0)\right| < \varepsilon$$

if $0 < |x - x_0| < \delta_\varepsilon$. To begin, we note that if $x \in [a,b]$ and $x \ne x_0$, then

$$\begin{aligned} \left|\frac{f(x) - f(x_0)}{x - x_0} - \varphi(x_0)\right| &\le \left|\frac{f(x) - f(x_0)}{x - x_0} - \frac{f_n(x) - f_n(x_0)}{x - x_0}\right| \\ &\quad + \left|\frac{f_n(x) - f_n(x_0)}{x - x_0} - f_n'(x_0)\right| \\ &\quad + |f_n'(x_0) - \varphi(x_0)| \end{aligned} \tag{6.3.3}$$

for any $n \in \mathbb{N}$. It is therefore enough to show that there is an $n_\varepsilon \in \mathbb{N}$ and a $\delta_\varepsilon > 0$ such that the three terms on the right side are each bounded by $\varepsilon/3$ for $n = n_\varepsilon$ and $0 < |x - x_0| < \delta_\varepsilon$.

Given $\varepsilon > 0$, there is an $n_\varepsilon \in \mathbb{N}$ such that if $n > n_\varepsilon$, then by Theorem 5.1.3,

$$|f_n'(u) - f_{n_\varepsilon}'(u)| < \varepsilon/3 \tag{6.3.4}$$

for all $u \in [a,b]$. Since $f_n' \to \varphi$, this implies that $|f_{n_\varepsilon}'(u) - \varphi(u)| \le \varepsilon/3$ for all $u \in [a,b]$,[1] so

$$|f_{n_\varepsilon}'(x_0) - \varphi(x_0)| \le \varepsilon/3.$$

Next, it follows from the differentiability of f_{n_ε} that there is a $\delta_\varepsilon > 0$ such that if $0 < |x - x_0| < \delta_\varepsilon$, then

$$\left|\frac{f_{n_\varepsilon}(x) - f_{n_\varepsilon}(x_0)}{x - x_0} - f_{n_\varepsilon}'(x_0)\right| < \varepsilon/3.$$

[1]See **ex2.2.3**.

In other words, with $n = n_\varepsilon$ and $0 < |x - x_0| < \delta_\varepsilon$, the second and third terms on the right-hand side of (6.3.3) are both bounded by $\varepsilon/3$, and it remains to show that the same is true of the first term.

Fix $x \in [a, b]$ with $0 < |x - x_0| < \delta_\varepsilon$. For any $n \in \mathbb{N}$,

$$\frac{(f_n(x) - f_{n_\varepsilon}(x)) - (f_n(x_0) - f_{n_\varepsilon}(x_0))}{x - x_0} = f_n'(u_n) - f_{n_\varepsilon}'(u_n),$$

for some u_n between x and x_0, by the mean value theorem, whether $x > x_0$ or vice versa. Rearranging the terms on the left-hand side above, shows that

$$\frac{f_n(x) - f_n(x_0)}{x - x_0} - \frac{f_{n_\varepsilon}(x) - f_{n_\varepsilon}(x_0)}{x - x_0} = f_n'(u_n) - f_{n_\varepsilon}'(u_n),$$

and hence, if $n > n_\varepsilon$, then by (6.3.4)

$$\left|\frac{f_n(x) - f_n(x_0)}{x - x_0} - \frac{f_{n_\varepsilon}(x) - f_{n_\varepsilon}(x_0)}{x - x_0}\right| = |f_n'(u_n) - f_{n_\varepsilon}'(u_n)| < \varepsilon/3. \tag{6.3.5}$$

Finally, since $f_n \to f$ (uniformly) in $[a, b]$,[2] it follows that (with x fixed above),

$$\lim_{n\to\infty} \frac{f_n(x) - f_n(x_0)}{x - x_0} = \frac{f(x) - f(x_0)}{x - x_0},$$

and therefore the inequality (6.3.5) implies that, if $n > n_\varepsilon$ and $0 < |x - x_0| < \delta_\varepsilon$, then

$$\left|\frac{f(x) - f(x_0)}{x - x_0} - \frac{f_{n_\varepsilon}(x) - f_{n_\varepsilon}(x_0)}{x - x_0}\right| \le \varepsilon/3.$$

Thus, all three terms on the right-hand side of (6.3.3) are bounded by $\varepsilon/3$, so f is differentiable at x_0, with $f'(x_0) = \varphi(x_0)$. Since $x_0 \in [a, b]$ was arbitrary, the same is true for all $x \in [a, b]$. □

Exercises

ex6.3.1. Show that if f is differentiable in $\mathbb{R}$ and f' is a polynomial, then f is a polynomial.

ex6.3.2. Suppose that f is differentiable in (a, b). Show that f' cannot have jump discontinuities.

ex6.3.3. A function f is *Hölder continuous* on the interval I, with exponent $\alpha > 0$, if there is a constant $\lambda > 0$ such that $|f(x) - f(y)| < \lambda|x - y|^\alpha$ for all $x, y \in I$. E.g., Hölder continuity with exponent 1 is Lipschitz continuity.

Show that if f is Hölder continuous in I with exponent $\alpha > 1$, then f is constant in I.

ex6.3.4. Show that if f is Hölder continuous in I, with exponent $1 > \alpha > 0$, then f is uniformly continuous in I.

[2]Pointwise convergence would be sufficient here. The uniform convergence of $\{f_n\}$ is just an added bonus.

6.4. Derivatives of power series

Power series behave very much like polynomials in their intervals of convergence. This similarity extends to differentiation. The derivative of a polynomial is the sum of the derivatives of its terms, and the same is true of power series.

6.4.1. Differentiating term by term

Suppose that $f(x) = \sum_{n=0}^{\infty} a_n(x - x_0)^n$ has radius of convergence $R > 0$. Following the model of polynomials, it is natural to guess that f is differentiable and that its derivative is given by the series $\sum_{n=1}^{\infty} na_n(x - x_0)^{n-1}$. This is true, but does not follow from Theorem 6.2.1 which is a statement about sums with finitely many terms.

To differentiate a series "term by term", requires interchanging the order of two limits — the limit of the partial sums that defines the value of the series at a point and the limit of difference quotients that defines the derivative at a point.

We begin by showing that the candidate series for f' converges at every point that f does.

Lemma. *The series $\sum_{n=0}^{\infty} a_n(x - x_0)^n$ and $\sum_{n=1}^{\infty} na_n(x - x_0)^{n-1}$ have the same radius of convergence.*

Proof. By Theorem 2.3.5 (v), $\lim_{n\to\infty} \sqrt[n]{n} = 1$, so for any $\varepsilon > 0$, there is an N_ε, such that $1 < \sqrt[n]{n} < 1 + \varepsilon$ and hence

$$\sqrt[n]{|a_n|} \leq \sqrt[n]{n|a_n|} \leq (1 + \varepsilon)\sqrt[n]{|a_n|},$$

for all $n > N_\varepsilon$. Therefore $\limsup \sqrt[n]{n|a_n|} = \limsup \sqrt[n]{|a_n|}$, and the claim follows from Proposition 5.4.1. □

Theorem. *If $f = \sum_{n=0}^{\infty} a_n(x - x_0)^n$ has radius of convergence $R > 0$, then f is differentiable in $(x_0 - R, x_0 + R)$ and $f' = \sum_{n=1}^{\infty} na_n(x - x_0)^{n-1}$.*

Proof. To keep the exposition simple (and without loss of generality),[3] we assume that $x_0 = 0$, so $f = \sum_{n=0}^{\infty} a_n x^n$.

Let $x \in (-R, R)$, and denote by $\delta_x = ||x| - R|$ the distance of x to the boundary of the interval. Both f and the series $\tilde{f}(x) = \sum_{n=1}^{\infty} na_n x^{n-1}$ converge absolutely at x, by the preceding lemma. Moreover, if $|h| < \delta_x$, then f converges absolutely at $x + h$, in which case

$$\begin{aligned} f(x+h) - f(x) - h\tilde{f}(x) &= \sum_{n=0}^{\infty} a_n\left((x+h)^n - x^n\right) - \sum_{n=1}^{\infty} na_n x^{n-1}h \\ &= \sum_{n=1}^{\infty} a_n\left[(x+h)^n - x^n - nx^{n-1}h\right] \\ &= h^2 \sum_{n=2}^{\infty} a_n\left(\sum_{j=2}^{n} \binom{n}{j} x^{n-j}h^{j-2}\right), \end{aligned}$$

[3]To see that there is no loss of generality, use the substitution $u = x - x_0$.

as follows from the binomial theorem (and the vanishing of the a_0 and a_1 terms). Hence, if $0 < |h| < \delta_x$, then

$$\left|\frac{f(x+h)-f(x)}{h} - \tilde{f}(x)\right| \le |h| \sum_{n=2}^{\infty} |a_n| \left(\sum_{j=2}^{n} \binom{n}{j} |x|^{n-j}|h|^{j-2}\right). \tag{6.4.1}$$

Next, if $|h| \le \delta_x/2 = \epsilon_x$, then

$$\sum_{j=2}^{n} \binom{n}{j} |x|^{n-j}|h|^{j-2} \le \sum_{j=2}^{n} \binom{n}{j} |x|^{n-j}\epsilon_x^{j-2} \le \epsilon_x^{-2} \sum_{j=0}^{n} \binom{n}{j} |x|^{n-j}\epsilon_x^{j} = \frac{(|x|+\epsilon_x)^n}{\epsilon_x^2},$$

and it follows from this and (6.4.1) that

$$\left|\frac{f(x+h)-f(x)}{h} - \tilde{f}(x)\right| \le \frac{|h|}{\epsilon_x^2} \sum_{n=2}^{\infty} |a_n|(|x|+\epsilon_x)^n \le \frac{|h|}{\epsilon_x^2} \sum_{n=0}^{\infty} |a_n|(|x|+\epsilon_x)^n.$$

The series on the right converges because $|x| + \epsilon_x < R$, and denoting its value by S_x, if $\varepsilon > 0$ and

$$0 < |h| < \min\left(\epsilon_x, \frac{\varepsilon\epsilon_x^2}{S_x+1}\right),$$

then

$$\left|\frac{f(x+h)-f(x)}{h} - \tilde{f}(x)\right| < \varepsilon,$$

from which the theorem follows. □

Comment. If $f_n(x)$ is differentiable in (a, b) for all n and the series $f = \sum f_n(x)$ converges absolutely in (a, b), it is not necessarily the case that f is differentiable in (a, b). We will explore the additional conditions that might guarantee the differentiability of a series of differentiable functions after discussing *integrals*.

6.4.2. The exponential function and its inverse

It follows from Theorem 6.4.1 that the exponential function $\exp(x)$ is differentiable on the entire real line, and that

$$\frac{d}{dx}\exp(x) = \sum_{n=1}^{\infty} \frac{nx^{n-1}}{n!} = \sum_{n=0}^{\infty} \frac{x^n}{n!} = \exp(x).$$

Since $\exp(x) = e^x$ is strictly monotone increasing on $\mathbb{R}$, with range $(0, \infty)$, it follows that it has an inverse, denoted $\log x$ (as usual), defined for all $x > 0$. Moreover, $\log x$ is differentiable in $(0, \infty)$ by theorem 6.2.4, and if $x_0 = \exp(y_0)$, then

$$\log'(x_0) = \frac{1}{\exp'(y_0)} = \frac{1}{\exp(y_0)} = \frac{1}{x_0},$$

i.e., $\log'(x) = 1/x$ for all $x > 0$.

6.5. Taylor's theorem

6.5.1. Higher order derivatives

Definition. Suppose that $y = f(x)$ is differentiable. The derivative of $f'(x)$ (at the points where it exists) is called the *second derivative* of $f(x)$. Likewise, the derivative of $f''(x)$ is the *third derivative* of $f(x)$, and in general, we define the nth derivative of $f(x)$ recursively as the derivative of the $(n-1)^{st}$ derivative.

If the nth derivative of $f(x)$ is defined in the interval I, we say that f is *n-times differentiable* in I.

The first, second and third derivatives are typically denoted $f'(x)$, $f''(x)$ and $f'''(x)$, while for $n \geq 4$, the nth derivative is denoted by $f^{(n)}(x)$. Alternatively, we also write

$$f''(x) = \frac{d^2y}{dx^2},\ f'''(x) = \frac{d^3y}{dx^3}, \dots, f^{(n)}(x) = \frac{d^ny}{dx^n}.$$

Polynomials may be differentiated repeatedly, because the derivative of a polynomial is a polynomial. Similarly, if the power series $f(x) = \sum c_k(x-a)^k$ converges in the interval I around a, then it is differentiable in I and its derivative is also a power series that converges in I, so $f(x)$ may differentiated repeatedly. Polynomials and convergent power series are examples of *infinitely differentiable* functions.

6.5.2. Polynomials are characterized by their derivatives

Differentiating a power series with infinitely many nonzero coefficients results in another power series with infinitely many nonzero coefficients. On the other hand, differentiating a polynomial results in a polynomial of one degree lower, and repeated differentiation eventually reduces the degree to 0.

Proposition. *The function $f(x)$ is a polynomial if and only if there is a $m \in \mathbb{N}$ such that $f^{(m)}(x) = 0$ for all $x \in \mathbb{R}$.*

Proof. If $f(x)$ is a polynomial of degree n, then $f^{(n+1)}(x) = 0$ for all x, as can be shown by induction (exercise). To verify the other implication, we rely on part (b) of Corollary 6.3.1.

Suppose that $m > 0$ and $f^{(m)}(x) = 0$ for all x. Then it follows from Corollary 6.3.1 (a) that $f^{(m-1)}(x)$ is a constant, which is a polynomial (of degree 0). It now follows from part (b) of the same corollary that $f^{(m-2)}(x)$ is a polynomial, and repeating this argument m times, shows that $f(x)$ is a polynomial. □

This is one way that a polynomial may be said to be characterized by its derivatives. But a more complete characterization is available.

Theorem. *A polynomial $f(x)$ of degree $n > 0$ is completely determined by its value and the values of its first n derivatives at any point a.*

Proof. First note that if $f(x) = \sum_{k=0}^{n} c_k x^k$, then $c_0 = f(0)$ and for $1 \leq k \leq n$,

$$f^{(k)}(0) = k!\, c_k, \tag{6.5.1}$$

as can be verified by induction. Therefore,

$$f(x) = \sum_{k=0}^{n} \frac{f^{(k)}(0)}{k!} x^k,$$

where by convention $f^{(0)}(x) = f(x)$.

Now, for any $a \in \mathbb{R}$, let $f_a(x) = f(x+a)$, then $f^{(k)}(a) = f_a^{(k)}(0)$, and

$$f(x) = f_a(x-a) = \sum_{k=0}^{n} \frac{f_a^{(k)}(0)}{k!}(x-a)^k = \sum_{k=0}^{n} \frac{f^{(k)}(a)}{k!}(x-a)^k. \tag{6.5.2}$$ □

The polynomial on the right-hand side above is called the *Taylor polynomial* of $f(x)$ of degree n, centered at the point $x = a$.

6.5.3. Taylor polynomials

The fact that a polynomial of degree n is equal to its degree n Taylor polynomial centered at any point can be generalized to n-times differentiable functions, as long as we replace the phrase "equal to" with the phrase "approximated by".

To begin, If f is n-times differentiable in a neighborhood of a point a, then we can certainly construct the polynomial that appears on the right-hand side of (6.5.2) for f.

Definition. If $n \geq 1$ and $f(x)$ is n-times differentiable in an interval around $a \in \mathbb{R}$, then the *Taylor polynomial of degree n* for f, centered at a is the polynomial

$$T_n(x) = \sum_{k=0}^{n} \frac{f^{(k)}(a)}{k!}(x-a)^k.$$

When $n = 1$, the Taylor polynomial $T_1(x)$ is the best linear approximation to $f(x)$ at a as shown in Theorem 6.1.2, and hence

$$f(x) - T_1(x) = o(x-a).$$

The following result generalizes this result to Taylor polynomials of higher degree.

Theorem (Taylor's theorem). *Suppose that $f(x)$ is n-times differentiable in a neighborhood of a, and $T_n(x)$ is the degree n Taylor polynomial of f, centered at $x = a$, then*

$$f(x) - T_n(x) = o(x-a)^n.$$

Proof. We need to show that

$$\lim_{x \to a} \frac{f(x) - T_n(x)}{(x-a)^n} = 0,$$

and to do this, we apply the special case of L'Hôpital's rule that appeared as Corollary 6.3.2. Indeed, the functions $f(x) - T_n(x)$ and $(x-a)^n$ satisfy the conditions of the corollary, and hence

$$\lim_{x \to a} \frac{f(x) - T_n(x)}{(x-a)^n} = \lim_{x \to a} \frac{f'(x) - T_n'(x)}{n(x-a)}.$$

The functions $n(x-a)^{n-1}$ and

$$f'(x) - T_n'(x) = f'(x) - \left(f'(a) + f''(a)(x-a) + \cdots + \frac{f^{(n)}(a)}{(n-1)!}(x-a)^{n-1}\right)$$

likewise satisfy the conditions of the corollary, and applying it again, gives

$$\lim_{x\to a} \frac{f(x) - T_n(x)}{(x-a)^n} = \lim_{x\to a} \frac{f'(x) - T_n'(x)}{n(x-a)^{n-1}} = \lim_{x\to a} \frac{f''(x) - T_n''(x)}{n(n-1)(x-a)^{n-2}}.$$

Continuing, we note that for $k < n-1$, the pairs of functions $f^{(k)}(x) - T_n^{(k)}(x)$ and $\frac{n!}{(n-k)!}(x-a)^k$ all satisfy the conditions of the corollary, and applying it $n-1$ times we conclude that

$$\lim_{x\to a} \frac{f(x) - T_n(x)}{(x-a)^n} = \lim_{x\to a} \frac{f'(x) - T_n'(x)}{n(x-a)^{n-1}} = \cdots = \lim_{x\to a} \frac{f^{(n-1)}(x) - T_n^{(n-1)}(x)}{n!\,(x-a)}.$$

We now observe that $T_n^{(n-1)}(x) = f^{(n-1)}(a) + f^{(n)}(a)(x-a)$, so that

$$\begin{aligned}\lim_{x\to a} \frac{f(x) - T_n(x)}{(x-a)^n} &= \lim_{x\to a} \frac{f^{(n-1)}(x) - \left(f^{(n-1)}(a) + f^{(n)}(a)(x-a)\right)}{n!\,(x-a)} \\ &= \frac{1}{n!}\left(\lim_{x\to a} \frac{f^{(n-1)}(x) - f^{(n-1)}(a)}{(x-a)} - f^{(n)}(a)\right) \\ &= \frac{f^{(n)}(a) - f^{(n)}(a)}{n!} = 0,\end{aligned}$$

as follows from the definition of the nth derivative. □

With additional conditions on the function f, we can obtain more precise estimates for the remainder $R_n(x) = f(x) - T_n(x)$.

Proposition. *Suppose that $f(x)$ is $n+1$ times differentiable in an open interval I containing a, f^{n+1} is continuous in I, and $T_n(x)$ is the degree n Taylor polynomial for f centered at $x = a$, then for any $x \in I$*

$$R_n(x) = \frac{f^{n+1}(u)}{(n+1)!}(x-a)^{n+1}, \tag{6.5.3}$$

for some u between x and a.

Proof. Exercise. □

The expression on the right of (6.5.3) is called Lagrange's form of the remainder.

Comment. With the assumptions of the proposition above, if I_0 is a closed subinterval of I that contains a in its interior, then

$$|f(x) - T_n(x)| \leq \frac{M|x-a|^{n+1}}{(n+1)!}$$

for all $x \in I_0$, where $M = \max\{|f^{n+1}(u)| : u \in I_0\}$.

6.5.4. The Newton-Raphson method

In Section 2.4.2, we found a sequence of approximate solutions to the equation $x^2 - 2 = 0$ using a method that dates back to antiquity (the "Babylonian method"). The Newton-Raphson method (aka "Newton's method") is a generalization, that finds increasingly accurate approximations to the solution(s) of the equation $f(x) = 0$ for a differentiable function f, as long as certain conditions are satisfied.

Suppose that $f(a) = 0$ and that for some $r > 0$, $f'(x) \neq 0$ if $|x - a| < r$. We choose an initial approximation, x_0 that is sufficiently close to a, where "sufficiently close" will be quantified more precisely as we go along.

If $|x_0 - a| < \delta < r$, with δ sufficiently small, the linear approximation to $f(x)$ centered at x_0,

$$L_0(x) = f(x_0) + f'(x_0)(x - x_0)$$

should behave very much like f in the interval $|x - a| < \delta$. In particular, the equation $L_0(x) = 0$ will have a solution x_1 (because $f'(x_0) \neq 0$) that is very close to a (the solution of the equation $f(x) = 0$). Moreover, if "certain conditions" are met,

$$x_1 = x_0 - \frac{f(x_0)}{f'(x_0)}$$

will be a much better approximation to a than the initial approximation, x_0. In particular, x_1 will also be "sufficiently close" to a, and we can repeat this process to find the next, better approximation $x_2 = x_1 - \frac{f(x_1)}{f'(x_1)}$, and so on. In other words, if x_n is the nth approximation to a found in this way, then the x_{n+1} is given by

$$x_{n+1} = x_n - \frac{f(x_n)}{f'(x_n)}. \tag{6.5.4}$$

Conditions that guarantee that that this method produces a sequence of increasingly accurate approximations of a are summarized in the following theorem.

Theorem. *Suppose that $f(a) = 0$, and that there is an $r > 0$ such that $f'(x) \neq 0$ and $f''(x)$ is continuous for all $x \in (a - r, a + r)$. Then there exists a constant $\delta > 0$, such that if $|x_0 - a| < \delta$, then for $n \geq 0$*

$$|x_{n+1} - a| < \frac{1}{2}|x_n - a|,$$

for the sequence of approximations $\{x_0, x_1, \dots, x_n, \dots\}$ generated by the Newton-Raphson method, as described above.

Proof. If $|x - a| < r$, then Proposition 6.5.3 implies that

$$0 = f(a) = f(x) + f'(x)(a - x) + \frac{f''(u)}{2}(a - x)^2$$

for some u between a and x, from which it follows that

$$a - \left(x - \frac{f(x)}{f'(x)}\right) = a - x + \frac{f(x)}{f'(x)} = -\frac{f''(u)}{2f'(x)}(a - x)^2. \tag{6.5.5}$$

Next, let

$$M = 2 + \frac{\max\{|f''(u)| : a - r/2 \leq u \leq a + r/2\}}{2\min\{|f'(x)| : a - r/2 \leq x \leq a + r/2\}},$$

and observe that $2 \le M < \infty$ because $f'(x)$ is bounded away from 0 in the closed interval $[a - r/2, a + r/2]$ and $f''(u)$ is bounded there.

Now set $\delta = \min(1/M^2, r/2)$, choose $x_0 \in (a - \delta, a + \delta)$, and let

$$x_1 = x_0 - \frac{f(x_0)}{f'(x_0)},$$

as prescribed by the Newton-Raphson method. Then it follows from (6.5.5) that

$$|a - x_1| = \left|\frac{f''(u)}{2f'(x_0)}\right|(a - x_0)^2 < M\delta|a - x_0| \le \frac{1}{2}|a - x_0|,$$

(where, as before, u is some number between a and x_0). In particular, we have $|a-x_1| < |a - x_0| < \delta$, so we may repeat the argument above to conclude that

$$|a - x_2| < \frac{1}{2}|a - x_1|,$$

where x_2 is defined as in (6.5.4), and so on. I.e., if $|x_0 - a| < \delta$ and the sequence $\{x_n\}$ is defined recursively by (6.5.4), then the theorem follows by induction. □

Exercises

ex6.5.1. Prove the identity (6.5.1).

ex6.5.2. Prove Proposition 6.5.3.

Hint: Use Cauchy's mean value theorem.

ex6.5.3. Assume $f \in C^\infty([-1,1])$, and $\sup_{|x|\le 1} |f^{(n)}(x)| \le 2^n$.

(i) Prove that $f(x) = \sum_0^\infty \frac{f^{(j)}(0)}{j!} x^j$ for $-1 \le x \le 1$.

(ii) What is the radius of convergence of the series $\sum_0^\infty \frac{f^{(j)}(0)}{j!} x^j$?

ex6.5.4. Let f be an infinitely differentiable function on $[-1,1]$. Assume that $f^{(j)}(0) = 0$ for all j, and assume that $f^{(n_k)}(x) \le 1$ for all $x \in [-1,1]$, for some infinite sequence $\{n_k\}$. Prove that $f(x) = 0$ everywhere.

Hint: Use the result of the previous exercise.

ex6.5.5. Show that if the conditions of Theorem 6.5.4 are satisfied, and x_0 is chosen as in the its proof, then $|a - x_n| < 2^{-(n+1)}$ for $n \ge 0$.

6.6. Trigonometric functions (without circles)

In this section, we study the familiar trigonometric functions $\sin x$ and $\cos x$. They are traditionally (and naturally) defined as functions of angles, first in triangles, then on the unit circle, and then generalized to functions of real variables. In this section we define them as the power series

$$(6.6.1)\quad \begin{cases} \cos x = \sum_{n=0}^{\infty} \frac{(-1)^n x^{2n}}{(2n)!} = 1 - \frac{x^2}{2} + \frac{x^4}{24} - \frac{x^6}{720} + \cdots & \text{and} \\ \sin x = \sum_{n=0}^{\infty} \frac{(-1)^n x^{2n+1}}{(2n+1)!} = x - \frac{x^3}{6} + \frac{x^5}{120} - \frac{x^7}{5040} + \cdots \end{cases}$$

and explore the extent to which we can rediscover their familiar properties based on this definition.

The point here is not to reinvent the wheel,[4] but to practice using some of the analytic tools that we have developed.

6.6.1. Basic properties

To begin, we observe that both series have infinite radii of convergence (by comparison to the exponential function, for example). It is also evident that $\cos x$ is an *even function* (i.e., $\cos x = \cos(-x)$ for all $x \in \mathbb{R}$) because all the powers in its series are even. Similarly, $\sin x$ is an *odd function* ($\sin(-x) = -\sin(x)$ for all $x \in \mathbb{R}$), because all the powers in its series are odd.

Next, the series for $\sin x$ and $\cos x$ may both be differentiated term by term by Theorem 6.4.1 and therefore

$$(\sin x)' = \sum_{n=0}^{\infty} \left(\frac{(-1)^n x^{2n+1}}{(2n+1)!} \right)' = \sum_{n=0}^{\infty} \frac{(-1)^n x^{2n}}{(2n)!} = \cos x$$

and

$$(\cos x)' = \sum_{n=0}^{\infty} \left(\frac{(-1)^n x^{2n}}{(2n)!} \right)' = \sum_{n=1}^{\infty} \frac{(-1)^n x^{2n-1}}{(2n-1)!} = -\sum_{m=0}^{\infty} \frac{(-1)^m x^{2m+1}}{(2m+1)!} = -\sin x,$$

as follows from pulling out a factor of -1 and substituting $m = n-1$. As a consequence, both $\sin x$ and $\cos x$ satisfy the *differential equation* $y + y'' = 0$.[5]

Proposition. $\sin^2 x + \cos^2 x = 1$ *for all* $x \in \mathbb{R}$.

Proof. Using the chain rule, we have

$$(\sin^2 x + \cos^2 x)' = 2\sin x(\sin x)' + 2\cos x(\cos x)' = 2\sin x \cos x - 2\cos x \sin x = 0$$

for all x, so $\sin^2 x + \cos^2 x$ is constant on $\mathbb{R}$ by Corollary 6.3.1. The proposition follows from the observation that $\sin^2 0 + \cos^2 0 = 0 + 1 = 1$. □

Using a similar idea, we can recover the all-important addition formulas.

Theorem. *For all* $x, y \in \mathbb{R}$,

$$(6.6.2)\qquad \sin(x+y) = \sin x \cos y + \cos x \sin y$$

[4]Math-pun intended.

[5]In fact, $\cos x$ and $\sin x$ are characterized by this differential equation, together with the initial values $\sin(0) = 0$ and $\cos(0) = 1$. See **ex6.6.1**.

and

$$\cos(x + y) = \cos x \cos y - \sin x \sin y. \tag{6.6.3}$$

Proof. For arbitrary $y \in \mathbb{R}$, define $\varphi(x) = \sin(x + y) - \sin x \cos y - \cos x \sin y$. We claim that $\varphi(x) = 0$ for all x, the truth of which will establish (6.6.2). This identity together with the fact that

$$\frac{d}{dx} \sin(x + y) = \cos(x + y)$$

proves (6.6.3).

To prove the claim, we note that $\varphi'(x) = \cos(x + y) - \cos x \cos y + \sin x \sin y$, and hence $\varphi''(x) = -\sin(x + y) + \sin x \cos y + \cos x \sin y = -\varphi(x)$. It follows that for all x,

$$\frac{d}{dx} (\varphi(x)^2 + \varphi'(x)^2) = 2\varphi(x)\varphi'(x) + 2\varphi'(x)\varphi''(x) = 2\varphi'(x)(\varphi(x) + \varphi''(x)) = 0,$$

and hence $\varphi(x)^2 + \varphi'(x)^2$ is constant on $\mathbb{R}$. To find the value of the constant, we observe that $\varphi(0) = \sin y - \sin y = 0$ and $\varphi'(0) = \cos y - \cos y = 0$, which means that $\varphi(x)^2 + \varphi'(x)^2 = 0$ for all x, and hence $\varphi(x) = 0$ on $\mathbb{R}$, as claimed. □

Comment. It is also possible to recover the addition formulas "directly" from the power series,[6] using Theorem 5.4.2, the binomial theorem and quite a bit of finesse.

Corollary. *For all* $x \in \mathbb{R}$, $\cos x = \cos^2(x/2) - \sin^2(x/2)$.

Proof. Exercise. □

We use this special case of the addition formula for cos in the next section.

6.6.2. Periodicity

The fact that $\sin x$ and $\cos x$, are both periodic with the same period (i.e., $\sin(x + P) = \sin x$ and $\cos(x + P) = \cos x$, for some $P > 0$) follows immediately from their definition as functions on the unit-circle, as does the value, $P = 2\pi$, of their period. In this section we derive the periodicity directly from the power series, though we won't find the precise value of P.

Lemma. *There is an* $\alpha_0 > 0$ *such that* $\cos \alpha_0 = 0$ *and* $\cos x > 0$ *for* $0 \le x < \alpha_0$.

Proof. To begin, we note that if $\cos x < 0$ for some $x > 0$, then $\cos y = 0$ for some $0 < y < x$, by the intermediate value theorem, because $\cos 0 = 1 > 0$. The set $A = \{y > 0 : \cos y = 0\}$ is closed and bounded away from 0, so $\alpha_0 = \min A$ exists and satisfies the lemma's claim. In other words, it is enough to show that $\cos x < 0$ for some $x > 0$, and to finish the proof, we show that $\cos 2 < -1/3$.

Since the series for $\cos x$ converges absolutely for every x, it follows from Theorem 2.10.1 that the terms in the series may be grouped (and rearranged) without affecting

[6] I.e., without using derivatives.

the sum. Hence,

$$
\begin{aligned}
\cos 2 &= \sum_{n=0}^{\infty} \frac{(-1)^n 2^{2n}}{(2n)!} = 1 - 2 + \frac{2}{3} + \sum_{n=3}^{\infty} \frac{(-1)^n 2^{2n}}{(2n)!} \\
&= -\frac{1}{3} - \sum_{k=1}^{\infty} \left(\frac{2^{4k+2}}{(4k+2)!} - \frac{2^{4k+4}}{(4k+4)!} \right), \qquad (6.6.4)
\end{aligned}
$$

as we can see by pairing the terms corresponding to odd $n = 2k+1$ with the next even term, $n = 2k+2$. Next, for $k \geq 1$, we have

$$
\begin{aligned}
\frac{2^{4k+2}}{(4k+2)!} - \frac{2^{4k+4}}{(4k+4)!} &= \frac{2^{4k+2}}{(4k+2)!} \left(1 - \frac{4}{(4k+3)(4k+4)} \right) \\
&\geq \frac{13}{14} \cdot \frac{2^{2k+2}}{(2k+2)!} > 0,
\end{aligned}
$$

so the terms in the sum in (6.6.4) above are all positive. □

It follows from the proof that $0 < \alpha_0 < 2$.[7] A more detailed analysis along the same lines can produce more accurate estimates. Two examples appear in the exercises.

Theorem. *With α_0 as in the Lemma,*

$$\cos(x + 4\alpha_0) = \cos x \quad \textit{and} \quad \sin(x + 4\alpha_0) = \sin x$$

for all real x.

Proof. To begin, we note that $\sin \alpha_0 = \pm 1$, because $\sin^2 \alpha_0 = 1$. On the other hand, $\sin' x = \cos x > 0$ in the interval $[0, \alpha_0)$, so $\sin x$ is increasing in $[0, \alpha_0)$, and since $\sin 0 = 0$, it follows that $\sin \alpha_0 = 1$.

Next, from the addition formulas, we have

$$\cos(x + \alpha_0) = \cos x \cos \alpha_0 - \sin x \sin \alpha_0 = -\sin x,$$

and

$$\sin(x + \alpha_0) = \sin x \cos \alpha_0 + \sin \alpha_0 \cos x = \cos x$$

for all $x \in \mathbb{R}$. It follows that

$$
\begin{aligned}
\cos(x + 2\alpha_0) &= -\sin(x + \alpha_0) = -\cos x, \\
\cos(x + 3\alpha_0) &= -\cos(x + \alpha_0) = \sin x,
\end{aligned}
$$

so

$$\cos(x + 4\alpha_0) = \sin(x + \alpha_0) = \cos x.$$

Likewise, $\sin(x+2\alpha_0) = -\sin x$, $\sin(x+3\alpha_0) = -\cos x$ and so $\sin(x+4\alpha_0) = \sin x$. □

To conclude that $4\alpha_0$ is the (smallest) period of $\sin x$ and $\cos x$, we need to show that if $\cos(x + \beta) = \cos x$ or $\sin(x + \beta) = \sin x$ for all x, then $\beta \geq 4\alpha_0$. This follows from the next proposition.

Proposition. *If $\cos(\beta) = 1$ or $\sin(\alpha_0 + \beta) = 1$, then $\beta \geq 4\alpha_0$.*

[7]Remember: $\alpha_0 = \pi/2 \approx 1.5708$.

Proof. We first note that $\sin(\alpha_0 + \beta) = \sin\alpha_0 \cos\beta + \sin\beta\cos\alpha_0 = \cos\beta$, so it is enough to consider the case that $\cos\beta = 1$. In this case, it follows from Corollary 6.6.1 that

$$\cos^2(\beta/2) - \sin^2(\beta/2) = \cos\beta = \cos(0+\beta) = \cos 0 = 1,$$

which implies that $\cos^2(\beta/2) = 1$, so $\cos(\beta/2) = \pm 1$, and $\sin(\beta/2) = 0$. If $\cos(\beta/2) = 1$, we can repeat this argument, to conclude that $\cos(\beta/4) = \pm 1$ and $\sin(\beta/4) = 0$. If $\cos(\beta/4) = 1$, then $\cos(\beta/8) = \pm 1$ and $\sin(\beta/8) = 0$, etc.

If k is an integer such that $\beta/2^k < \alpha_0$, then $0 < \cos(\beta/2^k) < 1$. This implies that it is impossible that $\cos(\beta/2^m) = 1$ for all natural numbers m, and therefore $\cos(\beta/2^{m_0}) = -1$ for some $m_0 \geq 1$. Applying Corollary 6.6.1 again, shows that

$$\cos^2(\beta/2^{m_0+1}) - \sin^2(\beta/2^{m_0+1}) = -1,$$

so that $\cos(\beta/2^{m_0+1}) = 0$, and hence $\alpha_0 \leq \beta/2^{m_0+1} \leq \beta/4$, by the minimality of α_0. □

Exercises

ex6.6.1. Show that if $f(x)$ is twice differentiable in a neighborhood of 0, and $f + f'' = 0$, then $f(x) = a\cos x + b\sin x$, where a and b depend on $f(0)$ and $f'(0)$.

ex6.6.2. Show that $\sqrt{2} < \alpha_0 < \sqrt{3}$.

ex6.6.3. Show that $\alpha_0 < \sqrt{6 - \sqrt{12}}$.

Chapter 7

The Riemann Integral

Addition on $\mathbb{R}$ is formally defined for sums of two terms, and can be extended recursively to sums of finitely many terms. Furthermore, by introducing limits of partial sums, we extended addition to (certain) sums with countably many terms (i.e., series).

Integrals can be viewed as an even broader generalization of addition that extends the operation to weighted-sums over sets with a possibly uncountable number of terms. These generalizations come in several (closely related) varieties, but once again all of them define integrals in terms of limiting values of certain sets of sums with finitely many terms.

The limiting processes that lead to integrals go back at least to Eudoxus of Cnidus and Archimedes of Syracus in the third century BCE. They used *the method of exhaustion* to calculate the areas of certain regions by approximating the regions by polygons (whose areas are given by a finite sum of areas of triangles) with an increasingly large number of sides. The method was rediscovered by the Chinese mathematician Liu Hui in the third century CE, and by others in the centuries that followed, leading eventually to the integral calculus of Leibniz and Newton.

The definitions of Newton and Leibniz combined with the *Fundamental Theorem of Calculus* were quite general and very useful, but somewhat lacking in mathematical rigor. The first rigorous definition of the integral was given by Bernhard Riemann, in the middle of the nineteenth century. This, or the equivalent formulation by Gaston Darboux, are what most students are taught in introductory calculus classes.[1]

Later mathematicians generalized his work in different ways, most notably the Dutch mathematician Thomas Stieltjes introduced one generalization in 1894, leading to what is commonly called the *Riemann-Stieltjes integral*, and Henri Lebesgue introduced a different generalization, the *Lebesgue integral* in 1904. Lebesgue's integral is the one most commonly used in real (and complex) analysis.

[1] Darboux's definition is generally easier to use in theoretical settings.

To keep things simple (and familiar), this chapter will focus primarily on the Riemann integral, leaving the Stieltjes integral for the last section. In the next chapter we will introduce *Lebesgue measure* and the Lebesgue integral.

7.1. The Riemann integral

The Riemann integral of a function f on a (closed) interval $[a, b]$ is defined as a limit of *Riemann sums*, which are defined in terms of *partitions* of $[a, b]$.

7.1.1. Partitions

Conceptually, a *partition* of the interval $[a, b] \subset \mathbb{R}$ is a finite set of closed subintervals of $[a, b]$ whose union is $[a, b]$ and whose pairwise intersections are either empty or consist of a shared endpoint. More formally, we define partitions in terms the endpoints of the subintervals that comprise them.

Definition. A *partition* of $[a, b]$ is a finite set of points, $P = \{t_0, t_1, \ldots, t_n\}$, where

$$a = t_0 < t_1 < t_2 < \cdots < t_n = b,$$

and the *intervals of the partition* are $I_j = [t_{j-1}, t_j]$, for $1 \le j \le n$.

The *mesh* of the partition $P = \{t_0, t_1, \ldots, t_n\}$ is the length of its longest subinterval,

$$\text{mesh}(P) = \max_{1 \le j \le n}(t_j - t_{j-1}).$$

If P and Q are partitions of the same interval, with $P \subseteq Q$, we say that Q is a *refinement* of P. If P_1 and P_2 are partitions of the same interval, then $P_1 \wedge P_2$ is the partition given by the set $P_1 \cup P_2$ renumbered in the natural ascending order.

For any two partitions P_1 and P_2 of $[a, b]$, $P_1 \wedge P_2$ is the "coarsest" common refinement of P_1 and P_2, in the following sense.

Proposition. *If Q is a common refinement of P_1 and of P_2, then Q is a refinement of $P_1 \wedge P_2$.*

Proof. Exercise. □

7.1.2. Riemann sums and the definition of the Riemann integral

Let $f(x)$ be defined on $[a, b]$, let $P = \{t_0, t_1, \cdots, t_n\}$ be a partition of $[a, b]$, and let $x_j \in [t_{j-1}, t_j]$ be an arbitrary point from each subinterval of the partition,[2] then the sum

$$R(f, P, \{x_1, \ldots, x_n\}) = \sum_{j=1}^{n} f(x_j)(t_j - t_{j-1}), \tag{7.1.1}$$

is called a *Riemann sum* of f on the partition P of I.

Intuitively, the Riemann integral of f (if it exists) is a "limit of Riemann sums" for f on partitions whose meshes are approaching 0. We make this intuition precise in the following definition.

[2]A partition P of the interval I, together with a choice of a point in every part of P is often called a *tagged partition* of I.

Definition. A function f defined on $[a, b]$ is *Riemann-integrable* if there is an $\mathcal{J} \in \mathbb{R}$ with the property that for every $\varepsilon > 0$, there exists a $\delta > 0$ such that

$$|R(f, P, \{x_1, \dots, x_n\}) - \mathcal{J}| < \varepsilon,$$

for every partition P with mesh less than δ and every Riemann sum of f on P. In this case $\mathcal{J}$ is the *Riemann integral* of f on $[a, b]$ and we write

$$\mathcal{J} = \int_a^b f(x)\,dx.$$

As an immediate consequence of the definition, it follows that unbounded functions are not Riemann-integrable.

Theorem. *If f is Riemann-integrable on $[a, b]$, then f is bounded on $[a, b]$.*

Proof. If f is unbounded on $[a, b]$, then without loss of generality, we may assume that there is a sequence $\{x_k\}_{k=1}^{\infty} \subset [a, b]$ such that $f(x_k) > k$ for all k.

For any $\delta > 0$, choose $n > (b - a)/\delta$ and let $u_j = a + j \cdot (b - a)/n$, for $0 \leq j \leq n$. The partition $P = \{u_0, \dots, u_n\}$ of $[a, b]$ has mesh $(b - a)/n < \delta$, and one of the subintervals $[u_{m-1}, u_m]$ of P must contain infinitely many points of the sequence $\{x_k\}_{k=1}^{\infty}$. In particular, setting

$$\rho = \sum_{j \neq m} f(u_j)(u_j - u_{j-1}),$$

there is a $k > 2n|\rho| + n^2$ such that $u_{m-1} \leq x_k \leq u_m$. It follows that the Riemann sum

$$R(f, P, \{u_1, \dots, u_{m-1}, x_k, u_{m+1}, \dots, u_n\}) = \rho + \frac{f(x_k)}{n} > \rho + \frac{k}{n} > |\rho| + n.$$

It follows that f is not Riemann-integrable on $[a, b]$ because there is an unbounded sequence of Riemann sums of f on partitions of $[a, b]$ whose meshes are approaching 0. ◻

Comment. In the definition of the Riemann integral $\int_a^b f(x)\,dx$, it is assumed that $b > a$. We can define the Riemann integral $\int_b^a f(x)\,dx$ analogously by defining the "opposite" Riemann sums

$$R^-(f, P, \{x_j\}) = \sum_{j=1}^{n} f(x_j)(t_{j-1} - t_j).$$

Doing this leads immediately to two conclusions. First, the integral $\int_a^b f\,dx$ is defined if and only if the integral $\int_b^a f\,dx$ is defined and second

$$\int_b^a f(x)\,dx = -\int_a^b f(x)\,dx. \tag{7.1.2}$$

The details are left as an exercise, and going forward, we assume this identity when needed.

7.1.3. Darboux sums

The definition of the Riemann integral in terms of Riemann sums is not always convenient for demonstrating properties of the integral, and in particular, for characterizing functions that are Riemann integrable. To address this, we introduce another definition of the integral, due to Gaston Darboux, and show that the two definitions are equivalent. Darboux's idea was to replace the factors $f(x_j)$ appearing in a Riemann sum by the supremum and infimum of f on each part of the partition in question, producing upper and lower bounds for all Riemann sums of f on the given partition.

From Theorem 7.1.2, we may assume that any potentially Riemann integrable functions that we consider are necessarily bounded. If f is bounded on $[a, b]$, then f is bounded on every subinterval of any partition $P = \{t_1, t_1, \ldots, t_n\}$ of $[a, b]$, so the supremum and infimum of f

$$M_j = \sup\{f(x) : x \in [t_{j-1}, t_j]\} \quad \text{and} \quad m_j = \inf\{f(x) : x \in [t_{j-1}, t_j]\}$$

both exist for each subinterval of P.

Definition. The *upper and lower Darboux sums* of f on a partition P of $[a, b]$ are given by

$$U(f, P) = \sum_{j=1}^{n} M_j(t_j - t_{j-1}) \quad \text{and} \quad L(f, P) = \sum_{j=1}^{n} m_j(t_j - t_{j-1}), \tag{7.1.3}$$

It is clear from the definition that $U(f, P) \geq R(f, P, \{x_j\} \geq L(f, P)$ for any partition P of $[a, b]$ and any Riemann sum of f on P. Moreover, upper Darboux sums decrease and lower Darboux sums increase with refinements of partitions.

Theorem. *If f is bounded on $[a, b]$, P is a partition of $[a, b]$ and Q is a refinement of P, then*

$$U(f, P) \geq U(f, Q) \geq L(f, Q) \geq L(f, P).$$

Proof. For each subinterval $I_j = [t_{j-1}, t_j]$ of P, let $I_{j,1}, \ldots, I_{j,n_j}$ be the subintervals of Q that comprise I_j. We denote the length of I_j by Δ_j and by $\Delta_{j,k}$ the length of $I_{j,k}$, for $1 \leq k \leq n_j$, from which it follows that $\Delta_j = \sum_{k=1}^{n_j} \Delta_{j,k}$. If $M_{j,k} = \sup\{f(x) : x \in I_{j,k}\}$, then $M_j = \sup\{f(x) : x \in I_j\} \geq M_{j,k}$ for $1 \leq k \leq n_j$, and hence

$$U(f, P) = \sum_{j=1}^{n} M_j \Delta_j = \sum_{j=1}^{n} M_j \sum_{k=1}^{n_j} \Delta_{j,k} \geq \sum_{j=1}^{n} \sum_{k=1}^{n_j} M_{j,k} \Delta_{j,k} = U(f, Q).$$

This proves the left-hand inequality in the statement. The middle inequality follows from the definition of Darboux sums, and we leave the right-hand inequality as an exercise. □

Corollary. *If f is bounded on $[a, b]$, then for any partition P of $[a, b]$*

$$U(f, P) \geq \inf\{f(x) : a \leq x \leq b\} \cdot (b - a)$$

and

$$L(f, P) \leq \sup\{f(x) : a \leq x \leq b\} \cdot (b - a).$$

Proof. This follows from the theorem above because every partition of $[a,b]$ is a refinement of $P_0 = \{a,b\}$. □

7.1.4. The Darboux integral

Not surprsingly, Darboux's definition of the integral uses upper and lower Darboux sums. While Darboux's and Riemann's definitions of the integral are equivalent (as we will soon see), Darboux's definition is easier to apply in practice.

Definition. The *upper Darboux integral* of a function f on the interval $[a,b]$ is

$$\overline{\int_a^b} f(x)\,dx = \inf_P U(f,P)$$

and the *lower Darboux integral* of f on $[a,b]$ is

$$\underline{\int_a^b} f(x)\,dx = \sup_P L(f,P),$$

where the infimum and supremum above are taken over all partitions P of $[a,b]$. For any bounded function on $[a,b]$, both the upper and lower Darboux integrals exist because the upper Darboux sums are bounded below and the lower Darboux sums are bounded above.

A function f is *Darboux-integrable* on $[a,b]$ if and only if the upper and lower Darboux integrals of f on $[a,b]$ are equal, in which case the Darboux integral of f is equal to this common value:

$$\int_a^b f(x)\,dx = \underline{\int_a^b} f(x)\,dx = \overline{\int_a^b} f(x)\,dx.$$

As can be expected, the upper Darboux integral is at least as big as the lower Darboux integral.

Proposition. *For any function f defined on $[a,b]$,*

$$\overline{\int_a^b} f(x)\,dx \geq \underline{\int_a^b} f(x)\,dx.$$

Proof. It follows from Theorem 7.1.3 that for any partitions P and Q of $[a,b]$,

$$U(f,P) \geq U(f, P \wedge Q) \geq L(f, P \wedge Q) \geq L(f,Q).$$

Hence,

$$U(f,P) \geq \sup L(f,Q) = \underline{\int_a^b} f(x)\,dx,$$

for all partitions P of $[a,b]$, so $\inf U(f,P) \geq \underline{\int_a^b} f(x)\,dx$, from which the proposition follows. □

The following theorem gives a basic condition for Darboux integrability.

Theorem. *The bounded function f is Darboux-integrable on $[a,b]$ if and only if for any $\varepsilon > 0$, there exists a partition P of $[a,b]$ such that $U(f,P) - L(f,P) < \varepsilon$.*

Proof. Suppose first that for any $\varepsilon > 0$, there exists a partition P_ε of $[a,b]$ such that $U(f,P_\varepsilon) - L(f,P_\varepsilon) < \varepsilon$. Then for any $\varepsilon > 0$,

$$\varepsilon > U(f,P_\varepsilon) - L(f,P_\varepsilon) \geq \overline{\int_a^b} f(x)\,dx - \underline{\int_a^b} f(x)\,dx \geq 0,$$

so $\overline{\int_a^b} f(x)\,dx = \underline{\int_a^b} f(x)\,dx$ and f is Darboux-integrable.

Next suppose that f is Darboux-integrable on $[a,b]$. Given $\varepsilon > 0$, there are partitions $P_{U,\varepsilon}$ and $P_{L,\varepsilon}$ of $[a,b]$ such that

$$U(f,P_{U,\varepsilon}) - \overline{\int_a^b} f(x)\,dx < \varepsilon/2 \quad \text{and} \quad \underline{\int_a^b} f(x)\,dx - L(f,P_{L,\varepsilon}) < \varepsilon/2,$$

as follows from the definitions. Setting $P = P_{U,\varepsilon} \wedge P_{L,\varepsilon}$, it follows from Theorem 7.1.3 that $U(f,P) - \overline{\int_a^b} f(x)\,dx < \varepsilon/2$ and $\underline{\int_a^b} f(x)\,dx - L(f,P) < \varepsilon/2$, and adding these inequalities shows that

$$U(f,P) - L(f,P) = U(f,P) - L(f,P) + \underline{\int_a^b} f(x)\,dx - \overline{\int_a^b} f(x)\,dx < \varepsilon,$$

since $\underline{\int_a^b} f(x)\,dx - \overline{\int_a^b} f(x)\,dx = 0$ by assumption. □

7.1.5. The equivalence of the Riemann and Darboux integrals

To show that Riemann's and Darboux's definitions of the integral are equivalent, we begin with the observation that the Darboux integral can be defined in a manner similar to the definition of the Riemann integral.

Lemma. *The function f is Darboux-integrable on $[a,b]$ if and only if for every $\varepsilon > 0$, there is a $\delta > 0$ such that $U(f,P) - L(f,P) < \varepsilon$ for all partitions P of $[a,b]$ with* $\operatorname{mesh}(P) < \delta$.

Proof. Suppose that f satisfies the second condition. Given any $\varepsilon > 0$, if P is a partition of $[a,b]$ with $\operatorname{mesh}(P) < \delta_\varepsilon$, then

$$0 \leq \overline{\int_a^b} f(x)\,dx - \underline{\int_a^b} f(x)\,dx \leq U(f,P) - L(f,P) < \varepsilon.$$

It follows that $\overline{\int_a^b} f(x)\,dx = \underline{\int_a^b} f(x)\,dx$, so f is Darboux-integrable on $[a,b]$.

Now suppose that f is Darboux-integrable on $[a,b]$ and let $\varepsilon > 0$ be given. By Theorem 7.1.4, there exists a partition $P_\varepsilon = \{s_0, s_1, \dots, s_N\}$ of $[a,b]$ for which $U(f,P_\varepsilon) - L(f,P_\varepsilon) < \varepsilon/2$. Let $M = \sup_{x\in[a,b]} |f(x)|$ and set $\delta = \frac{\varepsilon}{4NM}$. We claim that $U(f,P) - L(f,P) < \varepsilon$ for any partition P, with $\operatorname{mesh}(P) < \delta$.

Let $P = \{t_0, \ldots, t_n\}$ be such a partition. For $1 \le j \le n$, we set $\Delta_j = (t_j - t_{j-1})$ and $D_j = \sup_{I_j} f(x) - \inf_{I_j} f(x)$. We have $\Delta_j < \delta$ and $D_j \le 2M$, for $1 \le j \le n$, and $U(f,P) - L(f,P) = \sum_{j=1}^{n} D_j \cdot \Delta_j$.

Let $Q = P \wedge P_\varepsilon$, so that $U(f,Q) - L(f,Q) < \varepsilon/2$, by Theorem 7.1.3. We divide the set $\{j : 1 \le j \le n\}$ into two sets A and B: $j \in A$ if $[t_{j-1}, t_j]$ is not subdivided in Q by points of P_ε, and $j \in B$ otherwise. Since P_ε contains $N-1$ points not equal to a or b, it follows that $|B| < N$.

We observe that if $j \in A$, then $D_j \cdot \Delta_j$ appears as a term in the sum for $U(f,Q) - L(f,Q)$, so $\sum_{j \in A} D_j \cdot \Delta_j \le U(f,Q) - L(f,Q)$, because all the terms in the sum for $U(f,Q) - L(f,Q)$ are nonnegative. Therefore

$$\begin{aligned} U(f,P) - L(f,P) &= \sum_{j=1}^{n} D_j \cdot \Delta_j = \sum_{j \in A} D_j \cdot \Delta_j + \sum_{j \in B} D_j \cdot \Delta_j \\ &< U(f,Q) - L(f,Q) + \sum_{j \in B} 2M\delta \\ &< \varepsilon/2 + 2NM\delta = \varepsilon. \end{aligned}$$

□

Theorem. *A function f is Riemann-integrable on $[a,b]$ if and only if it is Darboux-integrable on $[a,b]$, in which case the values of its Riemann and Darboux integrals on $[a,b]$ are the same.*

Proof. Suppose that f is Riemann-integrable on $[a,b]$, with Riemann integral $\mathcal{I}_R$, and let $\varepsilon > 0$ be given. Then there is a $\delta_\varepsilon > 0$ such that

$$|R(f,P,\{x_1,\ldots,x_n\}) - \mathcal{I}_R| < \varepsilon/4,$$

for all partitions P of $[a,b]$ with $\text{mesh}(P) < \delta$ and any choice of representative points $x_j \in [t_{j-1}, t_j]$.

If $P = \{t_0, \ldots, t_n\}$ is a partition with mesh less than δ_ε, then for $1 \le j \le n$ let u_j and v_j be points in $[t_{j-1}, t_j]$ such that

$$\eta_j = M_j - f(u_j) < \frac{\varepsilon}{4n\delta_\varepsilon} \quad \text{and} \quad \xi_j = f(v_j) - m_j < \frac{\varepsilon}{4n\delta_\varepsilon},$$

where M_j and m_j are the supremum and infimum of f in $[t_{j-1}, t_j]$, respectively. Setting $\Delta_j = t_j - t_{j-1}$, as before, we have

$$|U(f,P) - \mathcal{I}_R| = \left| R(f,P,\{u_1,\ldots,u_n\}) - \mathcal{I} + \sum_{j=1}^{n} \eta_j \Delta_j \right| < \frac{\varepsilon}{2}, \tag{7.1.4}$$

by the triangle inequality, and the bounds on η_j, Δ_j and $|R(f,P,\{u_j\} - \mathcal{I}|$, and by the same argument, we have

$$|L(f,P) - \mathcal{I}_R| = \left| R(f,P,\{v_1,\ldots,v_n\}) - \mathcal{I} - \sum_{j=1}^{n} \xi_j \Delta_j \right| < \frac{\varepsilon}{2}. \tag{7.1.5}$$

It follows that $U(f,P) - L(f,P) < \varepsilon$ for any partition with $\text{mesh}(P) < \delta_\varepsilon$, so f is Darboux-integrable by the previous lemma. Moreover, $U(f,P) \ge \mathcal{I}_D \ge L(f,P)$ for all partitions P of $[a, b]$, so $|\mathcal{I}_D - \mathcal{I}_R| < \varepsilon/2$ for any $\varepsilon > 0$, by (7.1.4) and (7.1.5), whence $\mathcal{I}_D = \mathcal{I}_R$.

Now suppose that f is Darboux-integrable on $[a, b]$, with Darboux integral $\mathcal{I}_D$. Then $U(f, P) \geq \mathcal{I}_D \geq L(f, P)$ for all partitions P of $[a, b]$, and given $\varepsilon > 0$ there is a $\delta_\varepsilon > 0$ such that $U(f, P) - L(f, P) < \varepsilon$ if $\operatorname{mesh}(P) < \delta_\varepsilon$. If $R(f, P, \{x_j\})$ is a Riemann sum of f on such a partition, then $U(f, P) \geq R(f, P, \{x_j\}) \geq L(f, P)$, so $|R(f, P, \{x_j\}) - \mathcal{I}_D| < \varepsilon$. Hence f is Riemann integrable with Riemann integral $\mathcal{I}_R = \mathcal{I}_D$. □

Corollary. *The function f is integrable on $[a, b]$ if and only if for every $\varepsilon > 0$ there is a $\delta > 0$ such that if P is a partition of $[a, b]$ with* $\operatorname{mesh}(P) < \delta$, *then $|R(f, P, \{x_j\}) - R(f, P, \{y_j\})| < \varepsilon$ for any two Riemann sums of f on P.*

Proof. Exercise. □

Exercises

ex7.1.1. Prove Proposition 7.1.1.

ex7.1.2. Prove the right-hand inequality in the statement of Theorem 7.1.3.

ex7.1.3. Show that if f is integrable in $[a, b]$, then for every $\varepsilon > 0$ there is a $\delta > 0$ such that

$$U(f, P) - \int_a^b f(x)\,dx < \varepsilon \quad \text{and} \quad \int_a^b f(x)\,dx - L(f, P) < \varepsilon$$

for all partitions P of $[a, b]$ with mesh $< \delta$.

ex7.1.4. Suppose that f is bounded on $[a, b]$ and $R(f, P)$ is the set of all Riemann sums for f on a partition P of $[a, b]$. Show that $U(f, P) = \sup R(f, P)$ and $L(f, P) = \inf R(f, P)$.

ex7.1.5. Prove Corollary 7.1.5.

7.2. Properties of Riemann integrals

The Riemann integral $\int_a^b f(x)\,dx$ is defined in terms of finite sums and shares many properties with finite sums — both with respect to the terms of the sum (the integrand) and with respect to the range of summation (the interval).

7.2.1. Integrals of sums and products

The set of functions that are Riemann integrable on an interval $[a, b]$ form an *algebra* over the real numbers: it includes all constants and is closed under addition and multiplication.

Proposition. *For any constant c, $\int_a^b c\,dx = c(b - a)$.*

Proof. Exercise. □

Theorem. *If f and g are both integrable in $[a,b]$, then $f+g$ and $f\cdot g$ are integrable in $[a,b]$.*

Proof. For any partition P of $[a,b]$ and any subinterval $[t_{j-1},t_j]$ of P, we have

$$\sup_{[t_{j-1},t_j]}(f(x)+g(x)) \le \sup_{[t_{j-1},t_j]} f(x) + \sup_{[t_{j-1},t_j]} g(x),$$

from which it follows that $U(f+g,P) \le U(f,P)+U(g,P)$. Likewise,

$$\inf_{[t_{j-1},t_j]}(f(x)+g(x)) \ge \inf_{[t_{j-1},t_j]} f(x) + \inf_{[t_{j-1},t_j]} g(x),$$

so $L(f+g,P) \ge L(f,P)+L(g,P)$. Given $\varepsilon>0$, if P is a partition of $[a,b]$ with mesh small enough so that $U(f,P)-L(f,P)<\varepsilon/2$ and $U(g,P)-L(g,P)<\varepsilon/2$, then

$$U(f+g,P)-L(f+g,P) \le (U(f,P)+U(g,P))-(L(f,P)+L(g,P)) < \varepsilon.$$

Hence $f+g$ is integrable.

Now, let $M_f = \sup_{[a,b]}|f(x)|$ and $M_g = \sup_{[a,b]}|g(x)|$. Given $\varepsilon>0$, let P be any partition of $[a,b]$ with mesh small enough that

$$U(f,P)-L(f,P) < \frac{\varepsilon}{M_f+M_g} \text{ and } U(g,P)-L(g,P) < \frac{\varepsilon}{M_f+M_g}.$$

Let $R_1 = R(fg,P,\{x_0,\dots,x_n\})$ and $R_2 = R(fg,P,\{y_0\dots,y_n\})$ be any two Riemann sums of fg on P. If we can show that $|R_1-R_2|<\varepsilon$, then it will follow from Corollary 7.1.5 that fg is integrable.

We observe that

$$\begin{aligned}
R_1-R_2 &= \sum_{j=1}^n [f(x_j)g(x_j)-f(y_j)g(y_j)](t_j-t_{j-1})\\
&= \sum_{j-1}^n [f(x_j)g(x_j)-f(y_j)g(x_j)+f(y_j)g(x_j)-f(y_j)g(y_j)](t_j-t_{j-1})\\
&= \sum_{j=1}^n [g(x_j)(f(x_j)-f(y_j))+f(y_j)(g(x_j)-g(y_j))](t_j-t_{j-1}),
\end{aligned}$$

from which it follows that

$$\begin{aligned}
|R_1-R_2| &\le M_g\sum_{j=1}^n |f(x_j)-f(y_j)|(t_j-t_{j-1}) + M_f\sum_{j=1}^n |g(x_j)-g(y_j)|(t_j-t_{j-1})\\
&\le M_g(U(f,P)-L(f,P)) + M_f(U(g,P)-L(g,P)) < \varepsilon.
\end{aligned}$$

□

Corollary. *If f and g are both integrable on $[a,b]$, then*

$$\int_a^b \alpha f(x)+\beta g(x)\,dx = \alpha\int_a^b f(x)\,dx + \beta\int_a^b g(x)\,dx.$$

Proof. The integrability of $\alpha f+\beta g$ follows from the previous theorem and the proposition before it.

For any partition P of $[a,b]$, we have

$$R(\alpha f+\beta g,P,\{x_j\}) = \alpha R(f,P,\{x_j\}) + \beta R(g,P,\{x_j\}),$$

so that

$$\begin{aligned}
&\left|\int_a^b \alpha f(x)+\beta g(x)\,dx-\alpha\int_a^b f(x)\,dx-\beta\int_a^b g(x)\,dx\right|\\
&\quad\le\left|\int_a^b \alpha f(x)+\beta g(x)\,dx-R(\alpha f+\beta g,P,\{x_j\})\right|\\
&\quad\quad+\left|\left(\alpha R(f,P,\{x_j\}-\alpha\int_a^b f(x)\,dx\right)\right|\\
&\quad\quad+\left|\left(\beta R(g,P,\{x_j\}-\beta\int_a^b g(x)\,dx\right)\right|.
\end{aligned}$$

The equality claimed by the corollary follows from the observation that given any $\varepsilon > 0$, there is a $\delta > 0$ such that for any partition with mesh $< \delta$, the terms in the second, third and fourth lines above are all less than $\varepsilon/3$. □

7.2.2. Integration on subintervals

It can be convenient to split an integral over one interval into a sum of integrals over some of its subintervals.

Lemma. *If f is integrable on $[a,b]$, then for any $a \le c < d \le b$, f is integrable on $[c,d]$.*

Proof. Let $\varepsilon > 0$ be given and let $\delta > 0$ be such that $U(f,P) - L(f,P) < \varepsilon$, for all partitions P of $[a,b]$ with mesh$(P) < \delta$. Furthermore, let P_1 and P_2 be fixed partitions of $[a,c]$ and $[d,b]$ with meshes $< \delta$, and define the function f_0 by setting $f_0(x) = f(x)$ for $c \le x \le d$ and $f_0(x) = 0$ otherwise.

Now, for any partition P_0 of $[c,d]$ with mesh$(P_0) < \delta$, let $P = P_0 \cup P_1 \cup P_2$, so P is a partition of $[a,b]$ with mesh $< \delta$. Then

$$U(f,P_0) - L(f,P_0) = U(f_0,P) - L(f_0,P) \le U(f,P) - L(f,P) < \varepsilon,$$

so f is integrable on $[c,d]$ by Lemma 7.1.5. □

Theorem. *If f is integrable in $[a,b]$ and $a < c < b$, then*

$$\int_a^b f(x)\,dx = \int_a^c f(x)\,dx + \int_c^b f(x)\,dx.$$

Proof. By the lemma above, both integrals on the right are defined.

Given $\varepsilon > 0$, let $\delta > 0$ be such that if P_1, P_2 and P of $[a,c]$, $[c,b]$ and $[a,b]$, respectively, all with meshes $< \delta$, then

$$U(f,P_1) - \int_a^c f(x)\,dx < \varepsilon/3, \quad U(f,P_2) - \int_c^b f(x)\,dx < \varepsilon/3$$
$$\text{and}\quad U(f,P) - \int_a^b f(x)\,dx < \varepsilon/3. \tag{7.2.1}$$

Choose any two partitions P_1 and P_2 of $[a,c]$ and $[c,b]$, with mesh$(P_1) < \delta$ and mesh$(P_2) < \delta$, and set $P = P_1 \cup P_2$. Then $U(f,P) = U(f,P_1) + U(f,P_2)$ and mesh$(P) < \delta$.

Hence, subtracting the third difference in (7.2.1) from the sum of the first two, shows that

$$\left|\int_a^b f(x)\,dx - \int_a^c f(x)\,dx + \int_c^b f(x)\,dx\right| < \frac{\varepsilon}{3} + \frac{\varepsilon}{3} + \frac{\varepsilon}{3} = \varepsilon,$$

from which the theorem follows, since ε was arbitrary. □

7.2.3. Inequalities

Proposition. *If $f(x) \geq g(x)$ and both f and g are integrable in $[a,b]$, then*

$$\int_a^b f(x)\,dx \geq \int_a^b g(x)\,dx.$$

Proof. By Corollary 7.2.1,

$$\int_a^b f(x)\,dx - \int_a^b g(x)\,dx = \int_a^b f(x) - g(x)\,dx \geq 0,$$

where the inequality on the right follows from the fact that $R(f-g, P, \{x_j\}) \geq 0$ for all Riemann sums of $f - g$ in $[a,b]$. □

Theorem. *If f is integrable in $[a,b]$, then $|f|$ is integrable in $[a,b]$ and*

$$\int_a^b |f(x)|\,dx \geq \left|\int_a^b f(x)\,dx\right|. \tag{7.2.2}$$

This is the *triangle inequality* for integrals.

Proof. Let $f^+(x) = \max(f(x), 0)$, then for any partition P of $[a,b]$,

$$U(f^+, P) - L(f^+, P) \leq U(f, P) - L(f, P),$$

from which it follows that f^+ is integrable in $[a,b]$. Hence, $f^- = f^+ - f$ is also integrable in $[a,b]$ by Theorem 7.2.1, and therefore $|f| = f^+ + f^-$ is likewise integrable on $[a,b]$. Given any Riemann sum $R(f, P, \{x_j\})$ on $[a,b]$, we have

$$|R(f, P, \{x_j\})| = \left|\sum_{j=1}^{n} f(x_j)(t_j - t_{j-1})\right| \leq \sum_{j=1}^{n} |f(x_j)|(t_j - t_{j-1}) = R(|f|, P, \{x_j\}),$$

which implies (7.2.2). □

7.2.4. The fundamental theorem of calculus

The integral and the derivative are related by the concept of the *primitive* function.

Definition. If $f(x)$ is integrable on the interval $[a,b]$, then the function

$$F(x) = \int_a^x f(t)\,dt.$$

is the *primitive* of f in $[a,b]$, and it is defined for all $x \in [a,b]$ by Lemma 7.2.2.

It is common to refer to F as "the integral of f".

If F is the primitive of f in $[a,b]$, then if $a \le x < y \le b$,

$$|F(y) - F(x)| = \left| \int_x^y f(t)\,dt \right| \le \int_x^y |f(t)|\,dt \le M|y-x|,$$

where M is an upper bound for the (Riemann integrable) function $|f|$ in $[a,b]$. Hence, given $\varepsilon > 0$, if $|y-x| < \varepsilon/M$, then $|F(x) - F(y)| < \varepsilon$, so F is uniformly continuous in $[a,b]$. In fact, F is *absolutely continuous* in $[a,b]$.

Definition. A function f is *absolutely continuous* on an interval $I \subset \mathbb{R}$, if given any $\varepsilon > 0$ there is a $\delta > 0$ such that for any $m \in \mathbb{N}$ and points $x_1 < y_1 \le x_2 < y_2 \le \cdots \le x_m n y_n$ in I, if $\sum_{k=1}^n (y_k - x_k) < \delta$ then $\sum_{k=1}^n |f(y_k) - f(x_k)| < \varepsilon$.

Note that absolute continuity in I implies uniform continuity I (by taking $n = 1$).

Proposition. *If f is Riemann integrable on $[a,b]$, then $F(x) = \int_a^x f(t)\,dt$ is absolutely continuous in $[a,b]$.*

Proof. Let $M = \sup\{|f(x)| : x \in [a,b]\}$, then $M < \infty$ because f is Riemann integrable. Given $\varepsilon > 0$, set $\delta = \varepsilon/M$. If $a \le x_1 < y_1 \le \cdots \le x_n < y_n \le b$ and $\sum_{k=1}^n (y_k - x_k) < \delta$, then

$$\sum_{k=1}^n |F(y_k) - F(x_k)| = \sum_{k=1}^n \left| \int_{x_k}^{y_k} f(t)\,dt \right| \le \sum_{k=1}^n \int_{x_k}^{y_k} |f(t)|\,dt \le M \sum_{k=1}^n (y_k - x_k) \le \varepsilon.$$

□

More is true.

Theorem. *If $f(x)$ is integrable in $[a,b]$, then its privimitive, $F(x)$ in $[a,b]$ is differentiable at every point of continuity of f in (a,b), and at such a point, x_0, $F'(x_0) = f(x_0)$.*

In other words,

$$\frac{d}{dx} \int_a^x f(t)\,dt = f(x),$$

at every point of continuity of f in $[a,b]$.

Proof. Suppose that $a < x_0 < b$ is a point of continuity of f, let $\varepsilon > 0$ be given, and let $\delta > 0$ be such that $|f(x) - f(x_0)| < \varepsilon$ if $|x - x_0| < \delta$.

If $|x - x_0| < \min(\delta, b - x_0, x_0 - a)$, then

$$\begin{aligned} \left| \frac{F(x) - F(x_0)}{x - x_0} - f(x_0) \right| &= \left| \frac{1}{x - x_0} \int_{x_0}^x f(t)\,dt - \frac{1}{x - x_0} \int_{x_0}^x f(x_0)\,dt \right| \\ &= \left| \frac{1}{x - x_0} \int_{x_0}^x f(t) - f(x_0)\,dt \right|, \end{aligned}$$

by Proposition 7.2.1 and Theorem 7.2.1, and so by the triangle inequality,

$$\left| \frac{F(x) - F(x_0)}{x - x_0} - f(x_0) \right| \le \frac{1}{x - x_0} \int_{x_0}^x |f(t) - f(x_0)|\,dt < \frac{1}{x - x_0} \int_{x_0}^x \varepsilon\,dt = \varepsilon.$$

□

Corollary. *If $f'(x)$ is integrable on $[a,b]$, then*

$$\int_a^b f'(x)\,dx = f(b) - f(a).$$

Proof. Exercise. □

Proposition. *If g is differentiable, g' is integrable in $[a,b]$ and f is integrable in $[\alpha,\beta] = g([a,b])$, then*

$$\int_a^b f(g(x))g'(x)\,dx = F(g(b)) - F(g(a)),$$

where F is the primitive of f in $[\alpha,\beta]$.

Proof. Exercise. □

Exercises

ex7.2.1. Prove Corollary 7.2.4.

ex7.2.2. Show by example that the integral of a product of two integrable functions is not necessarily equal to the product of their integrals.

ex7.2.3. *Integration by parts:* Show that if f and g are both differentiable in $[a,b]$, and f' and g' are both integrable in $[a,b]$, then

$$\int_a^b f(x)g'(x)\,dx = f(b)g(b) - f(a)g(a) - \int_a^b g(x)f'(x)\,dx.$$

ex7.2.4. Show that if f is Lipschitz continuous in $[a,b]$, the f is absolutely continuous there.

ex7.2.5. Show that if f is differentiable in $[a,b]$ and $|f'|$ is bounded there, then f is absolutely continuous in $[a,b]$.

ex7.2.6. Let $l(x) = \displaystyle\int_1^x \frac{1}{t}\,dt$. Prove the following statements:

(i) $l(x)$ is strictly increasing, tends to $-\infty$ as $x \to 0$ and tends to $+\infty$ as $x \to \infty$.

(ii) $l(ab) = l(a) + l(b)$ for all positive a, b. In particular: $l(1/x) = -l(x)$.

(iii) The inverse function of l, denoted $E(x)$, exists and satisfies the functional equation $E(a+b) = E(a)E(b)$.

(iv) $E(x)$ is differentiable and satisfies: $E'(x) = E(x)$.

(v) $E(x)$, as defined here, is the exponential function, $\exp(x)$.

7.3. Integrable functions

We have already seen that Riemann-integrable functions are bounded. On the other hand, not all bounded functions are integrable. For example, the indicator function of the set of rational numbers,

$$\chi_{\mathbb{Q}}(x) = \begin{cases} 1 & : x \in \mathbb{Q} \\ 0 & : x \notin \mathbb{Q} \end{cases}$$

is bounded but not Riemann integrable on any interval. This follows from the fact that both $\mathbb{Q}$ and $\mathbb{R} \setminus \mathbb{Q}$ are dense in every interval, so if P is any partition of $[a, b]$, then

$$U(\chi_{\mathbb{Q}}, P) - L(\chi_{\mathbb{Q}}, P) = b - a.$$

The reason that $\chi_{\mathbb{Q}}$ is not Riemann-integrable is that it is '*too discontinuous*,' by which we mean that the set of its points of discontinuity is too large. Indeed, for all $x \in \mathbb{R}$, the *oscillation* $\omega_{\chi_{\mathbb{Q}}}(x) = 1$, so *every* real number is a point of discontinuity of $\chi_{\mathbb{Q}}$.[3]

In 1907, Henri Lebesgue and Giuseppe Vitali proved independently that the function $f(x)$ is Riemann integrable on $[a, b]$ if and only if it is bounded on $[a, b]$ *and* the set of its points of discontinuity in $[a, b]$ has measure zero. Recall that a set $E \subset \mathbb{R}$ has measure zero if for every $\varepsilon > 0$, there is a sequence of intervals $\{I_n : n \in \mathbb{N}\}$ whose union contains E as a subset and the sum of whose lengths is less than ε.[4]

We sketch a proof of the Vitali-Lebesgue theorem in Section*7.3.4, but first we consider some simple, special cases.

7.3.1. Continuous functions are integrable

Continuous functions have no points of discontinuity.

Proposition. *If f is continuous on $[a, b]$, then f is integrable on $[a, b]$.*

Proof. By Theorem 4.4.2 f is uniformly continuous on $[a, b]$, because $[a, b]$ is compact. Therefore, for any $\varepsilon > 0$, there exists $\delta > 0$ such that if $t, s \in [a, b]$ and $|t - s| < \delta$, then $|f(t) - f(s)| < \varepsilon/(b - a)$. If $P = \{t_0, \ldots, t_n\}$ is a partition of $[a, b]$ and $\text{mesh}(P) < \delta$ then, for every subinterval $[t_{j-1}, t_j]$ of P, we have $M_j - m_j < \varepsilon/(b - a)$. Therefore

$$U(f, P) - L(f, P) = \sum_{j=1}^{n} (M_j - m_j)(t_j - t_{j-1}) < \frac{\varepsilon}{b - a} \sum_{j=1}^{n} (t_j - t_{j-1}) = \varepsilon.$$

□

7.3.2. Monotone functions are integrable

Monotone functions have at most countably many points of discontinuity, as proved in Theorem 4.3.1, and countable sets also have measure zero.[5] Furthermore, the phrase "f is monotone on $[a, b]$", implies that $f(a)$ and $f(b)$ are both defined, so f is bounded between $f(a)$ and $f(b)$.

Proposition. *If f is monotone on $[a, b]$, then f is integrable on $[a, b]$.*

[3]See Theorem 4.3.4.
[4]See §3.3.1.
[5]See Proposition 3.3.1.

Proof. With no loss of generality we may assume that f is monotone nondecreasing (and not constant) on $[a,b]$. Given $\varepsilon > 0$, set $\delta = \varepsilon/(f(b) - f(a))$ and suppose that $P = \{t_0, \dots, t_n\}$ is a partition of $[a,b]$, with $\text{mesh}(P) < \delta$. Then

$$\begin{aligned} U(f,P) - L(f,P) &= \sum_{j=1}^{n} (f(t_j) - f(t_{j-1}))(t_j - t_{j-1}) \\ &< \delta \cdot \sum_{j=1}^{n} (f(t_j) - f(t_{j-1})) = \delta \cdot (f(b) - f(a)) = \varepsilon. \end{aligned}$$

□

7.3.3. Functions of bounded variation

If g is monotone nondecreasing on $[a,b]$, then

$$\sum_{j=1}^{n} |g(t_j) - g(t_{j-1}| = \sum_{j=1}^{n} g(t_j) - g(t_{j-1}) = g(b) - g(a) < \infty,$$

for any partition $P = \{t_0, \dots, t_n\}$ of $[a,b]$. Note that if g is monotone nonincreasing, then $\sum_{j=1}^{n} |g(t_j) - g(t_{j-1}|$ is also bounded for any partition of $[a,b]$.

This property of monotone functions leads to the following definition.

Definition. The *total-variation* $V_f(a,b)$ of a real-valued function f on a closed interval $[a,b]$ is defined by

$$V_f(a,b) = \sup_P \sum_{j=1}^{n} |f(t_j) - f(t_{j-1})| \tag{7.3.1}$$

where the supremum is taken over all partitions $P = \{t_0, \dots, t_n\}$ of $[a,b]$.

A function f is of *of bounded variation* on $[a,b]$ if $V_f(a,b)$ is finite, in which case we write $f \in BV[a,b]$.

In **ex7.3.2** you are asked to prove the theorem that $f \in BV[a,b]$ if and only if there exist monotone nondecreasing functions g and h on $[a,b]$, such that $f = g - h$. If follows from this and Proposition 7.3.2 that if $f \in BV[a,b]$, then f is integrable in $[a,b]$.[6] On the other hand, we can also show that functions of bounded variation are integrable directly from the definition.

Proposition. *If $f \in BV[a,b]$, then f is bounded on $[a,b]$.*

Proof. Exercise. □

Lemma. *Suppose that $f \in BV[a,b]$, and let $P = \{t_0, \dots, t_n\}$ be a partition of $[a,b]$. Then*

$$V_f(a,b) = \sum_{j=1}^{n} V_f(t_{j-1}, t_j), \tag{7.3.2}$$

[6]It also follows from this that the set of points of discontinuity of a function of bounded variation has measure zero as well.

Proof. If $P_j = \{t_{j,0}, \dots, t_{j,n_j}\}$ is a partition of $[t_{j-1}, t_j]$, for $1 \le j \le n$, then $P = P_1 \cup \cdots \cup P_n$ is a partition of $[a, b]$, and therefore

$$\sum_{j=1}^{n} \sum_{k=1}^{n_j} |f(t_{j,k}) - f(t_{j,k-1})| \le V_f(a, b). \tag{7.3.3}$$

Note that it follows from this that $\sum_{k=1}^{n_j} |f(t_{j,k}) - f(t_{j,k-1})| \le V_f(a, b)$, for each j, so $f \in BV[t_{j-1}, t_j]$ and $V_f(t_{j-1}, t_j) \le V_f(a, b)$.

Given $\varepsilon > 0$, let Q be a partition of $[a, b]$ such that

$$V_f(a, b) < \sum_{Q} |f(t_i) - f(t_{i-1})| + \varepsilon.$$

The partition $P \wedge Q$ of $[a, b]$ is refinement of Q, so

$$\sum_{Q} |f(t_i) - f(t_{i-1})| \le \sum_{Q \wedge P} |f(t_j) - f(t_{j-1})|,$$

by the triangle inequality. Moreover, $P \wedge Q$ includes the points of P so it is a union of partitions $\{P_1, \dots, P_n\}$ of the subintervals of P as above. Using the notation of (7.3.3), we have

$$\begin{aligned} V_f(a, b) &< \sum_{Q \wedge P} |f(t_i) - f(t_{i-1})| + \varepsilon \\ &= \sum_{j=1}^{n} \sum_{k=1}^{n_j} |f(t_{j,k}) - f(t_{j,k-1})| + \varepsilon \le \sum_{j=1}^{n} V_f(t_{j-1}, t_j) + \varepsilon. \end{aligned}$$

Since ε was arbitrary, (7.3.2) follows from this and (7.3.3). □

Theorem. *If $f \in BV[a, b]$, then f is Riemann integrable on $[a, b]$.*

Proof. We first note that f is bounded in $[a, b]$ by Proposition 7.3.3.

Next, for any subinterval $I = [c, d]$ of $[a, b]$, let $M_I = \sup_I f$ and $m_I = \inf_I f$. Then for any $\varepsilon > 0$, there exist $t_m, t_M \in [c, d]$ such that $0 < M_I - f(t_M) < \varepsilon/2$ and $0 < f(t_m) - m_I < \varepsilon/2$, and therefore

$$M_I - m_I < |f(t_M) - f(t_m)| + \varepsilon \le \sum_{P} |f(t_j) - f(t_{j-1})| + \varepsilon \le V_f(c, d) + \varepsilon,$$

where P is any partition of $[c, d]$ than includes t_m and t_M. It follows that

$$M_I - m_I \le V_f(c, d). \tag{7.3.4}$$

Now, given $\varepsilon > 0$, let $P = \{t_0, \dots, t_n\}$ be a partition of $[a, b]$ with mesh less than $\varepsilon / V_f(a, b)$. Then, setting $M_j = \sup_{[t_{j-1}, t_j]}$ and $m_j = \inf_{[t_{j-1}, t_j]}$, as usual, we have

$$\begin{aligned} U(f, P) - L(f, P) &= \sum_{j=1}^{n} (M_j - m_j)(t_j - t_{j-1}) \\ &< \frac{\varepsilon}{V_f(a, b)} \cdot \sum_{j=1}^{n} V_f(t_{j-1}, t_j) < \varepsilon, \end{aligned}$$

by (7.3.4) and Lemma 7.3.3. Thus f is integrable in $[a, b]$ by Lemma 7.1.5. □

*7.3.4. The Vitali-Lebesgue theorem

Theorem. *The function f is integrable on $[a,b]$ if and only if f is bounded on $[a,b]$ and the set of points of discontinuity of f in $[a,b]$ has measure zero.*

The proof below relies on the *oscillation* of a function f at a point x, $\omega_x(f)$, and the fact that f is discontinuous at x if and only if $\omega_x(f) > 0$. See §4.3.4.

Proof. First, let f be integrable on $[a,b]$ and let E be the set of points of discontinuity of f in $[a,b]$. For each $n \in \mathbb{N}$, let

$$E_n = \{x \in [a,b] \, : \, \omega_x(f) \geq 1/n\},$$

so that $E = \bigcup_n E_n$. Then E has measure zero if and only if E_n has measure zero for each $n \in \mathbb{N}$, because a countable union of sets of measure zero has measure zero by Theorem 3.3.1.

Given $\varepsilon > 0$, there is a partition P of $[a,b]$ such that $U(f,P) - L(f,P) < \varepsilon/n$, because f is integrable. Denote by P_n the subset of subintervals of P that have nonempty intersection with E_n, i.e.,

$$P_n = \{I_j = [t_{k_j}, t_{k_j+1}] \, : \, t_{k_j} \in P \text{ and } E_n \cap I_j \neq \emptyset\}.$$

If $I_j \in P_n$, then $M_j(f) - m_j(f) \geq \frac{1}{n}$ (using the notation of §7.1.3) because I_j contains a point of E_n, and therefore

$$\frac{\varepsilon}{n} > U(f,P) - L(f,P) \geq \sum_{P_n} (M_j(f) - m_j(f)) \cdot |I_j| \geq \frac{1}{n} \sum_{P_n} |I_j|,$$

where $|I_j|$ is the length of I_j, and middle inequality follows from the fact that $P_n \subset P$. Thus E_n has measure zero for each n, because ε and n were arbitrary, and therefore E has measure zero.

Now suppose that E has measure zero and that $|f(x)| < B$ for all $x \in [a,b]$. To prove that f is Riemann-integrable, we rely on Theorem 7.1.4 and show that for any $\varepsilon > 0$, there is a partition P of $[a,b]$ such that $U(f,P) - L(f,P) < \varepsilon$.

We construct such a partition from two subsets of intervals. One set, $\{J_r\}$, is comprised of intervals such that $\sup_J f(x) - \inf_J f(x)$ is uniformly small, and the other set, $\{I_s\}$, includes intervals, where $\sup_I f(x) - \inf_I f(x)$ may be larger,[7] but the sum of whose lengths is very small.

Let $\varepsilon > 0$ be given. For any $\eta > 0$, the set $E_\eta = \{x \in [a,b] \, : \, \omega_x(f) \geq \eta\}$ is closed by Proposition 4.3.4, and therefore compact.[8] Since E_η has measure zero (as a subset of E) there is a countable collection, C_η, of open intervals whose union contains E_η and the sum of whose lengths is $< \varepsilon/4B$. From the compactness of E_η, it follows that there is a finite subset $\{\mathcal{I}_1, \ldots, \mathcal{I}_{n_1}\} \subseteq C_\eta$ of disjoint intervals that covers E_η. Denoting by I_s the closure of $\mathcal{I}_s \cap [a,b]$ for $1 \leq s \leq n_1$, we have

$$\sum_{s=1}^{n_1} (\sup_{I_s} f(x) - \inf_{I_s} f(x))|I_s| \leq 2B \sum_{s=1}^{n_1} |I_s| < \varepsilon/2. \tag{7.3.5}$$

[7] But no larger than $2B$.

[8] We will choose an appropriate value for η later, but for the first part of the construction, its value does not play a role.

The set $[a,b] \setminus \bigcup_{s=1}^{n_1} I_s$ is a finite, disjoint union of subintervals of $[a,b]$, and we denote by $\{K_1,\dots,K_{n_2}\}$ the closures of these intervals. The interiors of the set of intervals $\{I_1,\dots,I_{n_1}\}\cup\{K_1,\dots,K_{n_2}\}$ are pairwise disjoint, and their union is equal to $[a,b]$, so that $\{I_1,\dots,I_{n_1},K_1,\dots,K_{n_2}\}$ forms a partition of $[a,b]$ into closed subintervals. The next step is to divide each K_l into smaller subintervals on each of which $\sup f - \inf f < \eta$.

We note that $K_l \cap E_\eta = \emptyset$ for $1 \le l \le n_2$, because E_η is contained in the *interior* of the union $\bigcup I_j$. Thus $\omega_f(x) < \eta$ in $\bigcup K_l$, and so for each $x \in \bigcup K_l$, there is an open interval $\mathcal{J}_x$ containing x such that $\sup_{\mathcal{J}_x} f(y) - \inf_{\mathcal{J}_x} f(y) < \eta$. The set of intervals $\{\mathcal{J}_x\}$ covers (the compact set) $\bigcup K_l$, so there is a finite subset $\{\mathcal{J}_1,\dots,\mathcal{J}_{n_3}\}$ of these intervals that covers $\bigcup K_l$.

Let $\{t_0,\dots,t_{n_4}\}$ be set of endpoints of all the intervals obtained by taking the intersections $\mathcal{J}_s \cap K_l$, for $1 \le s \le n_3$ and $1 \le l \le n_2$, arranged in ascending order. Then the intervals $J_r = [t_{r-1}, t_r]$, $1 \le r \le n_4$, form a partition of $\bigcup K_l$, with the property that

$$\sup_{J_r} f(y) - \inf_{J_r} f(y) < \eta$$

for each r. Choosing $\eta = \varepsilon/2(b-a)$, we have

$$\sum_{r=1}^{n_4} (\sup_{J_r} f(y) - \inf_{J_r} f(y))|J_r| < \frac{\varepsilon}{2(b-a)} \sum_{r=1}^{n_4} |J_r| \le \varepsilon/2, \tag{7.3.6}$$

since $\sum_{r=1}^{n_4} |J_r| \le b-a$. Finally, the set of subintervals $P = \{I_1,\dots,I_{n_1},J_1,\dots,J_{n_4}\}$ forms a partition of $[a,b]$, and

$$U(f,P) - L(f,P) = \sum_{j=1}^{n_1} (\sup_{I_j} f(x) - \inf_{I_j} f(x))|I_j| + \sum_{r=1}^{n_4} (\sup_{J_r} f(y) - \inf_{J_r} f(y))|J_r| < \varepsilon,$$

by (7.3.5) and (7.3.6). □

7.3.5. Sequences of integrable functions

The pointwise limit of a sequence of continuous functions on the interval $[a,b]$ need not be continuous (see Example 5.1.1(i)), but a uniform limit of continuous functions is continous. The same is true for Riemann integrable functions.

Example. For each $n \in \mathbb{N}$, define

$$f_n = \begin{cases} 1/x & : 1 \ge x \ge 1/n \\ 0 & : 1/n > x \ge 0. \end{cases}$$

For each n, f_n is bounded and is continuous everywhere in $[0,1]$, except at $x = 1/n$, so f_n is integrable in $[0,1]$. On the other hand, the sequence $\{f_n\}$ is converging pointwise to the function

$$f(x) = \begin{cases} 1/x & : 1 \ge x > 0 \\ 0 & : x = 0, \end{cases}$$

which is not bounded in $[0,1]$ and hence not Riemann integrable there.

Theorem. *If f_n is integrable on $[a,b]$ for $n \in \mathbb{N}$, and the sequence $\{f_n : n \in \mathbb{N}\}$ converges uniformly to f in $[a,b]$, then f is integrable in $[a,b]$ and*

$$\int_a^b f(x)\,dx = \lim_{n\to\infty} \int_a^b f_n(x)\,dx. \tag{7.3.7}$$

Comment. This is often written as $\displaystyle\int_a^b \lim_{n\to\infty} f_n(x)\,dx = \lim_{n\to\infty}\int_a^b f_n(x)\,dx.$

Proof. Let $\varepsilon > 0$ and let n_0 be large enough that $|f_{n_0}(x) - f(x)| < \varepsilon/3(b-a)$ for all $x \in [a,b]$. Then for any for any subinterval $I = [u,v]$ of $[a,b]$,

$$|\sup_I f - \sup_I f_{n_0}| \le \frac{\varepsilon}{3(b-a)} \quad \text{and} \quad |\inf_I f - \inf_I f_{n_0}| \le \frac{\varepsilon}{3(b-a)}. \tag{7.3.8}$$

It follows that for any partition $P = \{t_0, \dots, t_m\}$ of $[a,b]$,

$$|U(f,P) - U(f_{n_0},P)| \le \sum_{j=1}^{m} |\sup_{I_j} f(x) - \sup_{I_j} f_{n_0}(x)| \cdot |I_j| \le \frac{\varepsilon}{3},$$

where $I_j = [t_{j-1}, t_j]$, and the left-hand inequality follows from the triangle inequality. Likewise, $|L(f,P) - L(f_{n_0},P)| \le \frac{\varepsilon}{3}$ by the same argument.

Since f_{n_0} is integrable on $[a,b]$, there is a partition P_0 of $[a,b]$ such that $U(f_{n_0},P_0) - L(f_{n_0},P_0) < \varepsilon/3$, which implies that

$$\begin{aligned} U(f,P_0) - L(f,P_0) &= U(f,P_0) - U(f_{n_0},P_0) + U(f_{n_0},P_0) - L(f_{n_0},P_0) \\ &\quad + L(f_{n_0},P_0) - L(f,P_0) \\ &\le |U(f,P_0) - U(f_{n_0},P_0)| + U(f_{n_0},P_0) - L(f_{n_0},P_0) \\ &\quad + |L(f_{n_0},P_0) - L(f,P_0)| < \varepsilon, \end{aligned}$$

so f is integrable on $[a,b]$.

Next, for each $m \in \mathbb{N}$, let N_m be such that $|f(x) - f_n(x)| < \frac{1}{\mu(B-a)}$ for all $n > N_m$. Then

$$\left|\int_a^b f(x)\,dx - \int_a^b f_n(x)\,dx\right| \le \int_a^b |f(x) - f_n(x)|\,dx < \frac{1}{m},$$

for all $n > N_m$, so the sequence $\left\{\int_a^b f_n(x)\,dx\right\}$ converges to $\int_a^b f(x)\,dx$. □

Corollary. *If $\{f_n\}$ is integrable in $[a,b]$ for each n, and the series $\sum_{j=1}^{\infty} f_n(x)$ converges uniformly in $[a,b]$, then $\sum_{j=1}^{\infty} f_n(x)$ is integrable in $[a,b]$ and*

$$\int_a^b \sum_{j=1}^{\infty} f_n(x)\,dx = \sum_{j=1}^{\infty}\int_a^b f_n(x)\,dx.$$

Proof. Apply Theorem 7.3.5 to the sequence of partial sums $\sum_{j=1}^{n} f_j(x)$. □

Exercises

ex7.3.1. A function $\varphi(x)$ *piecewise constant* on the interval $[a, b]$ if there is a partition $\{t_0, \dots, t_n\}$ of $[a, b]$, such that $\varphi(x) = y_j$ in (t_{j-1}, t_j) for $1 \le j \le n$. The values of φ at the endpoints of the partition may be set arbitrarily.

(i) Show that if φ is a piecewise constant function on $[a, b]$, as above, then φ is Riemann integrable on $[a, b]$ and

$$\int_a^b \varphi(x)\, dx = \sum_{j=1}^{n} y_j(t_j - t_{j-1}).$$

(ii) Show that $f(x)$ is Riemann integrable on $[a, b]$, if and only if there is a sequence of piecewise constant functions $\{\varphi_n(x)\}$, such that $\varphi_n \to f$ uniformly on $[a, b]$ and hence

$$\int_a^b f(x)\, dx = \lim_{n\to\infty} \int_a^b \varphi_n(x)\, dx.$$

ex7.3.2. Show that f is of bounded variation on $[a, b]$ if and only if there exist monotone nondecreasing functions g and h on $[a, b]$, such that $f = g - h$.

Hint: For $a < x < b$, show that $V_f(a, x)$ and $f(x) + V_f(a, x)$ are both monotone nondecreasing on $[a, b]$.

ex7.3.3. Without quoting the Vitali-Lebesgue theorem, show that if f is bounded on $[a, b]$ and has at most *countably* many points of discontinuity in the interval, then f is integrable in $[a, b]$.

Hint: The *proof* of the theorem may offer some ideas.

ex7.3.4. Let $\{r_k : k \in \mathbb{N}\}$ be an enumeration of $\mathbb{Q} \cap [0, 1]$, and let $Q_n = \{r_1, \dots, r_n\}$. For each $n \in \mathbb{N}$, let

$$f_n = \begin{cases} 1 & : x \in Q_n \\ 0 & : \text{otherwise}, \end{cases}$$

Show that

(i) f_n is integrable on $[0, 1]$.

(ii) $\int_0^1 f_n\, dx = 0$ for all n.

(iii) f_n converges pointwise to

$$f = \begin{cases} 1 & : x \in \mathbb{Q} \cap [0, 1] \\ 0 & : \text{otherwise}, \end{cases}$$

(iv) f is (bounded but) not integrable on $[0, 1]$.

ex7.3.5. Use the Vitali-Lebesgue theorem to prove Theorem 7.2.1.

7.4. Improper integrals

The Riemann integral is defined for "almost continuous", bounded functions over a compact interval. Under certain conditions, this definition can be extended to cases where the functions or the intervals (or both) are unbounded.

7.4.1. Unbounded functions

If $b > a > 0$, then the integral $\int_a^b x^{-1/2}\,dx$ is defined and

$$\int_a^b x^{-1/2}\,dx = 2\left(b^{1/2} - a^{1/2}\right), \tag{7.4.1}$$

by the fundamental theorem of calculus. For any $b > 0$, the integrand, $x^{-1/2}$ is unbounded in every interval $[0, b]$, so $f(x) = x^{-1/2}$ is not Riemann integrable in the interval $[0, b]$. On the other hand, the right-hand side of (7.4.1) is defined for $a = 0$, so it is reasonable to say that $\int_0^b x^{-1/2}\,dx = 2b^{1/2}$.

Justifying this intuition requires a new definition.

Definition. If $\lim_{x\to a^+} f(x) = \pm\infty$ and for every $a < c < b$, the function $f(x)$ is integrable in the interval $[c, b]$, then the *improper integral* of f on $[a, b]$ is defined

$$\int_a^b f(x)\,dx = \lim_{c\to a^+} \int_c^b f(x)\,dx,$$

if the limit on the right exists. Likewise, If $\lim_{x\to b^-} f(x) = \pm\infty$ and for every $a < c < b$, the function $f(x)$ is integrable in the interval $[c, b]$, then

$$\int_a^b f(x)\,dx = \lim_{c\to b^-} \int_a^c f(x)\,dx,$$

if the limit on the right exists. Finally, if $\lim_{x\to\eta} f(x) = \pm\infty$, for some $\eta \in [a, b]$ and f is integrable in the intervals $[a, c_1]$ and $[c_2, b]$ for any $c_1 \in (a, \eta)$ and $c_2 \in (\eta, b)$, then the improper integral of f on $[a, b]$ is

$$\int_a^b f(x)\,dx = \int_a^\eta f(x)\,dx + \int_\eta^b f(x)\,dx,$$

provided that both improper integrals on the right are defined.

7.4.2. Unbounded intervals

For any $b > 1$, the function x^{-2} is integrable and

$$\int_1^b x^{-2}\,dx = 1 - \frac{1}{b}. \tag{7.4.2}$$

Moreover, $\lim_{b\to\infty} 1 - \frac{1}{b} = 1$. Once again, it is natural to say that

$$\int_1^\infty x^{-2}\,dx = 1,$$

but this also requires a formal definition.

Definition. If f is Riemann integrable on $[a, b]$ for every $b > a$, then the *improper integral* of f on $[a, \infty)$ is

$$\int_a^\infty f(x)\,dx = \lim_{b\to\infty} \int_a^b f(x)\,dx,$$

provided that the limit on the right exists.

If f is Riemann integrable on $[a, b]$ for every $a < b$, then the improper integral of f on $(-\infty, b]$ is

$$\int_{-\infty}^b f(x)\,dx = \lim_{a\to-\infty} \int_a^b f(x)\,dx,$$

provided that the limit on the right exists.

If f is Riemann integrable on $[a, b]$ for every $-\infty < a < b < \infty$, then the improper integral of f on $(-\infty, \infty)$ is

$$\int_{-\infty}^\infty f(x)\,dx = \int_{-\infty}^0 f(x)\,dx + \int_0^\infty f(x)\,dx,$$

provided that both improper integrals on the right exist.

Exercises

ex7.4.1. Show that the improper integral $\int_0^1 x^{-\alpha}\,dx$ exists for any $\alpha \in (0, 1)$.

ex7.4.2. Show that the improper integral $\int_1^\infty x^{-\alpha}\,dx$ exists for any $\alpha \in (1, \infty)$.

7.5. The Riemann-Stieltjes Integral

Stieltjes generalized the Riemann integral by replacing the lengths of subintervals, $\Delta t_j = t_j - t_{j-1}$, appearing in Riemann or Darboux sums by more general factors of the form $\Delta g_j = g(t_j) - g(t_{j-1})$, for a bounded, nondecreasing function g (called an *integrator*) on the interval $[a, b]$. Thus, Stieltjes's formulation gives rise to a family of integrals, which includes the Riemann integral when the integrator is $g(t) = t$.

In §7.1.5 we saw that Darboux's definition of the integral (via upper and lower integrals) and Riemann's (via Riemann sums) are equivalent. This equivalence continues to hold in this more general setting in many, but *not all* cases. The Riemann-Stieltjes integral, defined in terms of upper and lower "Darboux-Stieltjes" sums, is more generally applicable than the Riemann-Stieltjes integral, defined in terms of Riemann-Stieltjes sums. Specifically, there are cases where the integral defined using upper and lower sums exists, but the one defined using limits of Riemann-Stieltjes sums does not. On the other hand, when both integrals are defined, they both give the same result.

In this text, to keep things simple, we define the Riemann-Stieltjes integral using upper and lower sums and integrals. Many of the properties of the Riemann integral generalize to the Riemann-Stieltjes integral and their proofs are analogous, so we leave many of these proofs as exercises.

7.5.1. Upper and lower sums and integrals

Definition. Suppose that f and g are both defined on $[a, b]$, with f bounded and g nondecreasing, let $P = \{t_0, \ldots, t_n\}$ be a partition of $[a, b]$ and, as indicated above, let $\Delta g_j = g(t_j) - g(t_{j-1})$. We call the sums

$$U(f, g, P) = \sum_{j=1}^{n} M_j \Delta g_j \quad \text{and} \quad L(f, g, P) = \sum_{j=1}^{n} m_j \Delta g_j$$

the *upper* and *lower Darboux-Stieltjes sums* for f with respect to g and P, where as before,

$$M_j = \sup\{f(t) : t_{j-1} \leq t \leq t_j\} \quad \text{and} \quad m_j = \inf\{f(t) : t_{j-1} \leq t \leq t_j\}.$$

Upper and lower Darboux-Stieltjes sums have the same boundedness and monotonicity properties as standard upper and lower Darboux sums that we described in Theorem 7.1.3 and its corollary. We summarize these properties here in a single statement.

Theorem. *Suppose that f is bounded on $[a, b]$ and g is nondecreasing there. Then for any partition P of $[a, b]$,*

$$(\sup_{[a,b]} f(x))(g(b) - g(a)) \geq U(f, g, P) \geq L(f, g, P) \geq (\inf_{[a,b]} f(x))(g(b) - g(a)),$$

and for any refinement Q of P,

$$U(f, g, P) \geq U(f, g, Q) \geq L(f, g, Q) \geq L(f, g, P).$$

Proof. Exercise. □

7.5.2. The Riemann-Stieltjes Integral (á la Darboux)

We define the Riemann-Stiltjes integral analogously to Darboux's definition of the Riemann integral (in §7.1.4).

Definition. For a bounded function f and a nondecreasing function g, both defined on the interval $[a, b]$, the *upper Riemann-Stieltjes integral* of f, with respect to g on $[a, b]$ is

$$\overline{\int_a^b} f \, dg = \inf U(f, g, P),$$

and the *lower Riemann-Stieltjes integral* of f, with respect to g on $[a, b]$ is

$$\underline{\int_a^b} f \, dg = \sup L(f, g, P),$$

where the infimum and supremum above are taken over all partitions P of $[a, b]$.

Both the upper and lower integrals exist because all upper sums are bounded below and all lower sums are bounded above, as follows from Theorem 7.5.1.

The function f is integrable with respect to g on $[a,b]$ if and only if the upper and lower Riemann-Stieltjes integrals of f with respect to g are equal, in which case the Riemann-Stieltjes integral is equal to this common value:

$$\int_a^b f\,dg = \overline{\int_a^b} f\,dg = \underline{\int_a^b} f\,dg.$$

Going forward, to save ink, we shorten the phrase *"Riemann-Stieltjes integrable with respect to g"* to *RS(g) integrable*, and we assume in this context that the function g is bounded and nondecreasing.

Lemma. *If f is RS(g) integrable on $[a,b]$, then for any $\varepsilon > 0$ there is a partition P of $[a,b]$ such that $U(f,g,P) - \int_a^b f\,dg < \varepsilon$ and $\int_a^b f\,dg - L(f,g,P) < \varepsilon$.*

Proof. This follows from the definition of the Riemann-Stieltjes integral and Theorem 7.5.1. We leave the details as an exercise. □

Theorem. *The bounded function f is RS(g) integrable on $[a,b]$ if and only if for every $\varepsilon > 0$, there is a partition P of $[a,b]$ such that $U(f,g,P) - L(f,g,P) < \varepsilon$.*

Proof. Exercise. □

We use this theorem to characterize functions that are RS(g) integrable on an interval $[a,b]$, as in the following corollary.

Corollary. *If f is continuous in $[a,b]$, then f is RS(g) integrable on $[a,b]$.*

Proof. Let f be continuous on $[a,b]$, and g be bounded and nondecreasing there. Given $\varepsilon > 0$ there is a $\delta > 0$ such that $|f(x) - f(y)| < \varepsilon/(g(b) - g(a))$ for all $x, y \in [a,b]$ with $|x - y| < \delta$, because f is uniformly continuous in $[a,b]$.

If $P = \{t_0, \ldots, t_n\}$ is a partition of $[a,b]$ with $\operatorname{mesh}(P) < \delta$, then

$$\sup_{[t_{j-1},t_j]} f - \inf_{[t_{j-1},t_j]} f < \frac{\varepsilon}{g(b) - g(a))}$$

for $1 \le j \le n$, and therefore

$$U(f,g,P) - L(f,g,P) < \frac{\varepsilon}{g(b) - g(a)} \sum_{j=1}^{n} g(t_j) - g(t_{j-1}) = \varepsilon. \qquad \square$$

If f is not continuous on $[a,b]$, then the RS(g) integrability of f depends on the behavior of the integrator g at points of discontinuity of f. In particular, if f and g 'misbehave' in the same way at some point, then f is not RS(g) integrable.

Proposition. *If neither the function f nor the integrator g are right-continuous at some point $\xi \in [a,b)$ (or neither are left-continuous at $\xi \in (a,b]$), then f is not RS(g) integrable in $[a,b]$.*

Proof. Suppose that neither function is right-continuous at $\xi \in [a, b)$. The proof in the *not-left-continuous* case is analogous.

Since g is nondecreasing in $[a, b]$, it follows that there is a $d_1 > 0$ such that $g(t) - g(s) \geq d_1$ for all $s \leq \xi < t \leq b$. Similarly, there is a $d_2 > 0$ such that for any $\delta > 0$, there is a $t \in (\xi, \xi + \delta)$ such that $|f(t) - f(\xi)| \geq d_2$.

Given any partition $P = \{t_0, \dots, t_n\}$ of $[a, b]$, there is some $1 \leq j_0 \leq n$, such that $t_{j_0-1} \leq \xi < t_{j_0}$ (since $\xi < b$). With M_i and m_i denoting the sup and inf of f in the ith subinterval of P, as usual, it follows that

$$\begin{aligned} U(f, g, P) - L(f, g, P) &= \sum_{i=1}^{n} (M_i - m_i)(g(t_i) - g(t_{i-1})) \\ &\geq (M_{j_0} - m_{j_0})(g(t_{j_0}) - g(t_{j_0-1})) \geq d_1 d_2. \end{aligned}$$

Hence f is not RS(g) integrable on $[a, b]$, by Theorem 7.5.2. □

7.5.3. The case of finitely many points of discontinuity

If the integrator g is continuous at a point ξ of discontinuity of f, then we can find intervals $[s, t]$ containing ξ on which $\Delta g = g(t) - g(s)$ can be made arbitrarily small. We take advantage of this to prove the next theorem.

Theorem. *Suppose that f is bounded on $[a, b]$ with finitely many points of discontinuity. If the integrator g is continuous at each point of discontinuity of f, then f is RS(g) integrable in $[a, b]$.*

Proof. Suppose that $|f| < B$ in $[a, b]$ and that $a < x_1 < x_2 < \dots x_n < b$ are the points of discontinuity of f there.[9]

Given $\varepsilon > 0$, there are intervals $I_j = [s_j, t_j] \subset [a, b]$ such that $x_j \in (s_j, t_j)$ and

$$g(t_j) - g(s_j) < \frac{\varepsilon}{4Bn},$$

for $1 \leq j \leq n$, since $\omega_g(x_j) = 0$ for each j. Furthermore, shrinking the intervals if necessary, we may also assume that $I_j \cap I_k = \varnothing$ if $j \neq k$.

The closure of $[a, b] \setminus \bigcup I_j$ is a finite collection of closed intervals, on all of which f is continuous, so f is uniformly continuous on their union, K. Hence, there is a $\delta > 0$ such that if $|x - y| < \delta$ then $|f(x) - f(y)| < \varepsilon/2(g(b) - g(a))$, for all $x, y \in K$. We partition each of the intervals comprising K into subintervals of length $< \delta$ to obtain a finite collection of closed intervals $\{J_l\}_{i=1}^{l}$.

The collection $P = \{I_j\} \cup \{J_i\}$ is a partition of $[a, b]$ and

$$U(f, g, P) - L(f, g, P) = \sum_{j=1}^{n} (M_j - m_j)\Delta g_j + \sum_{i=1}^{l} (M_i - m_i)\Delta g_i = S_1 + S_2.$$

The factors $M_j - m_j$ in the terms of the sum S_1 are all bounded by $2B$ and the Δg_j factors of these terms are all bounded by $\varepsilon/4Bn$. Since there are n terms in S_1, it follows that $S_1 < n \cdot 2B \cdot (\varepsilon/4Bn) = \varepsilon/2$. For the sum S_2, we have $M_i - m_i < \varepsilon/2(g(b) - g(a))$ by

[9]The case where f has a point of discontinuity at a or b (or both), is left as an exercise.

construction, and $\sum_i \Delta g_i \leq g(b) - g(a)$, by the monotonicity of g. Hence $S_2 < \varepsilon/2$, so $U(f, g, P) - L(f, g, P) = S_1 + S_2 < \varepsilon$, and the claim follows from Theorem 7.5.2. □

7.5.4. The case of monotone integrands

Theorem. *If f is monotone and bounded on $[a, b]$ and the integrator g is continuous at all points of discontinuity of f in $[a, b]$, then f is RS(g) integrable on $[a, b]$.*

Proof. To keep things simple, we will assume that f is nondecreasing. The proof in the case that f is nonincreasing is essentially the same. We recall that a monotone function has at most countably many points of discontinuity. If f has finitely many points of discontinuity, then the claim follows from Theorem 7.5.3, so we assume that f has a countably infinite set of points of discontinuity, and denote this set by $E = \{\xi_j : j \in \mathbb{N}\}$. We will again assume that f is continuous at a and b, to streamline the exposition.

It also follows from the monotonicity of f, that both the one-sided limits $\lim_{x \to \xi_j^-} f(x)$ and $\lim_{x \to \xi_j^+} f(x)$ exist at each point in E (see Lemma 4.1.5), and the discontinuities of f at these points are all "jumps". We denote by δ_j the (positive) jump at ξ_j, i.e., $\delta_j = \lim_{x \to \xi_j^+} f(x) - \lim_{x \to \xi_j^-} f(x) > 0$, and observe that

$$\sum_j \delta_j \leq f(b) - f(a).$$

Given $\varepsilon > 0$, the convergence of the sum above implies that there is an $N \in \mathbb{N}$ such that $\sum_{j>N} \delta_j < \varepsilon/4(g(b) - g(a))$. In particular, $\delta_j < \varepsilon/4(g(b) - g(a))$ for all $j > N$.

Similar to the proof of Theorem 7.5.3, for each $j \in \{1, \ldots, N\}$, there is an interval $I_j = [s_j, t_j] \subset [a, b]$ such that $\xi_j \in (s_j, t_j)$ and

$$g(t_j) - g(s_j) < \frac{\varepsilon}{2(f(b) - f(a))},$$

since g is continuous at ξ_j, where the intervals may be made smaller if necessary, to assure that $I_j \cap I_k = \varnothing$ if $k \neq j$.

The closure K of $[a, b] \setminus \bigcup_j I_j$ is a finite union of disjoint closed intervals, and hence compact. If $x \in K$, then either f is continuous at x or $x = \xi_j$, with $j > N$, in which case $\delta_j < \varepsilon/4(g(b) - g(a))$. It follows, from this and the monotonicity of f, that for every $x \in K$, there is an open interval (u_x, v_x) containing x, with

$$f(v_x) - f(u_x) < \frac{\varepsilon}{2(g(b) - g(a))}.$$

The union $\bigcup_x \mathcal{J}_x$ covers K and hence there is a finite subset $\{\mathcal{J}_{x_1}, \ldots, \mathcal{J}_{x_M}\}$ of these intervals that covers K.

Denoting the closure of $\mathcal{J}_{x_i} \cap K$ by $J_i = [u_i, v_i]$, for $1 \le i \le M$, it follows that $P = \{I_1, \dots, I_N\} \cup \{J_1, \dots, J_M\}$ is a partition of $[a, b]$ into closed subintervals, and

$$\begin{aligned} U(f, g, P) - L(f, g, P) &= \sum_{j=1}^{N} (f(t_j) - f(s_j))(g(t_j) - g(s_j)) \\ &\quad + \sum_{i=1}^{M} (f(v_i) - f(u_i))(g(v_i) - g(u_i)) \\ &< \frac{\varepsilon}{2(f(b) - f(a))} \sum_{j=1}^{N} f(t_j) - f(s_j) \\ &\quad + \frac{\varepsilon}{2(g(b) - g(a))} \sum_{i=1}^{M} g(v_i) - g(u_i) \le \frac{\varepsilon}{2} + \frac{\varepsilon}{2} = \varepsilon, \end{aligned}$$

because $\sum_{j=1}^{N} f(t_j) - f(s_j) \le f(b) - f(a)$ and $\sum_{i=1}^{M} g(v_i) - g(u_i) \le g(b) - g(a)$. □

7.5.5. Integration by parts

If both f and g are nondecreasing functions on $[a, b]$, then $\int_a^b f\, dg$ and $\int_a^b g\, df$ are both defined, by Theorem 7.5.4. The relation between the two integrals is referred to as "integration by parts."[10]

Theorem. *If f and g are nondecreasing on $[a, b]$, then*

$$\int_a^b f\, dg = f(b)g(b) - f(a)g(a) - \int_a^b g\, df.$$

Proof. If $P = \{t_0, \dots, t_n\}$ is any partition of $[a, b]$,

$$\begin{aligned} U(f, g, P) + L(g, f, P) &= \sum_{i=1}^{n} [f(t_i)(g(t_i) - g(t_{i-1}))] + [g(t_{i-1})(f(t_i) - f(t_{i-1}))] \\ &= \sum_{i=1}^{n} f(t_i)(g(t_i) - f(t_{i-1})g(t_{i-1}) \\ &= f(t_n)g(t_n) - f(t_0)g(t_0) = f(b)g(b) - f(a)g(a), \end{aligned}$$

since both functions are nondecreasing.

Now for any $\varepsilon > 0$, there is a partition P_ε of $[a, b]$ such that both

$$0 < U(f, g, P) - \int_a^b f\, dg < \varepsilon/2 \quad \text{and} \quad 0 < \int_a^b g\, df - L(g, f, P) < \varepsilon/2,$$

[10]Compare to exercises **ex7.2.3** and **ex7.5.7**.

as follows from Lemma 7.5.2 and Theorem 7.5.1. It follows from this and the identity above that

$$\left| \int_a^b f\, dg + \int_a^b g\, df - (f(b)g(b) - f(a)g(a)) \right| < \varepsilon,$$

so $\int_a^b f\, dg + \int_a^b g\, df = f(b)g(b) - f(a)g(a)$. □

Exercises

ex7.5.1. Complete the proof of Lemma 7.5.2.

ex7.5.2. Adapt the proof of Theorem 7.1.3 (and its corollary) to prove Theorem 7.5.1.

ex7.5.3. Adapt the proof of Proposition 7.1.4 to show that If f is bounded on $[a, b]$ and g in nondecreasing there, then

$$\overline{\int_a^b} f\, dg \geq \underline{\int_a^b} f\, dg.$$

ex7.5.4. Adapt the proof of Theorem 7.1.4 to prove Theorem 7.5.2.

ex7.5.5. Suppose that f is continuous on $[a, b]$, except at the point $\xi \in (a, b)$ where f is right-continuous, but not left-continuous. I.e.,

$$\lim_{x \to \xi+} f(x) = f(\xi) \neq \lim_{x \to \xi-} f(x).$$

Show that if the integrator g is left continuous at ξ, then f is RS(g) integrable on $[a, b]$.

ex7.5.6. Complete the proof of Theorem 7.5.3 by considering the case that f has a point of discontinuity at one or both of the endpoints of $[a, b]$.

ex7.5.7. Suppose that $g' \geq 0$ in $[a, b]$ (so g is nondecreasing in $[a, b]$) and Riemann integrable there. Show that if f is Riemann integrable in $[a, b]$, then f is RS(g) integrable in $[a, b]$ and

$$\int_a^b f\, dg = \int_a^b f \cdot g'\, dx.$$

Chapter 8

The Lebesgue Integral

If f is Riemann integrable on $[0,1]$ and $P = \{t_0, \ldots, t_n\}$ is a partition of $[0,1]$ with "tags" $\{x_1, \ldots, x_n\}$, then

$$\sum_{j=1}^{n} f(x_j)(t_j - t_{j-1})$$

can be thought of as a weighted average of the values $\{f(x_1), \ldots, f(x_n)\}$ on $[0,1]$, with weights $(t_j - t_{j-1})$. In this context, the Riemann integral

$$\int_0^1 f\,dx$$

may be interpreted as the average value of f on $[0,1]$, as the limit of such sums.[1]

If f is continuous on $[0,1]$, then this approach to defining an average value makes good sense. In this case, once the mesh of the partition is small enough, the function f will be almost constant on each part of the partition, so the choice of tags is like a "representative sample" from $[0,1]$. And continuing the analogy, as the sample size grows (and the mesh approaches zero), the "sample averages" approach the "population average" of f on $[0,1]$.

On the other hand if f is not continuous, then a value $f(x_j)$, for some $x_j \in [t_{j-1}, t_j]$, may not be 'representative' of the function in $[t_{j-1}, t_j]$.

Example. Let $\chi_{\mathbb{Q}}$ be the characteristic function of $\mathbb{Q}$, i.e.,

$$\chi_{\mathbb{Q}}(x) = \begin{cases} 1 & : x \text{ rational} \\ 0 & : x \text{ irrational.} \end{cases}$$

Then $\chi_{\mathbb{Q}}$ is not continuous at any point, because both the rationals and the irrationals are dense in $\mathbb{R}$, so $\chi_{\mathbb{Q}}$ is not Riemann integrable. On the other hand, $\mathbb{Q}$ has measure

[1] More generally, if f is Riemann integrable on $[a,b]$, then $\frac{1}{b-a}\int_a^b f\,dx$ may be interpreted as the average value of f on $[a,b]$.

zero in $[0,1]$, so the "weight" of the set of points where $\chi_{\mathbb{Q}} = 1$ has measure zero. It seems reasonable to conclude, therefore, that the average value of $\chi_{\mathbb{Q}}$ in $[0,1]$ is

$$1 \cdot 0 + 0 \cdot w_0 = 0,$$

where w_0 is the "weight" of the set of points in $[0,1]$ (the irrational points) where $\chi_{\mathbb{Q}} = 0$.

In his doctoral thesis, Henri Lebesgue introduced a new approach to integration that generalizes the idea illustrated in this example.

Suppose that f is bounded on a set $X \subset \mathbb{R}$, with $\min_X f = m$ and $\max_X f = M$. Let $m = s_0 < s_1 \cdots < s_n = M$, be a partition of the *range* of f, and let $y_j \in (s_{j-1}, s_j)$ for $1 \le j \le n$. Loosely speaking, the Lebesgue integral of f on X is defined in terms of sums of the form

$$\sum_{j=1}^{n} y_j \cdot \mu(E_j),$$

where $E_j = f^{-1}((s_{j-1}, s_j)) \cap X$, and $\mu(E_j)$ is the *Lebesgue measure* of E_j.[2] The Lebesgue measure of a set is a generalization of length to subsets of $\mathbb{R}$ that are not intervals. Such a generalization is necessary in this context, because while the sets E_j above are pairwise disjoint, they are generally *not* intervals (unless f is monotone on X).

The sums that appear in the constructions of the Lebesgue integral and Riemann integral are very similar, in that the terms in each are products of the form

$$(\text{value of function}) \cdot (\text{size of set}).$$

But starting with a "decomposition" of the range of the function, rather than the domain, Lebesgue's integral is a versatile generalization of Riemann's.

It is a generalization in that if f is Riemann integrable on an interval $I = [a,b]$, then f is also Lebesgue integrable there, and the two integrals of f on I agree (Theorem 8.4.2). On the other hand, Lebesgue's approach has some notable advantages. The Lebesgue integral is not limited to compact intervals, admits a larger class of integrable functions, and is more flexible with sequences of functions than the Riemann integral.

The first step is to define Lebesgue measure and Lebesgue measurable sets.

8.1. Lebesgue measure on the real line

The length $\ell(I)$ of a bounded interval I is the distance between its endpoints, and is the same whether the interval is open, half open or closed. If I is unbounded, we set $\ell(I) = \infty$. This notion of length can be extended to a disjoint unions of intervals, $E = \bigcup_n I_n$ (with $I_n \cap I_m = \emptyset$ for $n \ne m$), by setting

$$\ell(E) = \sum_n \ell(I_n),$$

where $\ell(E) = \infty$ if the sum diverges, including the case where one of the constituent intervals is unbounded. Lebesgue measure generalizes this notion of length to sets that are not disjoint unions of intervals.

[2] E.g., $\mu([0,1] \setminus \mathbb{Q}) = w_0$.

8.1.1. Outer measure

To define Lebesgue measure on $\mathbb{R}$, we begin with the outer measure. Both measure and outer measure (and inner measure) are *set functions* on $\mathbb{R}$. A set function η on $\mathbb{R}$ is a function from $\mathscr{P}(\mathbb{R})$ to the *extended* (nonnegative) real numbers,[3] where ∞ is not just shorthand for "unbounded", but represents a possible value of $\eta(A)$ for $A \subseteq \mathbb{R}$.

Definition. The *outer measure*, $\mu^*(E)$, for $E \subset \mathbb{R}$ is defined

$$\mu^*(E) = \inf\left(\sum_n \ell(I_n)\right),$$

where the infimum is taken over all countable collections $\{I_n\}$ of bounded intervals that cover E. If all such sums diverge, then $\mu^*(E) = \infty$.

Every subset of $\mathbb{R}$ may be covered by a countable union of bounded intervals, and since the sums appearing in the definition of outer measure are all nonnegative (or infinite), the outer measure is defined on all subsets of $\mathbb{R}$.

Comments.

(i) If $E \subset \mathbb{R}$ is bounded, then $\mu^*(E) < \infty$.

(ii) The definition of outer measure of $E \subset \mathbb{R}$ does not depend on the topological nature of the bounded intervals covering E — they may be open, half open or closed, or a mix of these. The infimum of the sums of lengths is the same in all cases. See **ex8.1.1**. This allows us to choose the types of intervals in a given cover that best suit our purposes.

(iii) Any bounded interval I may be divided into finitely many (shorter) intervals of bounded length. Hence, in the definition of outer measure, we may choose the intervals that cover a set E to have uniformly short lengths.

(iv) In §3.3.1 we defined sets of measure zero in this way. Technically, what we defined in Chapter 3 were sets of *outer* measure zero, but we will see in §8.1.3 that measure and outer measure are the same for sets with outer measure zero.

(v) The outer measure is *monotone*: if $F \subseteq E$ then $\mu^*(F) \leq \mu^*(E)$.

Proposition. *If $I \subseteq \mathbb{R}$ is an interval, then $\mu^*(I) = \ell(I)$.*

Proof. Suppose first that $I = [a, b]$, is closed and bounded. Then $\mu^*(I) \leq \ell(I)$, since $I \subseteq I$. By the compactness of I, every open cover $\{J_m\}$ of I has a finite subcover $\{J_k\}_{k=1}^n$, so $\mu^*(I) = \inf\left(\sum_{k=1}^n \ell(J_k)\right)$, where the infimum may be taken over all *finite* (open) covers of I. Without loss of generality, we may also assume that $J_k \cap I \neq \varnothing$, for each J_k in such a cover.[4] If $\{J_k = (a_k, b_k) : 1 \leq k \leq n\}$ is a cover of $[a, b]$, satisfying this assumption, then there can be no gaps between its constituent intervals, so

$$[a, b] \subset \bigcup_{k=1}^n (a_k, b_k) = (\alpha, \beta),$$

[3] See §1.5.6.

[4] Why?

where $\alpha = \min\{a_k : 1 \le k \le n\} < a$ and $\beta = \max\{b_k : 1 \le k \le n\} > b$. Hence $\sum_{k=1}^{n} \ell(J_k) = \beta - \alpha > b - a = \ell(I)$. Thus, $\mu^*(I) \ge \ell(I)$, so the claim is true for closed and bounded intervals.

If I is any bounded interval with endpoints $a < b$, then

$$[a + \varepsilon/2, b - \varepsilon/2] \subset I \subset [a - \varepsilon/2, b + \varepsilon/2],$$

for any $0 < \varepsilon < \ell(I)$, so $\ell(I) - \varepsilon < \mu^*(I) < \ell(I) + \varepsilon$. Thus, $\ell(I) = \mu^*(I)$ for all bounded intervals.

Finally, if I is not bounded, then I contains arbitrarily long bounded intervals, so $\mu^*(I) \ge B$ for all $B > 0$, whence $\mu^*(I) = \infty = \ell(I)$. □

Bounded sets have finite outer measure, but sets with finite outer measure need not be bounded, as illustrated by the set $G = \bigcup_{n=1}^{\infty} (n, n + 2^{-n})$. On the other hand, sets with finite outer measure are 'almost bounded'.

Theorem. *If $\mu^*(E) < \infty$, then for any $\varepsilon > 0$, there is an $\alpha > 0$ such that*

$$\mu^*(E \cap [-\alpha, \alpha]) \ge \mu^*(E) - \varepsilon.$$

Proof. Given $\varepsilon > 0$, it follows from the definition of outer measure that there exists a collection $\{I_n\}$ of bounded intervals such that $E \subset \bigcup_n I_n$ and

$$\mu^*(E) \le \sum_n \ell(I_n) < \mu^*(E) + \varepsilon/2.$$

If $\{I_n\}$ is a finite set, then E is bounded and there is nothing more to prove. Otherwise, the series of lengths above converges, so there is an $N \in \mathbb{N}$ such that $\sum_{n=1}^{N} \ell(I_n) > \mu^*(E) - \varepsilon/2$, from which it follows that

$$\sum_{n>N} \ell(I_n) < \varepsilon. \tag{8.1.1}$$

Let $\alpha = \sup\left(|x| : x \in \bigcup_{n=1}^{N} I_n\right)$. If $x \in E$ and $|x| > \alpha$, then $x \in \bigcup_{n>N} I_n$, so that $E \cap [-\alpha, \alpha]^c$ is covered by $\{I_n\}_{n>N}$. If $\{J_m\}$ is a set of bounded intervals whose union covers $E \cap [-\alpha, \alpha]$, then the set $\{I_n\}_{n>N} \cup \{J_m\}$ covers all of E, so $\sum_m \ell(J_m) + \sum_{n>N} \ell(I_n) \ge \mu^*(E)$. It follows from this and (8.1.1) that

$$\sum_m \ell(J_m) > \mu^*(E) - \varepsilon. \tag{8.1.2}$$

Hence, $\mu^*(E \cap [-\alpha, \alpha] \ge \mu^*(E) - \varepsilon$, because the lower bound (8.1.2) holds for any collection of intervals that covers $E \cap [-\alpha, \alpha]$. □

8.1.2. The countable subadditivity of the outer measure

Length is a *countably additive* function on the set of intervals and their disjoint unions.[5] This means that if $\{I_n\}$ is a countable (or finite) set of mutually disjoint intervals and $J = \bigcup_n I_n$, then, by definition $\ell(J) = \sum_n \ell(I_n)$. The outer measure does not have this property. There exist disjoint, bounded sets $E_1, \dots E_n \subset \mathbb{R}$ such that $\mu^*(\bigcup_{j=1}^{n} E_j) < \sum_{j=1}^{n} \mu^*(E_j)$ (we give an example in §8.1.5). On the other hand, outer measure is countably *subadditive*.

[5] Countable additivity is also called σ additivity.

Theorem. *If $E_n \subseteq \mathbb{R}$ for $n \geq 1$, and $E_j \cap E_k = \emptyset$ for $j \neq k$, then*

$$\mu^*\left(\bigcup_n E_n\right) \leq \sum_n \mu^*(E_n).$$

Proof. If $\sum_n \mu^*(E_n) = \infty$, then the claim is true (and not interesting), so we may assume that the sum converges, and hence $\mu^*(E_n) < \infty$ for all n.

For any $\varepsilon > 0$, there is, for each $n \geq 1$, a cover $\{I_{n,k}\}_{n=1}^{\infty}$ of E_n such that

$$\mu^*(E_n) \leq \sum_{k=1}^{\infty} \ell(I_{n,k}) < \mu^*(E_n) + \frac{\varepsilon}{2^n}.$$

The intervals $\{I_{n,k} : n, k \geq 1\}$ cover $\bigcup E_n$, and therefore

$$\mu^*\left(\bigcup_n E_n\right) \leq \sum_{n=1}^{\infty}\sum_{k=1}^{\infty} \ell(I_{n,k}) < \sum_{n=1}^{\infty}\left(\mu^*(E_n) + \frac{\varepsilon}{2^n}\right) = \varepsilon + \sum_{n=1}^{\infty} \mu^*(E_n),$$

from which the claim follows. □

8.1.3. Lebesgue measure

The outer measure of a set can be viewed as the least upper bound on its 'size'. The *inner measure* of a set provides the greatest lower bound for this size. If the two are equal, then the set is measurable.

Definition. The *inner measure*, $\mu_*(E)$, of a set $E \subseteq \mathbb{R}$ is defined by

$$\mu_*(E) = \sup(\mu^*(K)),$$

where the supremum is taken over all compact subsets $K \subseteq E$. It follows from the monotonicity of outer measure that $\mu_*(E) \leq \mu^*(E)$ for all $E \subset \mathbb{R}$.

If $\mu^*(E) < \infty$ and $\mu^*(E) = \mu_*(E)$, then E is *measurable* in the sense of Lebesgue, in which case, the (Lebesgue) measure of E is $\mu(E) = \mu^*(E)$.

If $\mu^*(E) = \infty$, then E is measurable in the sense of Lebesgue if and only if $E \cap [-a, a]$ is measurable for all $a > 0$, in which case, we set $\mu(E) = \infty$.

Comments.

(i) From this point forward, when we say that a set is measurable, we mean "measurable in the sense of Lebesgue".

(ii) If K is compact, then $\mu_*(K) = \mu^*(K)$ by the monotonicity of the outer measure, so all compact subsets of $\mathbb{R}$ are measurable.

(iii) Lebesgue measure is countably subadditive on disjoint unions of measurable sets, because the outer measure is.

(iv) If $E \subset \mathbb{R}$ has outer measure zero, then E is measurable and has measure zero.

(v) If a statement is true for all $x \in E$, with the possible exception of a subset of measure zero, we say that the statement is true for *almost all* $x \in E$, or *almost everywhere* (usually shortened to "a.e.") in E. For example, the conclusion of Theorem 7.3.4 can be stated as "a function is Riemann integrable in $[a, b]$, if and only if it is bounded there and continuous a.e. in $[a, b]$."

Proposition. *If $I \subset \mathbb{R}$ is an interval, then I is measurable and $\mu(I) = \ell(I)$.*

Proof. If I is bounded with endpoints $a < b$, then $\mu^*(I) = \ell(I)$. If $\varepsilon > 0$, (and $b-a > \varepsilon$) then $[a + \varepsilon/2, b - \varepsilon/2] \subset I$, so $\mu_*(I) > \ell(I) - \varepsilon$, so $\mu^*(I) = \mu_*(I)$ and the claim is true for bounded intervals. If I is unbounded, then $\mu^*(I) = \ell(I) = \infty$, and $I \cap [-a, a]$ is a bounded interval, and hence measurable for all $a > 0$, so I is measurable in this case too. □

Compact sets are closed, but not all closed sets are compact. However, all closed sets are measurable.

Theorem. *Closed sets in $\mathbb{R}$ are measurable.*

Proof. Let $F \subseteq \mathbb{R}$ be closed. If $\mu^*(F) = \infty$, then F is measurable because $F \cap [-a, a]$ is compact (and hence measurable) for all $a > 0$.

Now suppose that $\mu^*(F) < \infty$, and let $\varepsilon > 0$. It follows from Theorem 8.1.1 that there exists $a > 0$ such that $\mu^*(F \cap [-a, a]) > \mu^*(F) - \varepsilon$, and hence that $\mu_*(F) > \mu^*(F) - \varepsilon$, because $F \cap [-a, a]$ is compact. Since this is true for any $\varepsilon > 0$, it follows that $\mu_*(F) = \mu^*(F)$, and F is measurable. □

8.1.4. The countable additivity of Lebesgue measure

The outer measure is countably subadditive on the collection of all subsets of $\mathbb{R}$, but not necessarily countably additive. But if we restrict ourselves to measurable sets, then outer measure, and consequently Lebesgue measure, is countably additive.

Recall that in §3.2.1, we defined the distance between $E, F \subseteq \mathbb{R}$ as

$$d(E, F) = \inf(|x - y| : x \in E, \; y \in F).$$

Lemma. *If $E_1, \ldots E_N \subset \mathbb{R}$ and $d(E_i, E_j) > 0$ for $i \neq j$, then*

$$\mu^*\left(\bigcup_{i=1}^{n} E_i\right) = \sum_{i=1}^{N} \mu^*(E_i).$$

Proof. If $\mu^*(\bigcup_{i=1}^N E_i) = \infty$, then the claim follows from the subadditivity of the outer measure, so we may assume that $\mu^*(\bigcup_{i=1}^N E_i) < \infty$.

Let $\delta = \min(d(E_i, E_j) : i \neq j)$. Suppose that $\{I_n\}$ is any cover of $\bigcup_{i=1}^N E_i$ by intervals. Dividing some (or all) of these intervals into subintervals (if necessary) we may assume that $\ell(I_n) < \delta$ for all n, which implies that none of the intervals intersects more than one of the sets E_i. Thus, $\{I_n\} = \bigcup_{k=1}^N \{J_{k,m} : m \in \mathbb{N}\}$, where for $1 \leq k \leq N$, the intervals $\{J_{k,m} : m \in \mathbb{N}\}$ cover E_k, and $J_{k,m} \cap E_i = \varnothing$ if $i \neq k$. Thus, for every cover $\{I_m\}$ of $\bigcup_{i=1}^N E_i$, there are covers $\{J_{1,m}\}, \ldots \{J_{N,m}\}$ of $E_1, \ldots, E_N$, respectively, such that

$$\sum_n \ell(I_n) = \sum_{k=1}^{N} \sum_m \ell(J_{k,m}),$$

and the claim follows from this. □

Theorem. *If $\{E_n\}_{n=1}^{\infty}$ is a set of measurable subsets of $\mathbb{R}$, with $E_j \cap E_k = \varnothing$ for $j \neq k$, then $\bigcup_n E_n$ is measurable and*

$$\mu\left(\bigcup_n E_n\right) = \sum_n \mu(E_n). \tag{8.1.3}$$

Proof. We begin with the case that $\mu^*(\bigcup_n E_n) < \infty$, in which case it is sufficient to show that

$$\mu_*\left(\bigcup_n E_n\right) \geq \sum_n \mu(E_n), \tag{8.1.4}$$

because this inequality implies two things. First, that $\mu_*\left(\bigcup_n E_n\right) \geq \mu^*\left(\bigcup_n E_n\right)$, by the subadditivity of outer measure, so that $\bigcup_n E_n$ is measurable. Second, with $\bigcup_n E_n$ known to be measurable, it also follows from (8.1.4) that $\mu\left(\bigcup_n E_n\right) \geq \sum_n \mu(E_n)$, which together with subadditivity proves (8.1.3).

To justify (8.1.4), let $\varepsilon > 0$, and for every $n \in \mathbb{N}$, let $K_n \subseteq E_n$ be a compact subset such that $\mu(K_n) > \mu(E_n) - \varepsilon/2^n$. Such compact subsets exist by the definition of inner measure and the fact that $\mu(E_n) = \mu_*(E_n)$ for the measurable set E_n. Since the sets $\{E_n\}$ are mutually disjoint in pairs, the same is true of the compact subsets $\{K_n\}$, and this implies that $d(K_i, K_j) > 0$ for $i \neq j$ by Proposition 3.2.1.

For any $N \in \mathbb{N}$, $\bigcup_{n=1}^{N} K_n$ is a compact subset of $\bigcup_{n=1}^{\infty} E_n$, and therefore

$$\mu_*\left(\bigcup_{n=1}^{\infty} E_n\right) \geq \mu\left(\bigcup_{n=1}^{N} K_n\right) = \sum_{n=1}^{N} \mu(K_n) > \sum_{n=1}^{N} (\mu(E_n) - \varepsilon/2^n),$$

where the equality in the middle follows from the preceding lemma. It follows that

$$\mu_*\left(\bigcup_{n=1}^{\infty} E_n\right) > \sum_{n=1}^{N} \mu(E_n) \quad \varepsilon,$$

for every $N \in \mathbb{N}$ and every $\varepsilon > 0$, from which (8.1.4) follows.

In the case that $\mu^*(\bigcup_n E_n) = \infty$, let $a > 0$ and set $E_{a,n} = E_n \cap [-a, a]$ for each $n \in \mathbb{N}$. The sets $\{E_{a,n}\}_n$ are mutually disjoint in pairs and

$$\mu^*\left(\bigcup_n E_{a,n}\right) \leq 2a,$$

so we may repeat the argument above for the collection $\{E_{a,n}\}_n$. This shows that $(\bigcup_n E_n) \cap [-a, a]$ is measurable for every $a > 0$, and hence, $\bigcup_n E_n$ is measurable. Since $\mu(\bigcup_n E_n) = \infty$ in this case, the identity (8.1.3) follows from the subadditivity of Lebesgue measure. □

Corollary. *Open sets in $\mathbb{R}$ are measurable.*

Proof. Every open set in $\mathbb{R}$ is a union of countably many disjoint open intervals, by Proposition 3.1.8, and intervals are measurable. □

8.1.5. A nonmeasurable set

The outer measure and inner measure are defined on all subsets of $\mathbb{R}$, but they are not always equal. I.e., nonmeasurable sets exist. The standard example is is called a *Vitali set*,[6] which we describe below. The proof that a Vitali set is not measurable relies on the *translation invariance* of Lebesgue measure. If $E \subseteq \mathbb{R}$ and $t \in \mathbb{R}$, then $E_t = \{x + t : x \in E\}$ is the *translation* of E by t.

Lemma. *If $E \subseteq \mathbb{R}$, then E is measurable if and only if E_t is measurable for any $t \in \mathbb{R}$, in which case $\mu(E_t) = \mu(E)$.*

Proof. First note that if $I = (a, b)$, then $I_t = (a + t, b + t)$, and $\ell(I) = \ell(I_t)$. Also, if $E \subseteq \bigcup_n (a_n, b_n)$, then $E_t \subset \bigcup_n (a_n + t, b_n + t)$, and *vice versa*. It follows from these two observations that $\mu^*(E) = \mu^*(E_t)$.

Next, $K \subset \mathbb{R}$ is compact if and only if K_t is compact. This follows from the observation that every cover of K by open intervals $\{I_n\}$ yields a cover of K_t by the open intervals $\{I_{n_t}\}$ and *vice versa*. Hence, K is a compact subset of E if and only if K_t is a compact subset of E_t, and since $\mu^*(K) = \mu^*(K_t)$, it follows that $\mu_*(E) = \mu_*(E_t)$.

Hence $\mu_*(E) = \mu^*(E)$ if and only $\mu_*(E_t) = \mu^*(E_t)$, and since $\mu^*(E) = \mu^*(E_t)$, the claim is true in the case that $\mu^*(E) < \infty$. We leave the case that $\mu^*(E) = \infty$ as an exercise. □

To construct a Vitali set, we start with *equivalence classes* (mod)Q. For every $x \in [0, 1)$, let $S_x = \{y \in [0, 1) : x - y \in \mathbb{Q}\}$.[7] If $0 \leq x \leq 1/2$, then $x + 1/n \in S_x$ for all $n \geq 2$, and if $1/2 < x \leq 1$, then $x - 1/n \in S_x$ for $n \geq 2$. This means that S_x is infinite for all $x \in [0, 1]$. On the other hand, S_x is countable because $\mathbb{Q}$ is countable, so S_x is countably infinite for all $x \in [0, 1)$.

Proposition. *For $x, y \in [0, 1)$, either $S_x = S_y$ or $S_x \cap S_y = \varnothing$.*

In other words, the sets S_x for a *partition* of $[0, 1)$. See Appendix, Section 1.2.

Proof. If $S_x \cap S_y \neq \varnothing$ and $z \in S_x \cap S_y$, then $x - z \in \mathbb{Q}$ and $y - z \in \mathbb{Q}$. This implies that $x - y = x - y + y - z \in \mathbb{Q}$ and $y - x = -(y - x) \in \mathbb{Q}$. Therefore, if $S_x \cap S_y \neq \varnothing$, then $x \in S_y$ and $y \in S_x$. In this case, for any $u \in S_x$, we have $x - u \in \mathbb{Q}$, so $y - u = y - x + x - u \in \mathbb{Q}$, so $u \in S_y$, and hence $S_x \subseteq S_y$. By the same argument, if $S_x \cap S_y \neq \varnothing$, then $S_y \subseteq S_x$. In other words, if $S_x \cap S_y \neq \varnothing$, then $S_x = S_y$. □

It follows from the proposition that every $x \in [0, 1)$ belongs to exactly one set of type S_ξ. Assuming the *axiom of choice*,[8] there exists a subset $V \subset [0, 1)$ such that $[0, 1) = \bigcup_{\xi \in V} S_\xi$, and $S_\xi \cap S_\lambda = \varnothing$ for $\xi, \lambda \in V$ with $\xi \neq \lambda$. Equivalently, we can say that for every $x \in [0, 1)$, there is a unique $\xi \in V$ such that $x - \xi \in \mathbb{Q}$.

---

[6] First described by Giuseppe Vitali.

[7] For the reader with some background in group theory: S_x is the *coset* of x with respect to the additive subgroup $\mathbb{Q} \cap [0, 1)$, in the additive group $G = ([0, 1), \bigoplus)$. We define addition, $x \bigoplus y$, in this group by $x \bigoplus y = x + y$ if $x + y < 1$ and $x \bigoplus y = x + y - 1$ if $x + y \geq 1$.

[8] There are models of set theory that omit the axiom of choice, and it is possible in such (esoteric) settings to construct models of the real numbers in which every subset is Lebesgue measurable. In other words, the axiom of choice plays an important role in this example.

Theorem. *The set V is not Lebesgue measurable.*

Proof. If $x, y \in [0,1)$, then $-1 < x - y < 1$. Hence, if $x \in S_\xi$, then $x = \xi + q$ for some $q \in \mathbb{Q} \cap (-1,1)$. With this in mind, let $\{q_n : n \in \mathbb{N}\}$ be an enumeration of $\mathbb{Q} \cap (-1,1)$, and set $V_n = V_{q_n} = \{\xi + q_n : \xi \in V\}$. If $n \neq m$, then $V_n \cap V_m = \varnothing$, because $x = \xi_1 + q_n = \xi_2 + q_m$ implies $\xi_1 \in S_{\xi_2}$, so $\xi_1 = \xi_2$ and hence $q_n = q_m$. It also follows from the lemma above (translation invariance of Lebesgue measure) that $\mu^*(V_n) = \mu^*(V)$ for all $n \in \mathbb{N}$.[9]

Now consider the set $E = \bigcup_n V_n$. First, $E \subset (-1,2)$, because $V_n \subset (-1,2)$ for all n, and second, $[0,1) \subset E$, because for every $x \in [0,1)$, there is an n such that $x \in V_n$. These inclusions imply that $3 \geq \mu^*(E) \geq 1$. Now, if V were measurable, then V_n would be measurable, with $\mu(V_n) = \mu(V) = \mu^*(V)$, for each n. Furthermore, by Theorem 8.1.4, the set E would be measurable and

$$\mu^*(E) = \mu(E) = \sum_n \mu(V_n) = \sum_n \mu(V).$$

But this is impossible since the sum on the right is either zero or infinite. Hence V is not measurable. □

We can now show that outer measure is *not* additive on all subsets of $\mathbb{R}$.

Corollary. *With the sets V_n as defined above, there exists $N \in \mathbb{N}$ such that*

$$\sum_{n=1}^{N} \mu^*(V_n) > \mu^*\left(\bigcup_{n=1}^{N} V_n\right).$$

Proof. The set $V \subset [0,1]$ is not measurable, so $1 \geq \mu^*(V) > \mu_*(V) \geq 0$. If $N > 3/\mu^*(V)$ then

$$\sum_{n=1}^{N} \mu^*(V_n) = N \cdot \mu^*(V) > 3 \geq \mu^*\left(\bigcup_{n=1}^{N} V_n\right),$$

since $\mu^*(V_n) = \mu^*(V)$ for all n and $\bigcup_{n=1}^{N} V_n \subset [-1,2]$. □

8.1.6. The "sandwich" criterion

Open sets and closed sets are measurable, countable disjoint unions of measurable sets are measurable, and sets of (outer) measure zero are measurable. On the other hand, not all subsets of $\mathbb{R}$ are measurable, and it is useful to have criteria, beyond the definition, that we can use to identify measurable sets.

To begin, we observe that in the definition of the outer measure of a set E, we may assume that all the intervals in the various covers of E are open, so that their union is open. This means that we can reformulate the definition of outer measure to more closely align with the definition of inner measure:

$$\mu^*(E) = \inf(\mu(G) : E \subseteq G,\ G \text{ open}),$$

since every open set can be expressed as a countable disjoint union of open intervals, the sum of whose lengths equals the measure of the set.

[9] More precisely, it follows from the proof of the lemma.

If E is measurable and $\mu(E) < \infty$, then for any $\varepsilon > 0$, there is an open set G such that $E \subset G$ and $\mu(E) > \mu(G) - \varepsilon$. So in a sense, E is 'almost open.' Similarly, there is a compact set $K \subseteq E$ such that $\mu(E) < \mu(K) + \varepsilon$, so E is also "almost compact." These observations lead to the "sandwich" criterion.

Theorem. *The set $E \subseteq \mathbb{R}$ is measurable if and only if for every $\varepsilon > 0$ there is an open set G and a closed set F such that $F \subseteq E \subseteq G$ and $\mu(G \setminus F) < \varepsilon$.*

Recall that $G \setminus F = \{x \in G : x \notin F\} = G \cap F^c$, so if G is open and F is closed, then $G \setminus F$ is open and hence measurable. We refer to the condition that $F \subseteq E \subseteq G$, with F closed, G open and $\mu(G \setminus F) < \varepsilon$, as the *sandwich condition* for $\varepsilon > 0$.

Proof. Suppose first that E is measurable, and $\mu(E) < \infty$. Given $\varepsilon > 0$, it follows from the definitions of inner and outer measure that there exist a compact set K and an open set G, such that $K \subseteq E \subseteq G$, and

$$\mu(K) + \varepsilon/2 > \mu(E) > \mu(G) - \varepsilon/2.$$

Since $K \subset G$ and both sets are measurable, we have $G = (G \setminus K) \cup K$, and $\mu(G) = \mu(G \setminus K) + \mu(K)$ (by the additivity of Lebesgue measure), and hence

$$\mu(G \setminus K) = \mu(G) - \mu(K) < \varepsilon.$$

If E is measurable, but $\mu(E) = \infty$, then $E \cap [-n, n]$ is measurable with finite measure, for all $n \in \mathbb{N}$. Given $\varepsilon > 0$, it follows from the argument above, that for every $n \in \mathbb{N}$, there is an open set G_n and a compact set K_n such that $K_n \subseteq E \cap [-n, n] \subseteq G_n$ and

$$\mu(G_n \setminus K_n) < \frac{\varepsilon}{2^n}.$$

Let $G = \bigcup_n G_n$ and $F = \bigcup_n K_n$, so that $F \subseteq E \subseteq G$. The set G is open as a union of open sets. Any convergent sequence in F is contained in the compact set $F_N = \bigcup_{n=1}^N K_n$ for some N, because convergent sequences are bounded, and hence has its limit in $F_N \subset F$, so F is closed. If $x \in G \setminus F$, then $x \notin K_n$ for any n, but $x \in G_{n_0}$ for some n_0, so $x \in G_{n_0} \setminus K_{n_0}$. Hence, $G \setminus F \subseteq \bigcup_n (G_n \setminus K_n)$ and therefore

$$\mu(G \setminus F) \le \mu\left(\bigcup_n (G_n \setminus K_n)\right) \le \sum_n \mu(G_n \setminus K_n) < \sum_n \frac{\varepsilon}{2^n} = \varepsilon,$$

where the inequality in the middle follows from the subadditivity of Lebesgue measure.

Now, let $E \subseteq \mathbb{R}$ satisfy the sandwich condition for any $\varepsilon > 0$, and to begin, suppose that $\mu^*(E) < \infty$. Given $\varepsilon > 0$, let G be open, and F closed, such that $F \subseteq E \subseteq G$ with $\mu(G \setminus F) < \varepsilon/2$. It follows that $\mu^*(E \setminus F) < \varepsilon/2$ because $E \setminus F \subseteq G \setminus F$. Moreover, since $E = (E \setminus F) \cup F$, we have $\mu^*(F) > \mu^*(E) - \varepsilon/2$, by the subadditivity of outer measure. Repeating this argument for G, we also have $\mu^*(G) < \mu^*(E) + \varepsilon/2$.

If E is bounded, then F is compact, so $\mu_*(E) \ge \mu^*(F) > \mu^*(E) - \varepsilon/2$. Hence, $\mu_*(E) = \mu^*(E)$ (since ε was arbitrary) so E is measurable. If E is not bounded, then there is an $a > 0$ such that $\mu^*(E \cap [-a, a]) > \mu^*(E) - \varepsilon/2$, by Theorem 8.1.1, from which it follows that $\mu^*(G \cap [-a, a]) > \mu^*(E) - \varepsilon/2$.

The sets $G \cap [-a, a]^c$ and $G \cap (-a, a)$ are both open, and $\{-a, a\}$ has measure zero, so all three are measurable. It follows that

$$\mu(G) = \mu(G \cap [-a, a]^c) + \mu(G \cap (-a, a)),$$

since G is the union of these disjoint sets (and $\{-a, a\}$ has measure zero). Moreover, $\mu^*(G \cap (-a, a)) = \mu^*(G \cap [-a, a]) > \mu^*(E) - \varepsilon/2$, and therefore

$$\mu^*(F \cap [-a, a]^c) \leq \mu^*(G \cap [-a, a]^c) < (\mu^*(E) + \varepsilon/2) - (\mu^*(E) - \varepsilon/2) = \varepsilon,$$

since $F \cap [-a, a]^c \subset G \cap [-a, a]^c$. Since $F = (F \cap [-a, a]^c) \cup (F \cap [-a, a])$, it follows that

$$\mu^*(F \cap [-a, a]) \geq \mu^*(F) - \mu^*(F \cap [-a, a]^c) > \mu^*(E) - \varepsilon,$$

so $\mu_*(E) > \mu^*(E) - \varepsilon$, because $F \cap [-a, a]$ is compact. Once again, this shows that $\mu_*(E) = \mu^*(E)$, so E is measurable.

Finally, suppose that E satisfies the sandwich condition for any $\varepsilon > 0$ and $\mu^*(E) = \infty$. In this case, we have to show that $E \cap [-a, a]$ is measurable for every $a > 0$. Let $\varepsilon > 0$ be given, and let $F \subseteq E \subseteq G$, with F closed, G open and $\mu(G \setminus F) < \varepsilon/2$. Given $a > 0$, the set $F_a = F \cap [-a, a]$ is closed, the set $G_a = G \cap (-a - \varepsilon/4, a + \varepsilon/4)$ is open, and $F_a \subseteq E \cap [-a, a] \subseteq G_a$. If $x \in G_a \setminus F_a$, then either $x \in (G \setminus F)$ or $x \in (-a - \varepsilon/4, -a) \cup (a, a + \varepsilon/4)$, so

$$\mu(G_a \setminus F_a) \leq \mu(G \setminus F) + \varepsilon/2 < \varepsilon.$$

It follows that $E \cap [-a, a]$ is measurable by the previous paragraphs, and hence that E is measurable. □

The following corollary illustrates the usefulness of the sandwich criterion, but first we need a definition.

Definition. If X is a set, then a *σ–algebra* of subsets of X is a collection $\mathcal{S} \subseteq \mathcal{P}(X)$ that contains X and $\emptyset$, and is closed under complements, countable unions and countable intersections.

Corollary. *The set of Lebesgue measurable sets is a σ–algebra of subsets of $\mathbb{R}$.*

Proof. To begin, the empty set is measurable, being a set of measure zero, and $\mathbb{R}$ is measurable, being open (and closed).

If $F \subseteq A \subseteq G$, with F closed and G open, then $G^c \subseteq A^c \subseteq F^c$, where G^c is closed and F^c is open. Moreover, $F^c \setminus G^c = F^c \cap G = G \setminus F$, so if A is measurable, then so is A^c, by the sandwich criterion.

If $F_A \subseteq A \subseteq G_A$, and $F_B \subseteq B \subseteq G_B$, with F_A, F_B closed and G_A, G_B closed, then $F_A \cap F_B \subseteq A \cap B \subseteq G_A \cap G_B$, where $F_A \cap F_B$ is closed and $G_A \cap G_B$ is open. Moreover, $(G_A \cap G_B) \setminus (F_A \cap F_B) \subseteq (G_A \setminus F_A) \cup (G_B \setminus F_B)$, so

$$\mu((G_A \cap G_B) \setminus (F_A \cap F_B)) \leq \mu(G_A \setminus F_A) + \mu(G_B \setminus F_B),$$

by subadditivity. Hence, the sandwich criterion implies that $A \cap B$ is measurable if both A and B are measurable. It follows, by induction that the intersection of any finite number of measurable sets is measurable.

Next, if A and B are both measurable, then $A \cap B^c$, $A \cap B$ and $A^c \cap B$ are all measurable,

$$A \cup B = (A \cap B^c) \cup (A \cap B) \cup (A^c \cap B)$$

is measurable by Theorem 8.1.4, since the three sets on the right are disjoint. Once again, it follows by induction that the union of any finite number of measurable sets is measurable.

Now suppose that $\{A_n\}$ is a countable collection of measurable sets. Set $B_1 = A_1$, and for $n > 1$ let $B_n = A_n \cap (\bigcup_{j=1}^{n-1} A_j)^c$. Then B_n is measurable for each n, by the previous paragraphs, $B_i \cap B_k = \emptyset$ if $i \neq k$ and $\bigcup_n A_n = \bigcup_n B_n$. It follows that $\bigcup_n A_n$ is measurable by Theorem 8.1.4, and hence that $\bigcap_n A_n = (\bigcup_n A_n^c)^c$ is likewise measurable. □

8.1.7. Continuity of Lebesgue measure from below and from above

Suppose that $\{E_n\} \subset \mathcal{P}(\mathbb{R})$. If $E_n \subseteq E_{n+1}$ for all n, then we say that $\{E_n\}$ is *approaching* $\bigcup E_n$ *from below*. Likewise, if $E_n \supseteq E_{n+1}$ for all n, then $\{E_n\}$ is *approaching* $\bigcap_n E_n$ *from above*.

Theorem. *Suppose that* $E_n \subseteq \mathbb{R}$ *is measurable for every* n. *If* $\{E_n\}$ *is approaching* $\bigcup_n E_n$ *from below, then*

$$\mu\left(\bigcup_n E_n\right) = \lim_{n\to\infty} \mu(E_n),$$

where the limit on the right may be infinite. If $\{E_n\}$ *is approaching* $\bigcap E_n$ *from above and* $\mu(E_1) < \infty$, *then*

$$\mu\left(\bigcap_n E_n\right) = \lim_{n\to\infty} \mu(E_n).$$

In the first case, we say that μ is *continuous from below*, and in the second case that μ is *continuous from above*.

Proof. Suppose first that $\{E_n\}$ is approaching $\bigcup_n E_n$ from below. Let $F_1 = E_1$, and for $n \geq 2$, let $F_n = E_n \setminus E_{n-1}$. Then $F_i \cap F_j = \emptyset$ for $i \neq j$, $\bigcup_{n=1}^N F_n = E_N$ and $\bigcup_{n=1}^\infty F_n = \bigcup_{n=1}^\infty E_n$. By the additivity of Lebesgue measure, $\mu(E_N) = \sum_{n=1}^N \mu(F_n)$, so by the countable additivity of Lebesgue measure,

$$\mu\left(\bigcup_{n=1}^\infty E_n\right) = \mu\left(\bigcup_{n=1}^\infty F_n\right) = \sum_{n=1}^\infty \mu(F_n) = \lim_{N\to\infty} \sum_{n=1}^N \mu(F_n) = \lim_{N\to\infty} \mu(E_N).$$

It follows that μ is continuous from below.

Now assume that $\{E_n\}$ is approaching $\bigcap_n E_n$ from above, and that $\mu(E_1) < \infty$. For each n, we have $\mu(E_1 \setminus E_n) = \mu(E_1) - \mu(E_n)$, and more generally,

$$\mu\left(E_1 \setminus \bigcap_{n=1}^\infty E_n\right) = \mu(E_1) - \mu\left(\bigcap_{n=1}^\infty E_n\right). \tag{8.1.5}$$

Setting $G_n = E_1 \setminus E_n$ for $n \in \mathbb{N}$, we have $G_n \subseteq G_{n+1}$ for $n \geq 1$, so

$$\mu\left(\bigcup_{n=1}^{\infty} G_n\right) = \lim_{n\to\infty} \mu(G_n) = \mu(E_1) - \lim_{n\to\infty} \mu(E_n), \tag{8.1.6}$$

because μ is continuous from below.

Now, $x \in \bigcup_n G_n$ if and only if $x \in E_1$ and $x \in E_n^c$ for some n, so

$$\bigcup_{n=1}^{\infty} G_n = E_1 \cap \bigcup_{n=1}^{\infty} E_n^C = E_1 \setminus \left(\bigcup_{n=1}^{\infty} E_n^C\right)^C = E_1 \setminus \bigcap_{n=1}^{\infty} E_n.$$

It follows then from (8.1.5) and (8.1.6) that

$$\mu(E_1) - \mu\left(\bigcap_{n=1}^{\infty} E_n\right) = \mu(E_1) - \lim_{n\to\infty} \mu(E_n),$$

so $\mu\left(\bigcap_{n=1}^{\infty} E_n\right) = \lim_{n\to\infty} \mu(E_n)$ because $\mu(E_1) < \infty$. □

Comments. (i) The sets F_n and G_n constructed in the proof are measurable, because the collection of measurable sets is a σ-algebra. (ii) The condition that $\mu(E_1) < \infty$ is necessary for continuity from above. If $E_n = (n, \infty)$, for example, then $\mu(E_n) = \infty$ for all n, but $\bigcap_n E_n = \emptyset$, so the claimed equality fails in this case.

8.1.8. Measurable sets are almost Borel sets

Theorem 8.1.6 implies measurable sets are almost open, and almost closed. We can take this characterization one step further, by considering *Borel sets*.

The *Borel algebra* of a topological space X is the smallest σ-algebra of sets in X that contains all open sets in X. Elements of the Borel algebra of X are called the Borel sets of X. The Borel algebra of $\mathbb{R}$ is the smallest σ-algebra in $\mathcal{P}(\mathbb{R})$ that contains all open intervals.

Theorem. *A set $E \subset \mathbb{R}$ is Lebesgue measurable if and only if there is a Borel set B and a set A of measure zero, such that $E = A \cup B$.*

Proof. One direction is clear. If B is a Borel set, then it is measurable by Corollary 8.1.6, since the Borel algebra is generated by open sets. Hence, $A \cup B$ is measurable for every measurable set A (as also follows from the corollary), and in particular for every set A of measure zero.

Now suppose that E is measurable. Then for every $n \in \mathbb{N}$, there is a closed set F_n such that $F_n \subset E$ and $\mu(E \setminus F_n) < \frac{1}{n}$, by Theorem 8.1.6. The set $B = \bigcup_n F_n$ is a Borel set (being a union of closed sets), and $B \subseteq E$. We have

$$\mu(E \setminus B) \leq \mu(E \setminus F_n) < \frac{1}{n},$$

for all n, so $\mu(E \setminus B) = 0$, and the statement follows since $E = B \cup (E \setminus B)$. □

8.1.9. The Vitali covering theorem

Roughly speaking, the Vitali covering theorem tells us that if a bounded set $E \subset \mathbb{R}$ can be covered very *efficiently* by a collection $\mathcal{V}$ of intervals, then E can be covered arbitrarily well by a finite subset of pairwise disjoint intervals from $\mathcal{V}$. This conclusion does not require that E be measurable. This theorem is an important tool in Lebesgue theory, as we illustrate in §8.5.

By 'efficiently', we mean that $\mathcal{V}$ is a Vitali covering of E. To describe and study such coverings, we use the following notation. We denote the center of a bounded interval I by u_I and its radius by r_I, whether I is open, closed or half open. If I is a bounded interval interval and $c > 0$, then cI is the interval with the same center as I, with radius c times the radius of I. I.e., $u_{cI} = u_I$ and $r_{cI} = cr_I$.

Definition. Given $E \subset \mathbb{R}$, a collection $\mathcal{V}$ of bounded, intervals is a *Vitali covering* of E if for any $x \in E$ and any $\delta > 0$, there is an $I \in \mathcal{V}$ such that $x \in I$ and $0 < r_I < \delta$. In other words, E is covered by arbitrarily short intervals in $\mathcal{V}$. This is what we mean by 'efficiently'.

To prove the covering theorem, we rely on the covering lemma.

Lemma (Vitali covering lemma). *Let $\mathcal{V}$ be a collection of bounded intervals in $\mathbb{R}$, with $r_I \leq 1$ for all $I \in \mathcal{V}$, and $\bigcup_{I \in \mathcal{V}} \subset (-R, R)$ for some $R > 0$. Then there exists a countable subset $\mathcal{W} \subseteq \mathcal{V}$ of pairwise disjoint intervals such that*

$$\bigcup_{\mathcal{V}} I \subset \bigcup_{\mathcal{W}} 5J.$$

Comments. The condition that $r_I \leq 1$ for all $I \in \mathcal{V}$ can be replaced by $r_I \leq \beta$ for all $I \in \mathcal{V}$, where $\beta > 0$ is any fixed constant, and the proof of the claim remains identical, with one small change in notation. We choose $\beta = 1$ for the sake of concreteness. The claim also remains true if we drop the condition that $\bigcup_{I \in \mathcal{V}}$ is bounded in $\mathbb{R}$. In this case the idea of the proof remains the same, but the execution requires a little more work. In fact, the general claim is true in any separable metric space X, with intervals replaced by metric balls, $B(x, r)$.

Proof. To begin, we divide $\mathcal{V}$ into mutually disjoint subsets according to the radii of the intervals. Setting $V_n = \{I \in \mathcal{V} : 1/2^{n+1} < r_I \leq 1/2^n\}$ for $n \geq 0$, then $V_j \cap V_k = \emptyset$ if $k \neq j$, and $\mathcal{V} = \bigcup_{n \geq 0} V_n$.

If $V_0 = \emptyset$, set $W_0 = \emptyset$. Otherwise choose $J_{0,1} \in V_0$. If $I \cap J_1 \neq \emptyset$ for all $I \in V_0$, let $W_0 = \{J_{0,1}\}$. Otherwise choose $J_{0,2} \in V_0$ such that $J_1 \cap J_2 = \emptyset$. If $I \cap (J_{0,1} \cup J_{0,2}) \neq \emptyset$ for all $I \in V_0$, set $W_0 = \{J_{0,1}, J_{0,2}\}$, otherwise continue. Suppose that we have found $J_{0,1}$, $\dots, J_{0,m} \in V_0$ such that $J_{0,i} \cap J_{0,k} = \emptyset$ for $i \neq k$. If $I \cap (\bigcup_{k=1}^{m} J_{0,k}) \neq \emptyset$ for all $I \in V_0$, set $W_0 = \{J_{0,1}, \dots, J_{0,m}\}$, otherwise continue. This process must terminate after no more than $4R$ steps, because $r_I > 1/2$ for all $I \in V_0$, and $\bigcup_{V_0} I \subset (-R, R)$.[10] This leaves us with a (finite) set $W_0 = \{J_{0,1}, \dots, J_{0,n_0}\} \subseteq V_0$ of pairwise disjoint intervals, such that for every $I \in V_0$ there is a $J \in W_0$ such that $I \cap J \neq \emptyset$.

[10]It is this part of the proof that would require a little more work if $\bigcup_{V_0} I$ were not bounded.

Next, let V_1' be the subset of all intervals in V_1 that are disjoint from the union of intervals in W_0. Once again, if $V_1' = \emptyset$, set $W_1 = \emptyset$, otherwise, choose $J_{1,1} \in V_1'$. Repeating the process we used for W_0, we can find a set $W_1 = \{J_{1,1}, \dots, J_{1,n_1}\} \subset V_1' \subset V_1$ of pairwise disjoint intervals such that for all $I \in V_0 \cup V_1$, $I \cap J \neq \emptyset$ for some $J \in W_0 \cup W_1$.

We repeat this process recursively. After $n+1$ steps, we have $n+1$ sets $W_0, \dots, W_n$, each of which is finite, such that (i) $\bigcup_{k=1}^n W_k \subset \bigcup_{k=1}^n V_k$, (ii) any two distinct intervals in $\bigcup_{k=1}^n W_k$ are disjoint, and (iii) for any $I \in \bigcup_{k=1}^n V_k$, there is a $J \in \bigcup_{k=1}^n W_k$ such that $I \cap J \neq \emptyset$. Once again, we denote by V_{n+1}' the subset of all intervals in V_{n+1} that are disjoint from $\bigcup_{k=1}^n W_k$. As above, we can find a finite set $W_{n+1} \subset V_{n+1}'$ of pairwise disjoint intervals whose union intersects every interval in V_{n+1}', from which it follows that for every $I \in \bigcup_{k=0}^{n+1} V_k$ there is a $J \in \bigcup_{k=1}^{n+1} W_k$ such that $I \cap J \neq \emptyset$.

Having constructed the sequence $\{W_n\}_{n\geq 0}$ as above, we let $\mathcal{W} = \bigcup_n W_n$, which is countable as a countable union of finite sets. If $I \in \mathcal{V}$, then $I \in V_n$ for some n, so there is a $J \in \bigcup_{k=0}^n W_k$, such that $I \cap J \neq \emptyset$. It follows that $r_I < 2r_J$, so $|x - u_J| < 2r_I + r_J < 5r_J$, for any $x \in I$. Hence, $I \subset 5J$, and therefore $\bigcup_{\mathcal{V}} I \subset \bigcup_{\mathcal{W}} 5I$, as claimed. $\square$

Theorem. *Let $E \subset \mathbb{R}$ be bounded, and let $\mathcal{V}$ be a Vitali covering of E. Then there exists a countable subset $\mathcal{W}$ of $\mathcal{V}$ of pairwise disjoint intervals such that*

$$\mu\left(E \setminus \bigcup_{I\in\mathcal{W}} I\right) = 0.$$

Comment. The claim is also true for unbounded $E \subset \mathbb{R}$, and we leave the proof of this case as an exercise.

Proof. We may assume without loss of generality that for all $I \in \mathcal{V}$, $I \cap E \neq \emptyset$ and that the radius of I is bounded by 1, since the subset of a Vitali covering of E satisfying these conditions is still a Vitali covering of E. With these assumptions, it follows from the boundedness of E that the set $\bigcup_{I\in\mathcal{V}} I$ is likewise bounded so that $\mathcal{V}$ satisfies the conditions of the covering lemma.

Using the same notation as in the statement and proof of the covering lemma, let $\mathcal{W} = \bigcup_{n\geq 0} W_n$ be a subset of pairwise disjoint intervals from $\mathcal{V}$, satisfying $\bigcup_{\mathcal{V}} I \subset \bigcup_{\mathcal{W}} 5J$. We enumerate the countable set $\mathcal{W} = \{J_k : k \in \mathbb{N}\}$, and note that $\bigcup_k J_k \subseteq \bigcup_{I\in\mathcal{V}} I$ is bounded, so by the countable additivity of Lebesgue measure

$$\sum_{k=1}^{\infty} \mu(J_k) = \mu\left(\bigcup_{k=1}^{\infty} J_k\right) < \infty, \tag{8.1.7}$$

because the intervals in $\mathcal{W}$ are disjoint.

For each $m \in \mathbb{N}$, let $E_m = E \setminus \bigcup_{k=1}^m \bar{J}_k$ (where $\bar{J}_k$ is the closure of J_k, as usual). Then

$$E \setminus \bigcup_{k=1}^{\infty} \bar{J}_k = \bigcap_{m=1}^{\infty} E_m$$

differs from $E \setminus \bigcup_{k=1}^{\infty} J_k$ by the set of endpoints of the intervals in $\mathcal{W}$ that are not already closed, which is a countable set, and so has measure zero. Hence, to show that

$E \setminus \bigcup_{k=1}^{\infty} J_k$ has measure zero, it is enough to show that $\mu^*(E_m) \to 0$, since $\bigcap_{m=1}^{\infty} E_m \subset E_m$ for every m.

Given $\varepsilon > 0$, it follows from (8.1.7) that there is an $N_1 \in \mathbb{N}$ such that

$$\sum_{k>N_1} \mu(J_k) < \varepsilon/5.$$

Furthermore, since W_n is finite for every n, there is an $N_2 \geq N_1$ such that if $n \geq N_2$ and $J_k \in W_n$, then $k > N_1$i. If $m \geq N_2$, and $x \in E_m$, then $x \notin \bigcup_{k=1}^{N_2} \bar{J}_k$, so the distance of x from $\bigcup_{k=1}^{N_2} \bar{J}_k$,

$$\delta_x = \min\left\{|x - y| \, : \, y \in \bigcup_{k=1}^{N_2} \bar{J}_k\right\} > 0,$$

because $\bigcup_{k=1}^{N_2} \bar{J}_k$ is compact, as a finite union of compact sets. Since $\mathcal{V}$ is a Vitali covering of E, there must be an $I_x \in \mathcal{V}$ such that $x \in I_x$ and $r_{I_x} < \delta_x$. Hence, $I_x \cap \bigcup_{k=1}^{N_2} J_k = \emptyset$, and it follows from the construction of $\mathcal{W}$ that there is an $n > N_2$ and a $J_k \in W_n$ such that $I_x \subset 5J_k$. Since $x \in E_m$ was arbitrary, it follows that if $m \geq N_2$, then $E_m \subset \bigcup_{k>N_1} 5J_k$, and

$$\mu^*(E_m) \leq \sum_{k>N_1} \mu(5J_k) < \varepsilon,$$

so $\lim_{m\to\infty} \mu^*(E_m) = 0$. □

Corollary. *If $E \subset \mathbb{R}$ is bounded, and $\mathcal{V}$ is a Vitali covering of E, then for any $\varepsilon > 0$, there is a finite set of pairwise disjoint intervals $\{J_1, \ldots, J_N\}$ such that*

$$\mu^*\left(E \setminus \bigcup_{k=1}^{N} J_k\right) < \varepsilon \tag{8.1.8}$$

and

$$\mu\left(\bigcup_{k=1}^{N} J_k\right) < \mu^*(E) + \varepsilon. \tag{8.1.9}$$

The inequalities (8.1.8) and (8.1.9) explain what we meant by "E can be covered arbitrarily well", in the opening paragraph of this subsection.

Proof. Let $\varepsilon > 0$, then by the definition of outer measure, there is an open set $G = G_\varepsilon$ containing E such that $\mu^*(E) \leq \mu(G) < \mu^*(E) + \varepsilon$. If $x \in E$, then $x \in G$, so there is a $\delta_x > 0$ such that $(x - \delta_x, x + \delta_x) \subset G$, because G is open. Since $\mathcal{V}$ is a Vitali covering of E, for any $\delta \in (0, \delta_x/2)$, there is an $I \in \mathcal{V}$ such that $x \in I$ and $r_I < \delta$, and for any such interval, $I \subset (x - \delta_x, x + \delta_x) \subset G$. It follows that the set $\mathcal{V}_G = \{I \in \mathcal{V} : I \subset G\}$ is also a Vitali covering of E.

If follows from the proof of the covering theorem applied to $\mathcal{V}_G$ that there is a set of pairwise disjoint intervals $\mathcal{W}_G = \{J_k : k \in \mathbb{N}\} \subset \mathcal{V}_G$, such that

$$\lim_{n\to\infty} \mu^*\left(E \setminus \bigcup_{k=1}^{n} J_k\right) = 0.$$

Hence there is an $N \in \mathbb{N}$ such that $\mu^*\left(E \setminus \bigcup_{k=1}^{N} J_k\right) < \varepsilon$, and since $\bigcup_{k=1}^{N} J_k \subset G$, it also follows that $\mu^*\left(\bigcup_{k=1}^{N} J_k\right) \le \mu(G) < \mu^*(E) + \varepsilon$. □

Exercises

ex8.1.1. Justify Comment (ii) in §8.1.1. Consider the cases $\mu^*(E) = \infty$ and $\mu^*(E) < \infty$ separately. In the second case, show that

$$\inf\left\{\sum_n \ell(I_n) : E \subset \bigcup_n I_n\right\} = \inf\left\{\sum_n \ell(J_n) : E \subset \bigcup_n J_n\right\},$$

where the infimum on the right is taken over all countable covers of E by bounded intervals, and the infimum on the left is taken over all countable covers of E by bounded *open* intervals.

Hint: If J is open, then adding one or both endpoints to J doesn't change its length. If I is closed or half open, then for any $\varepsilon > 0$, there is an open interval J, such that $I \subset J$ and $\ell(J) < \ell(I) + \varepsilon$.

ex8.1.2. Justify Comment (iv) in §8.1.1.

ex8.1.3. Justify Comment (iii) in §8.1.3.

ex8.1.4. Show that if K is a compact subset of the Vitali set V, then $\mu(K) = 0$, and conclude that $\mu_*(V) = 0$.

Hint: Show that $K_r \cap K_s = \varnothing$ for any rational numbers $r \neq s$.

ex8.1.5. Show that if $\mu(A) < \infty$ and $\mu(B) = 0$, then $\mu(A \cup B) = \mu(A)$. (What if $\mu(A) = \infty$?)

ex8.1.6. In §3.3.2 and §3.3.3, we showed that the Cantor set is uncountable, compact, nowhere dense, and has measure zero. Show that for $0 < \alpha < 1$, there is a subset $C_\alpha \subset [0, 1]$ that is nowhere dense and compact, with $\mu(C_\alpha) = \alpha$. Show that this is not possible if $\alpha = 1$.

Hint: $\sum_{n=1}^{\infty} \frac{1-\alpha}{2^n} = 1 - \alpha$.

ex8.1.7. Construct a measurable set $C_\infty \subset \mathbb{R}$ that is nowhere dense in $\mathbb{R}$, with $\mu(C_\infty) = \infty$.

ex8.1.8. For a sequence of sets $\{E_n\}$, the set

$$\limsup_{n\to\infty} E_n = \bigcap_{N=1}^{\infty} \bigcup_{n=N}^{\infty} E_n$$

consists of all points that belong to infinitely many of these sets.

Show that if $\{E_n\}$ is a sequence of measurable sets, and $\sum_n \mu(E_N) < \infty$, then

$$\mu\left(\limsup_{n\to\infty} E_n\right) = 0.$$

This is the *Borel-Cantelli* Lemma.

ex8.1.9. Prove the Vitali covering theorem in the case that $E \subset \mathbb{R}$ is not bounded.

***ex8.1.10.** Show that if $E \subset \mathbb{R}$ and $\mu(E) > 0$, then the set $E - E = \{x - y : x, y \in E\}$ contains an interval around 0.

8.2. Measurable functions

Recall from the introduction that Lebesgue's idea was to define the integral of f in terms of sums of the form $\sum y_j \cdot \mu(E_j)$, where roughly speaking, $E_j = f^{-1}((s_{j-1}, s_j))$ with and $s_{j-1} < y_j < s_j$. Such sums only make sense for a function f for which the pre-images of open intervals (and open sets more generally) are measurable.

8.2.1. Measurability

A function $f : \mathbb{R} \to \mathbb{R}$ is continuous if $f^{-1}(U)$ is open for every open set $U \subset \mathbb{R}$. Measurability is defined analogously.

Definition. The function f is *measurable* if

$$f^{-1}(U) = \{x \in \mathbb{R} : f(x) \in U\}$$

is measurable for every open $U \subseteq \mathbb{R}$.

Note that if f is measurable, then its domain $E = f^{-1}(\mathbb{R})$ must be measurable. Hence, going forward, any set $E \subset \mathbb{R}$ mentioned without qualification will be assumed to be measurable (to avoid overuse of the word "measurable").

The condition that $f^{-1}(U)$ is measurable for every open set $U \subseteq \mathbb{R}$ can be simplified.

Theorem. *The function f is measurable if and only if $\{x : f(x) < \alpha\}$ is measurable for every $\alpha \in \mathbb{R}$.*

Comment. The condition that $\{x : f(x) < \alpha\}$ in the statement of the theorem can be replaced by $\{x : f(x) \le \alpha\}$, $\{x : f(x) > \alpha\}$ or $\{x : f(x) \ge \alpha\}$ and the proofs in those cases are analogous to the one below (see Exercise **ex8.2.4**), and we use these conditions interchangeably.

Proof. One direction is clear, since $(-\infty, \alpha)$ is open for all $\alpha \in \mathbb{R}$. In the other direction, we begin by noting that

$$f^{-1}\left(\bigcup_{n=1}^{\infty} A_n\right) = \bigcup_{n=1}^{\infty} f^{-1}(A_n),$$

because $x \in f^{-1}\left(\bigcup_{n=1}^{\infty} A_n\right)$ if and only if $f(x) \in A_n$ for some n;

$$f^{-1}\left(\bigcap_{n=1}^{\infty} A_n\right) = \bigcap_{n=1}^{\infty} f^{-1}(A_n),$$

because $x \in f^{-1}\left(\bigcap_{n=1}^{\infty} A_n\right)$ if and only if $f(x) \in A_n$ for all n; and

$$f^{-1}(A^c) = \left(f^{-1}(A)\right)^c,$$

because $x \in \left(f^{-1}(A)\right)^c$ if and only if $f(x) \notin A$.

From our assumption that $f^{-1}((-\infty, \alpha))$ is measurable for every $\alpha \in \mathbb{R}$ it follows that $f^{-1}([\alpha, \infty))$ is measurable for every $\alpha \in \mathbb{R}$ and that if $\alpha < \beta$, then

$$f^{-1}([\alpha, \beta)) = f^{-1}([\alpha, \infty)) \cap f^{-1}((-\infty, \beta))$$

is measurable. Hence, if $\alpha < \beta$, then

$$f^{-1}((\alpha, \beta)) = \bigcup_{n=1}^{\infty} f^{-1}\left([\alpha + \tfrac{1}{n}, \beta)\right)$$

is measurable. Thus, the inverse image of any open interval is measurable, and since every open set in $\mathbb{R}$ is a countable union of open intervals, it follows that $f^{-1}(U)$ is measurable for all open $U \subseteq \mathbb{R}$. □

As it is with Lebesgue measure, it is useful to allow the functions we study to take values in the extended real numbers, $\mathbb{R} \cup \{\pm\infty\}$. For example $f(x) = 1/x^2$ may now be thought to be defined at $x = 0$ by setting $f(0) = \infty$. This extension has no effect on the definition of measurability, since $\{x : f(x) < \infty\} = \mathbb{R}$ and $\{x : f(x) < -\infty\} = \emptyset$ are both measurable.

The functions that we frequently encounter in practice are measurable.

Proposition. *Continuous functions are measurable, and if f is Riemann-integrable on the compact interval $[a, b]$, then f is measurable there.*

Proof. If f is continuous, then $f^{-1}(U)$ is open and therefore measurable, for every open set $U \subseteq \mathbb{R}$.

If f is Riemann-integrable on the compact interval $[a, b]$, then by the Vitali-Lebesgue theorem, the set $D \subset [a, b]$ of points of discontinuity of f in $[a, b]$, has measure 0. It follows that for any $\alpha \in \mathbb{R}$, $\mu(\{x \in D : f(x) < \alpha\}) = 0$. On the other hand, the set $\{x \in [a, b] \setminus D : f(x) < \alpha\}$ is open, so $\{x \in [a, b] : f(x) < \alpha\}$ is the union of an open set and a set of measure zero, so it is measurable. □

8.2.2. Extending the range of measurable functions

A function that is Riemann integrable must be bounded in the domain of the integral, but this is not the case for the Lebesgue integral as we shall see. In fact, it is useful to allow the measurable functions to take the "values" $\pm\infty$, and we denote the set $\mathbb{R} \cup \{-\infty, \infty\}$ by $\overline{\mathbb{R}}$. In other words, the general set of functions we consider are functions $f : \mathbb{R} \to \overline{\mathbb{R}}$. If the range of the function f on the domain $E \subseteq \mathbb{R}$ is contained in $\mathbb{R}$, we say that f is *finite* on E.

The definition of measurability for a function $f : \mathbb{R} \to \overline{\mathbb{R}}$ remains the same, i.e., f is measurable if and only if the set $\{x : f(x) \leq \alpha\}$ is measurable for every $\alpha \in \overline{\mathbb{R}}$.

If $\{f_n\}$ is a sequence of functions on some set $E \subset \mathbb{R}$, then $f_n(x) \to f$ pointwise in E, now means that for all $x \in E$, $|f_n(x) - f(x)| \to 0$, if $|f(x)| < \infty$, and $f_n(x) \to \pm\infty$ if $f(x) = \pm\infty$. Conversely if $\lim_{n\to\infty} f_n(x) = f(x)$ exists in $\overline{\mathbb{R}}$ for all $x \in E$, then $f_n \to f$ pointwise in E.

Proposition. *If f is measurable, then $\{x : f(x) = \infty\}$ and $\{x : f(x) = -\infty\}$ are both measurable.*

Proof. Exercise. □

8.2.3. The positive and negative parts of a function

Much of what we discuss in this chapter is easier to define or understand initially for nonnegative measurable functions, and generalize to all measurable functions in terms of their *positive* and *negative* parts.

Definition. The positive and negative parts of the function $f(x)$ are defined as $f^+(x) = \max(f, 0)$ and $f^-(x) = -\min(f, 0)$.

We note that $f(x) = f^+(x) - f^{-1}(x)$ and $|f(x)| = f^+(x) + f^-(x)$.

Theorem. *The function f is measurable if and only if f^+ and f^- are both measurable.*

Proof. It follows from Theorem 8.2.1 that f^+ and f^- are measurable if f is measurable. On the other hand, If f^+ and f^- are measurable and $\alpha \geq 0$, then

$$\{x : f(x) \leq \alpha\} = \{x : f^+(x) \leq \alpha\} \cup \{x : f^-(x) \geq 0\}$$

is measurable as a union of measurable sets, and if $\alpha \leq 0$, then

$$\{x : f(x) \leq \alpha\} = \{x : f^-(x) \geq -\alpha\}$$

is measurable. □

8.2.4. Simple functions

Another tool we use to develop the theory of the Lebesgue integral is the *simple function*.

Definition. The *indicator function* of a set $E \subseteq \mathbb{R}$ is

$$\chi_E(x) = \begin{cases} 1 & : x \in E \\ 0 & : x \notin E \end{cases}$$

A function $\varphi(x)$ is a *simple function* if there is a collection $\{E_1, \ldots, E_n\}$ of pairwise disjoint subsets of $\mathbb{R}$, and real numbers such that

$$\varphi(x) = \sum_{j=1}^{n} a_j \cdot \chi_{E_j}(x), \tag{8.2.1}$$

where $a_j \in \mathbb{R}$ for each j.

We observe that the positive and negative parts of a simple function are themselves (nonnegative) simple functions.

It is also clear that φ is measurable if the sets E_j appearing in the sum (8.2.1) are all measurable.

Going forward, if a simple function is given in the form of (8.2.1) without qualification, it will be understood that the sets $\{E_j\}$ are pairwise disjoint and measurable. In particular, we will assume that every simple function we consider is measurable.

Theorem. *Sums and products of simple functions are simple.*

Proof. Suppose that $\varphi = \sum_{j=1}^{n} a_j \chi_{A_j}$ and $\psi = \sum_{i=1}^{m} b_i \chi_{B_i}$. Setting $C_{ij} = A_i \cap B_j$ for $1 \leq i \leq m$ and $1 \leq j \leq n$, we have

$$\varphi(x) + \psi(x) = \sum_{i,j} (a_i + b_j)\chi_{C_{ij}}(x) \quad \text{and} \quad \varphi(x) \cdot \psi(x) = \sum_{i,j} (a_i \cdot b_j)\chi_{C_{ij}}(x).$$

Moreover, C_{ij} is measurable for all i and j because A_i and B_j are both measurable. □

8.2.5. Sequences of measurable functions

Theorem. *If $\{f_n(x)\}$ is a sequence of measurable functions, then the functions*

$$\inf_n f_n(x), \ \sup_n f_n(x), \ \liminf_{n\to\infty} f_n(x) \quad \textit{and} \quad \limsup_{n\to\infty} f_n(x)$$

are all measurable.

Proof. Let $g(x) = \sup_n f_n(x)$ and $\alpha \in \overline{\mathbb{R}}$, then

$$\{x : g(x) \leq \alpha\} = \bigcap_n \{x : f_n(x) \leq \alpha\},$$

is measurable, as an intersection of measurable sets.

Next, if $h(x) = \limsup f_n(x)$, then $h(x) \leq \alpha$, for $\alpha \in \overline{\mathbb{R}}$, if and only if for all $n \in \mathbb{N}$ and all $m > n$, $f_m(x) \leq \alpha$. Therefore

$$\{x : h(x) \leq \alpha\} = \bigcap_{n\in\mathbb{N}} \bigcap_{m>n} \{x : f_m(x) \leq \alpha\},$$

is measurable as an intersection of measurable sets.

The proof of the claims for $\inf f_n$ and $\liminf f_n$ is left as an exercise. □

Corollary. *If $\{f_n(x)\}$ is a sequence of measurable functions and $f_n(x)$ converges pointwise to $f(x)$, then $f(x)$ is measurable.*

Proof. If $f_n \to f$ pointwise, then $\limsup f_n = f$. □

8.2.6. Measurable functions as limits of simple functions

If f is the pointwise limit of a sequence of simple functions, then f is measurable, as follows from Corollary 8.2.5. The converse is also true.

Theorem. *If $f(x)$ is measurable on $E \subseteq \mathbb{R}$, then there is a sequence of simple functions $\{\varphi_n(x)\}$ that converges pointwise to $f(x)$.*

Proof. To begin, we note that the sets

$$E_+ = \{x \in E : f(x) = \infty\} \quad \text{and} \quad E_- = \{x \in E : f(x) = -\infty\}$$

are measurable by Proposition 8.2.2. and for $n, m \in \mathbb{N}$, the sets

$$E_{n,m} = f^{-1}\left(\left[\frac{m}{n}, \frac{m+1}{n}\right)\right),$$

are likewise measurable. Let φ_n be the (measurable) simple function

$$\varphi_n = n \cdot \chi_{E_+} - n \cdot \chi_{E_-} + \sum_{m=-n^2}^{n^2-1} \frac{m}{n} \cdot \chi_{E_{n,m}}.$$

For any $x \in E$, if $-n \leq f(x) < n$, then $f(x) \in [m/n, m+1/n)$ for some integer m in $[-n^2, n^2)$, and hence

$$|\varphi_n(x) - f(x)| = f(x) - \frac{m}{n} < \frac{1}{n}.$$

Thus, if $|f(x)| < \infty$, then $|\varphi_n(x) - f(x)| < 1/n$ for all $n > |f(x)|$, so $\{\varphi_n(x)\}$ converges pointwise $f(x)$ in $E \setminus (E_+ \cup E_-)$. On the other hand, if $x \in E_+$, then $\varphi_n(x) = n$, and if $x \in E_-$, then $\varphi_n(x) = -n$, so $\{\varphi_n(x)\}$ converges pointwise $f(x)$ at these points too. □

Note that, if $f \geq 0$, then the sequence of simple functions that we constructed in the proof above, is monotone increasing (pointwise) to f, and if f is bounded, then the convergence is uniform.

Corollary. *Sums and products of measurable functions are measurable.*

Proof. By Theorem 8.2.4, this is true for simple functions, and hence it is true for all measurable functions by the theorem above and Theorem 2.3.2. □

8.2.7. Egoroff's theorem

Theorem. *Suppose that $\mu(E) < \infty$ and the sequence of measurable functions $\{f_n\}$ converges pointwise to the finite function f a.e. in E. Then for any $\varepsilon > 0$, there is a measurable set $F \subseteq E$ such that $\mu(E \setminus F) < \varepsilon$ and $f_n \to f$ uniformly on F.*

Proof. The function f is measurable in E, by Corollary 8.2.5, so for each $n, m \in \mathbb{N}$, the set $D_{n,m} = \bigcup_{k \geq n} \{x \in E : |f(x) - f_k(x)| \geq 1/m\}$ is measurable.

For fixed $m \in \mathbb{N}$, we have $\bigcap_n D_{n,m} = \{x \in E : \limsup_n |f(x) - f_n(x)| \geq 1/m\}$ so $\mu(\bigcap_n D_{n,m}) = 0$, because $f_n \not\to f$ in this set. Furthermore, $D_{n+1,m} \subset D_{n,m}$ for all n, and $\mu(D_{1,m}) < \mu(E) < \infty$, so it follows from Theorem 8.1.7 that

$$\lim_{n \to \infty} \mu(D_{n,m}) = 0.$$

Let $\varepsilon > 0$, then for each $m \in \mathbb{N}$, there is an $n_m \in \mathbb{N}$ such $\mu(D_{n_m,m}) < \varepsilon/2^m$, by the limit above. Setting $D = \bigcup_m D_{n_m,m}$, we have

$$\mu(D) \leq \sum_{m=1}^{\infty} \mu(D_{n_m,m}) < \varepsilon.$$

If $x \notin D$, then $x \notin D_{n_m,m}$, and hence $|f(x) - f_n(x)| < 1/m$ for all $n \geq n_m$, and so $f_n \to f$ uniformly on $F = E \setminus D$. □

8.2.8. Lusin's theorem

If f is Riemann integrable on $I = [a, b]$ then f is continuous almost everywhere in I. Another way of saying this is that there exists a measurable subset $E \subset I$, such that $\mu(I \setminus E) = 0$ and f is continuous on E. Lusin's theorem is a slightly weaker generalization of this to measurable functions.

Theorem. *If $\mu(E) < \infty$, and f is finite and measurable on E, then for any ε there is a compact set $K \subseteq E$ such that $\mu(E \setminus K) < \varepsilon$ and f is continuous on K.*

Proof. We first prove the claim for simple functions and then use Egoroff's theorem to extend it to measurable functions in general.

Let $\varphi = \sum_{j=1}^{n} a_j \chi_{E_j}$ be a simple function on E, so $E_i \cap E_j = \varnothing$ for $i \neq j$. We may assume that $E_j \subset E$ for $1 \leq j \leq n$ (by replacing E_j by $E \cap E_j$ if necessary), and that $\bigcup_j E_j = E$ (by adding the term $0 \cdot \chi_{E_{n+1}}$ to φ where $E_{n+1} = E \setminus \bigcup_{j=1}^{n} E_j$ if necessary).

For any $\varepsilon > 0$, there are compact sets $K_1, \dots, K_n$ such that $K_j \subseteq E_j$ for each j, and $\mu(E_j \setminus K_j) < \varepsilon/n$. The set $K = \bigcup_j K_j \subseteq E$ is compact, and if $i \neq j$, then $d(K_i, K_j) > 0$ because $K_i \cap K_j = \varnothing$, so K_i and K_j have no limit points in common. It follows that φ is continuous on K, since φ is constant on K_j for each j. Since

$$\mu(E \setminus K) = \mu\left(\bigcup_j (E_j \setminus K_j)\right) = \sum_{j=1}^{n} \mu(E_j \setminus K_j) < \varepsilon,$$

the claim is true for simple functions.

Now let f be measurable on E and let $\varepsilon > 0$. By Theorem 8.2.4 there is a sequence of simple functions $\{\varphi_n\}$ converging pointwise to f in E, and by Egoroff's theorem, there is a measurable $F \subseteq E$ such that the convergence is uniform in F, with $\mu(E \setminus F) < \varepsilon/4$. It follows that there is a compact set $K_0 \subseteq F \subseteq E$, such that $\mu(F \setminus K_0) < \varepsilon/4$, so that $\mu(E \setminus K_0) < \varepsilon/2$.

For each $n \in \mathbb{N}$ there is a compact set $K_n \subset K_0$, such that φ_n is continuous on K_n and $\mu(K_0 \setminus K_n) < \varepsilon/2^{n+1}$, since the theorem is true for simple functions and $\mu(K_0) \leq \mu(E) < \infty$. The set $K = \bigcap_{n=1}^{\infty} K_n$ is compact, φ_n is continuous in K for all n, and

$$\mu(K_0 \setminus K) = \mu\left(\bigcup_{n=1}^{\infty} (K_0 \setminus K_n)\right) < \sum_{n=1}^{\infty} \mu(K_0 \setminus K_n) < \varepsilon/2.$$

It follows that $\mu(E \setminus K) = \mu(E \setminus K_0) + \mu(K_0 \setminus K) < \varepsilon$ and f is continuous in $K \subseteq F \subseteq E$ as the uniform limit of continuous functions. □

Exercises

ex8.2.1. Show that continuous functions are measurable.

ex8.2.2. Show that monotone functions are measurable.

ex8.2.3. Show that if f is measurable, then $f^{-1}(B)$ is measurable for every Borel set $B \subseteq \mathbb{R}$.

ex8.2.4. Show that the measurability of f is equivalent to any of the following:

- $\{x : f(x) \leq \alpha\}$ is measurable for all $\alpha \in \mathbb{R}$;
- $\{x : f(x) > \alpha\}$ is measurable for all $\alpha \in \mathbb{R}$;
- $\{x : f(x) \geq \alpha\}$ is measurable for all $\alpha \in \mathbb{R}$.

ex8.2.5. Prove Proposition 8.2.2.

ex8.2.6. Complete the proof of Theorem 8.2.5.

ex8.2.7. Show that if $\varphi_1, \ldots, \varphi_n$ are simple functions, then any linear combination of these functions, $\alpha_1\varphi_1 + \cdots + \alpha_n\varphi_n$, is a simple function.

ex8.2.8. Show that if f is a nonnegative measurable function on $\mathbb{R}$, then the simple functions $\{\varphi_n\}$ constructed in Theorem 8.2.4 to converge to f pointwise are nonnegative and that their convergence to f is monotone nondecreasing.

8.3. The Lebesgue integral

The (proper) Riemann integral of a function f is defined only on compact intervals. In contrast, the Lebesgue integral of a (measurable) function f, may be defined on any *measurable* set $E \subseteq \mathbb{R}$.

We begin by defining the *sum* of a simple function on a set $E \subseteq \mathbb{R}$, and then define the Lebesgue integral of general measurable function in terms of these sums.

8.3.1. Summing simple functions

Let $E \subseteq \mathbb{R}$ and $\varphi = \sum_{j=1}^{n} a_j \chi_{A_j}$ be a simple function.

Definition. If $\mu(A_j \cap E) < \infty$ for all $a_j \neq 0$, then the *sum* of φ on E, $\mathcal{S}_E(\varphi)$, defined by

$$\mathcal{S}_E(\varphi) = \sum_{j=1}^{n} a_j \cdot \mu(A_j \cap E),$$

is finite, and we say that φ is *summable* on E. If $\mu\left(\bigcup_{a_j<0}(A_j \cap E)\right) < \infty$, but $\mu\left(\bigcup_{a_j>0}(A_j \cap E)\right) = \infty$, we set $\mathcal{S}_E(\varphi) = \infty$, and if $\mu\left(\bigcup_{a_j>0}(A_j \cap E)\right) < \infty$, but $\mu\left(\bigcup_{a_j<0}(A_j \cap E)\right) = \infty$, we set $\mathcal{S}_E(\varphi) = -\infty$. In both these cases, $\mathcal{S}_E(\varphi)$ is defined, but φ is *not* summable on E.

Finally, if $\mu\left(\bigcup_{a_j>0}(A_j \cap E)\right) = \mu\left(\bigcup_{a_j<0}(A_j \cap E)\right) = \infty$, then $\mathcal{S}_E(\varphi)$ is not defined.

Once again, when we write $\mathcal{S}_E(\varphi) = \pm\infty$, this is its value in the extended real numbers.

Proposition. *If φ is summable on E, then both φ^+ and φ^- are summable on E and $\mathcal{S}_E(\varphi) = \mathcal{S}_E(\varphi^+) - \mathcal{S}_E(\varphi^-)$.*

Proof. Exercise. □

Theorem. *Let φ and ψ be summable simple functions on E.*

(i) *For any $\alpha, \beta \in \mathbb{R}$, $\alpha\varphi + \beta\psi$ is summable on E, and*

$$\mathcal{S}_E(\alpha\varphi + \beta\psi) = \alpha\mathcal{S}_E(\varphi) + \beta\mathcal{S}_E(\psi).$$

(ii) *If $\varphi \geq \psi$ in E, then*

$$\mathcal{S}_E(\varphi) \geq \mathcal{S}_E(\psi).$$

Proof.

(i) Write $\varphi(x) = \sum_{j=1}^{n} a_j \chi_{A_j}(x)$ and $\psi(x) = \sum_{i=1}^{m} b_i \chi_{B_i}(x)$, and let $C_{ij} = A_j \cap B_i$ for $1 \leq j \leq n$ and $1 \leq i \leq m$. For each i, $\sum_{j=1}^{n} \mu(C_{ij} \cap E) = \mu(B_i \cap E)$, and for each j, $\sum_{i=1}^{m} \mu(C_{ij} \cap E) = \mu(A_j \cap E)$, since the sets $\{A_1, \ldots, A_n\}$ are pairwise disjoint, as are the sets $\{B_1, \ldots, B_m\}$. Moreover, since

$$\mu(C_{ij} \cap E) \leq \min(\mu(A_j \cap E), \mu(B_i \cap E))$$

is finite for each pair i, j, it follows that

$$\alpha\varphi(x) + \beta\psi(x) = \sum_{i=1}^{m}\sum_{j=1}^{n}(\alpha a_j + \beta b_i)\chi_{C_{ij}},$$

is summable on E, and

$$\begin{aligned}
\mathcal{S}_E(\alpha\varphi + \beta\psi) &= \sum_{i=1}^{m}\sum_{j=1}^{n}(\alpha a_j + \beta b_i)m(C_{ij} \cap E) \\
&= \alpha \sum_{j=1}^{n} a_j \sum_{i=1}^{m} \mu(C_{ij} \cap E) + \beta \sum_{i=1}^{m} b_i \sum_{j=1}^{n} \mu(C_{ij} \cap E) \\
&= \alpha \sum_{j=1}^{n} a_j \mu(A_j \cap E) + \beta \sum_{i=1}^{m} b_i \mu(B_i \cap E) \\
&= \alpha\mathcal{S}_E(\varphi) + \beta\mathcal{S}_E(\psi).
\end{aligned}$$

(ii) If $\varphi \geq \psi$ on E, then it follows from (i) that

$$\mathcal{S}_E(\varphi) - \mathcal{S}_E(\psi) = \mathcal{S}_E(\varphi - \psi) \geq 0,$$

because all the terms in the sum $\mathcal{S}_E(\varphi - \psi)$ are nonnegative. □

8.3.2. The Lebesgue integral

Let f be a measurable function on $E \subseteq \mathbb{R}$. If $f \geq 0$ on E, we denote by $\Sigma_E(f)$ the set of all simple functions φ such that $f \geq \varphi \geq 0$ on E.

Definition. If $f \geq 0$, then the *Lebesgue integral* of f on E is defined

$$\int_E f\, d\mu = \sup_{\Sigma_E(f)} \mathcal{S}_E(\varphi).$$

This includes the possibility that $\int_E f\,d\mu = \infty$. If f is any measurable function, with both $\int_E f^+\,d\mu < \infty$ and $\int_E f^-\,d\mu < \infty$, then the Lebesgue integral of f on E is defined.

$$\int_E f\,d\mu = \int_E f^+\,d\mu - \int_E f^-\,d\mu.$$

If $\int_E f^+\,d\mu < \infty$ or $\int_E f^-\,d\mu < \infty$, but not both, then $\int_E f\,d\mu = -\infty$ in the first case, and $\int_E f\,d\mu = \infty$ in the second case, and if $\int_E f^+\,d\mu = \int_E f^-\,d\mu = \infty$, then $\int_E f\,d\mu$ is *not* defined. The function f is *Lebesgue integrable* (or *summable*) on E, if $\int_E f\,d\mu < \infty$. To be clear, $\int_E f\,d\mu$ may still be defined, even if f is not integrable on E.

If there is no possibility of confusion (with Riemann integrability, for example), we will write "integrable" instead of "Lebesgue integrable".

Proposition. *If the simple function φ is summable on $E \subset \mathbb{R}$ in the sense of Definition* 8.3.1, *then φ is integrable on E and*

$$\int_E \varphi\,d\mu = \mathcal{S}_E(\varphi).$$

I.e., the two definitions of summability for simple functions agree.

Proof. If $\varphi \geq 0$ on E, then it follows that $\varphi \in \Sigma_E(\varphi)$ and $\mathcal{S}_E(\varphi) \geq \mathcal{S}_E(\psi)$ for all $\psi \in \Sigma_E(\varphi)$, by Theorem 8.3.1 (ii). Hence $\int_E \varphi\,d\mu = \mathcal{S}_E(\varphi)$ in this case. The general case follows from this and Proposition 8.3.1. □

Theorem. *The measurable function f is integrable on E if and only if $|f|$ is integrable on E, in which case*

$$\left|\int_E f\,d\mu\right| \leq \int_E |f|\,d\mu. \tag{8.3.1}$$

Proof. If $|f| = f^+ + f^-$ is integrable on E, then both f^+ and f^- must both have finite integrals, so f is integrable on E.

If f is integrable on E, then $\int_E |f|\,d\mu$ is defined because $|f| = f^+ + f^-$ is both measurable and nonnegative. Moreover, the integrability of f means that both $\int_E f^+\,d\mu$ and $\int_E f^-\,d\mu$ are finite (and nonnegative), so the integrability of $|f|$ follows from the identity

$$\int_E |f|\,d\mu = \int_E f^+\,d\mu + \int_E f^-\,d\mu, \tag{8.3.2}$$

which we prove below. The inequality (8.3.1) follows from this and the triangle inequality.

To prove (8.3.2), we first note that for any $\varphi^+ \in \Sigma_E(f^+)$ and $\varphi^- \in \Sigma_E(f^-)$, we have $\varphi^+ + \varphi^- \in \Sigma_E(|f|)$. Moreover,

$$\mathcal{S}_E(\varphi^+ + \varphi^-) = \mathcal{S}_E(\varphi^+) + \mathcal{S}_E(\varphi^-),$$

because for any $x \in E$, $f^+(x) = 0$ if $f^-(x) > 0$ and vice versa, so the same must be true for φ^+ and φ^-. It follows that

$$\sup_{\Sigma_E E(|f|)} \mathcal{S}_E(\varphi) \geq \sup_{\Sigma_E E(f^+)} \mathcal{S}_E(\varphi^+) + \sup_{\Sigma_E E(f^-)} \mathcal{S}_E(\varphi^-). \tag{8.3.3}$$

Conversely, let $E^+ = \{x \in E : f^+(x) > 0\}$ and $E^- = \{x \in E : f^-(x) > 0\}$. These sets are measurable and disjoint from each other, $|f(x)| = f^+(x)$ in E^+, $|f(x)| = f^-(x)$ in E^-, and $|f(x)| = 0$ for all other $x \in E$. It follows that for any $\varphi \in \Sigma_E(|f|)$, if $\varphi^+ = \varphi \cdot \chi_{E^+}$ and $\varphi^- = \varphi \cdot \chi_{E^-}$, then $\varphi^+ \in \Sigma_E(f^+)$, $\varphi^- \in \Sigma_E(f^-)$, and once again $\mathcal{S}_E(\varphi) = \mathcal{S}_E(\varphi^+ + \varphi^-) = \mathcal{S}_E(\varphi^+) + \mathcal{S}_E(\varphi^-)$. Hence, $\sup_{\Sigma_E E(|f|)} \mathcal{S}_E(\varphi) \leq \sup_{\Sigma_E E(f^+)} \mathcal{S}_E(\varphi^+) + \sup_{\Sigma_E E(f^-)} \mathcal{S}_E(\varphi^-)$, which together with (8.3.3) shows that

$$\sup_{\Sigma_E E(|f|)} \mathcal{S}_E(\varphi) = \sup_{\Sigma_E E(f^+)} \mathcal{S}_E(\varphi^+) + \sup_{\Sigma_E E(f^-)} \mathcal{S}_E(\varphi^-),$$

from which (8.3.2) follows. □

8.3.3. "Countable additivity" of the Lebesgue integral

The countable additivity of Lebesgue measure says that if $\{E_n\}$ are all measurable and $E_i \cap E_j = \varnothing$ when $i \neq j$, then

$$\mu\left(\bigcup_n E_n\right) = \sum_n \mu(E_n).$$

An analogous property holds for the Lebesgue integral of a nonnegative measurable function, and more generally for the Lebesgue integral of an integrable function.

Lemma. *If E_1 and E_2 are disjoint measurable sets, and $\int f\,d\mu$ is defined on both, then $\int f\,d\mu$ is defined on $E = E_1 \cup E_2$ and*

$$\int_E f\,d\mu = \int_{E_1} f\,d\mu + \int_{E_2} f\,d\mu. \tag{8.3.4}$$

Proof. To begin, the assumptions imply that $E = E_1 \cup E_2$ is measurable and that f is measurable on E.

Consider first the case that $f \geq 0$ on E. In this case $\int_E f\,d\mu$ is defined, and for any $\varphi_1 \in \Sigma_{E_1}(f)$ and any $\varphi_2 \in \Sigma_{E_2}(f)$,

$$\varphi = \varphi_1 \cdot \chi_{E_1} + \varphi_2 \cdot \chi_{E_2} \in \Sigma_E(f),$$

and $\mathcal{S}_E(\varphi) = \mathcal{S}_{E_1}(\varphi_1) + \mathcal{S}_{E_2}(\varphi_2)$. Conversely, if $\varphi \in \Sigma_E(f)$, then $\varphi \in \Sigma_{E_1}(f)$ and $\varphi \in \Sigma_{E_2}(f)$, and again, $\mathcal{S}_E(\varphi) = \mathcal{S}_{E_1}(\varphi) + \mathcal{S}_{E_2}(\varphi)$. It follows that

$$\int_E f\,d\mu = \sup_{\Sigma_E(f)} \mathcal{S}_E(\varphi) = \sup_{\Sigma_{E_1}} \mathcal{S}_{E_1}(\varphi_1) + \sup_{\Sigma_{E_2}} \mathcal{S}_{E_2}(\varphi_2) = \int_{E_1} f\,d\mu + \int_{E_2} f\,d\mu,$$

so the claim is true for nonnegative measurable functions.

If f is integrable on E_1 and E_2, then (8.3.4) is true for f, because it is true for f^+ and f^-. □

Theorem. *If $\{E_n\}_{n=1}^{\infty}$ are all measurable and $E_i \cap E_j = \varnothing$, and f is measurable and nonnegative on E_n for every n, then $\int f\,d\mu$ is defined on $E = \bigcup_n E_n$, and*

$$\int_E f\,d\mu = \sum_{n=1}^{\infty} \int_{E_n} f\,d\mu. \tag{8.3.5}$$

This identity is also true if f is integrable on E.

Proof. Let $F_n = \bigcup_{i=1}^n E_i$, then for any $n \in \mathbb{N}$,

$$\int_{F_n} f\,d\mu = \sum_{i=1}^{n} \int_{E_i} f\,d\mu,$$

by the lemma above and induction on n. It also follows from the lemma that $\int_E f\,d\mu = \int_{F_n} f\,d\mu + \int_{E\setminus F_n} f\,d\mu$, so $\int_E f\,d\mu \geq \sum_{i=1}^n \int_{E_i} f\,d\mu$, for every $n \in \mathbb{N}$ because $\int_{E\setminus F_n} f\,d\mu \geq 0$, and therefore

$$\int_E f\,d\mu \geq \sum_{n=1}^{\infty} \int_{E_n} f\,d\mu.$$

On the other hand, if $\varphi \in \Sigma_E(f)$, then $\varphi \in \Sigma_{E_n}(f)$ for each n, and it follows from the pairwise disjointness of the $\{E_n\}$ that

$$\mathcal{S}_E(\varphi) = \sum_{n=1}^{\infty} \mathcal{S}_{E_n}(\varphi) \leq \sum_{n=1}^{\infty} \int_{E_n} f\,d\mu,$$

and therefore

$$\int_E f\,d\mu = \sup_{\Sigma_E(f)} \mathcal{S}_E(\varphi) \leq \sum_{n=1}^{\infty} \int_{E_n} f\,d\mu.$$

Thus the claim is true for nonnegative measurable functions, and hence it is true if f is integrable on E, since it is true for f^+ and f^-. □

Corollary. *Suppose that the sets $\{E_n\}$ are all measurable, and for each n, and $E_n \subseteq E_{n+1}$. If f is measurable and nonnegative, or f is integrable, on $E = \bigcup_n E_n$, then*

$$\int_E f\,d\mu = \lim_{n\to\infty} \int_{E_n} f\,d\mu.$$

Proof. Let $A_1 = E_1$ and for $n \geq 2$, let $A_n = E_n \setminus E_{n-1}$. The sets $\{A_n\}$ are measurable and pairwise disjoint, $E_n = \bigcup_{k\leq n} A_k$ for each n, and in particular, $E = \bigcup_n A_n$. If f is measurable and nonnegative on E, then it follows from the Theorem above that

$$\int_E f\,d\mu = \sum_{n=1}^{\infty} \int_{A_n} f\,d\mu = \lim_{n\to\infty} \left(\sum_{k=1}^{n} \int_{A_k} f\,d\mu \right) = \lim_{n\to\infty} \int_{E_n} f\,d\mu,$$

and the same is true if f is integrable on E. □

8.3.4. The (in)significance of sets of measure zero

When it comes to integrals, the behavior of functions on sets of measure zero can be ignored.

Proposition. *If f_n is measurable on E for every n, and $f_n \to f$ pointwise a.e. in E, then f is measurable in E.*

Proof. Let $E_1 = \{x \in E : f_n(x) \to f(x)\}$ and $E_2 = E \setminus E_1$, so that $\mu(E_2) = 0$. The set E_1 is measurable and it follows from Corollary 8.2.5 that f is measurable in E_1. Hence, the set $\{x \in E_1 : f(x) \leq \alpha\}$ is measurable for any $\alpha \in \overline{\mathbb{R}}$, and therefore

$$\{x \in E : f(x) \leq \alpha\} = \{x \in E_1 : f(x) \leq \alpha\} \cup \{x \in E_2 : f(x) \leq \alpha\}$$

is likewise measurable because the set on the right has measure zero. □

Lemma. *Suppose that $E = E_1 \cup E_2$, $E_1 \cap E_2 = \varnothing$, and $\mu(E_2) = 0$. If f is measurable in E, then $\int_{E_2} f\,d\mu = 0$ and $\int_E f\,d\mu$ is defined if and only if $\int_{E_1} f\,d\mu$ is defined, in which case*

$$\int_E f\,d\mu = \int_{E_1} f\,d\mu.$$

In other words, the behavior of functions on sets of measure zero has no effect on their integrability.

Proof. First, if $f \geq 0$ on E_2, then $\mathcal{S}_{E_2}(\varphi) = 0$ for all $\varphi \in \Sigma_{E_2}(f)$, and hence $\int_{E_2} f\,d\mu = 0$. The same is therefore true for any measurable function h, since it is true for h^+ and h^-.

Next, $E_1 = E \setminus E_2$ is measurable, and if $f \geq 0$ on E, then

$$\int_E f\,d\mu = \int_{E_1} f\,d\mu + \int_{E_2} f\,d\mu = \int_{E_1} f\,d\mu,$$

by Lemma 8.3.3. Thus the second claim is true for nonnegative functions, and therefore true in general. □

It follows from this that if E has measure zero, and $f(x) = \infty$ for all $x \in E$, then $\int_E f\,d\mu = 0$. This could be interpreted as saying that $0 \cdot \infty = 0$,[11] but what it actually says in this case is that $\sup_n\{x_n \cdot 0\} = 0$, for any sequence $\{x_n\} \subset \mathbb{R}$, even if $x_n \to \infty$.

Theorem. *Suppose that $f(x)$ and $g(x)$ are measurable functions on E.*

(i) *If $f = g$ a.e. on E, then $\int_E f\,d\mu$ is defined if and only if $\int_E g\,d\mu$ is defined, in which case*

$$\int_E f\,d\mu = \int_E g\,d\mu.$$

(ii) *If $f \geq g$ a.e. on E and $\int_E f\,d\mu$ and $\int_E g\,d\mu$ are both defined, then*

$$\int_E f\,d\mu \geq \int_E g\,d\mu.$$

Proof.

(i) If $f = g$ a.e. in E then $f^+ = g^+$ and $f^- = g^-$ a.e. in E, so it is enough to prove the claim when f and g are nonnegative a.e. on E. Let $E_1 = \{x \in E : f(x) = g(x) \geq 0\}$ and $E_2 = E \setminus E_1$. It then follows from Lemma 8.3.4 that

$$\int_E f\,d\mu = \int_{E_1} f\,d\mu = \int_{E_1} g\,d\mu = \int_E g\,d\mu,$$

since $\mu(E_2) = 0$.

(ii) Suppose first that $0 \leq g \leq f$ a.e. on E, and let $E_1 = \{x \in E : g(x) \leq f(x)\}$. Then $\Sigma_{E_1}(g) \subseteq \Sigma_{E_1}(f)$, so

$$\int_E f\,d\mu = \int_{E_1} f\,d\mu \geq \int_{E_1} g\,d\mu = \int_E g\,d\mu,$$

[11] And sometimes it is.

by part (i). In general, if $f \geq g$ a.e. in E, then $f^+ \geq g^+$ and $f^- \leq g^-$ a.e. in E, so

$$\int_E f\,d\mu = \int_E f^+\,d\mu - \int_E f^-\,d\mu \geq \int_E g^+\,d\mu - \int_E g^-\,d\mu = \int_E g\,d\mu. \qquad \square$$

8.3.5. Essentially bounded functions

Definition. A measurable function f is *essentially bounded* in $E \subseteq \mathbb{R}$ if there is some $B > 0$ such that the set $\{x \in E : |f(x)| > B\}$ has measure zero. I.e., f is essentially bounded in E if f is equal a.e. in E to a bounded function.

Lemma. *If f is measurable and essentially bounded on E and $\mu(E) < \infty$ then f is integrable on E.*

Proof. If f is measurable and $|f| < B$ a.e. in E for some $B < \infty$, then both $f^+ < B$ and $f^- < B$, so $\mathcal{S}_E(\varphi) < B \cdot \mu(E) < \infty$ for any φ in $\Sigma_E(f^+) \cup \Sigma_E(f^-)$. Hence $\int_E f^+\mu$ and $\int_E f^-\mu$ are both bounded by $B \cdot \mu(E)$. $\square$

The converse is not true, as illustrated in Example 8.4.2. On the other hand, integrable functions are *almost* essentially bounded in the following sense.

Proposition. *If f is integrable on $E \subseteq \mathbb{R}$, then $\mu(\{x \in E : f(x) = \pm\infty\}) = 0$.*

Proof. If f is integrable on E, then $\int_E f^+\,d\mu = \beta$ for some $\beta \in [0, \infty)$. Let $E_+ = \{x \in E : f^+(x) = \infty\}$, then $\varphi = (\beta + 1) \cdot \chi_{E_+} \in \Sigma_E(f^+)$. Hence,

$$(\beta + 1) \cdot \mu(E_+) \leq \int_E f^+\,d\mu = \beta,$$

from which it follows that $\mu(E_+) = 0$, and $\mu(\{x \in E : f(x) = -\infty\}) = 0$, by the analogous argument applied to f^-. $\square$

Exercises

ex8.3.1. Prove Proposition 8.3.1.

ex8.3.2. Show that if f is integrable on E, then for all $\alpha \in \mathbb{R}$, αf is integrable on E and $\int_E \alpha f\,d\mu = \alpha \int_E f\,d\mu$.

ex8.3.3. Show that if f is integrable on E, $\alpha \leq f \leq \beta$ a.e. on E, and $\mu(E) < \infty$, then $\alpha\mu(E) \leq \int_E f\,d\mu \leq \beta\mu(E)$.

ex8.3.4. Show that if f is integrable on E, then for any $\varepsilon > 0$, there is a B_ε such that $\mu(\{x \in E : |f(x)| \geq B_\varepsilon\}) < \varepsilon$.

8.4. Convergence theorems

For the Riemann integral, the claim that $\int_a^b f(x)\,dx = \lim_{n\to\infty} \int_a^b f_n(x)\,dx$ is only generally true if the Riemann integrable functions $\{f_n\}$ converge uniformly to f in $[a, b]$. For the Lebesgue integral, the identity

$$\int_E \lim_{n\to\infty} f_n\,d\mu = \lim_{n\to\infty} \int_E f_n\,d\mu$$

holds under less restrictive conditions: that the sequence $\{f_n\}$ of measurable functions converges pointwise in E, and the convergence isn't "too wild". The resulting convergence theorems can then be used to establish other important properties of the Lebesgue integral.

8.4.1. Lebesgue's monotone convergence theorem

Theorem. *If $\{f_n\}$ is a sequence of measurable functions on E, $0 \le f_n \le f_{n+1}$ for all n, and $f_n \to f$ pointwise a.e. in E, then*

$$\lim_{n\to\infty} \int_E f_n\,d\mu = \int_E f\,d\mu.$$

Proof. We first note that f is measurable on E, by Proposition 8.3.4, and $f \ge 0$ in E, so $\int_E f\,d\mu$ is defined. Moreover, for all $n \in \mathbb{N}$,

$$\int_E f_n\,d\mu \le \int_E f_{n+1}\,d\mu \le \int_E f\,d\mu,$$

by Theorem 8.3.4 (ii), since $f_n \le f_{n+1} \le f$. Hence $\lim_n \int_E f_n\,d\mu$ exists, though it may be infinite, and in any case,

$$\lim_{n\to\infty} \int_E f_n\,d\mu \le \int_E f\,d\mu.$$

For $0 < \delta < 1$, let $E_{n,\delta} = \{x \in E : f_n(x) \ge \delta \cdot f(x)\}$. Then $E_{n,\delta}$ is measurable for all n (because f and f_n are), $E_{n,\delta} \subseteq E_{n+1,\delta}$ for all n and $E \setminus \bigcup_n E_n$ has measure zero. Therefore,

$$\lim_{n\to\infty} \int_E f_n\,d\mu \ge \lim_{n\to\infty} \int_{E_n} f_n\,d\mu \ge \delta \lim_{n\to\infty} \int_{E_n} f\,d\mu = \delta \int_E f\,d\mu,$$

by Lemma 8.3.3, Theorem 8.3.4 (ii) and Corollary 8.3.3. Since this inequality holds for any $\delta < 1$, it follows that $\lim_{n\to\infty} \int_E f_n\,d\mu \ge \int_E f\,d\mu$, and the claim follows. □

Corollary. *If f and g are both integrable on E, then $f + g$ is integrable on E and*

$$\int_E f + g\,d\mu = \int_E f\,d\mu + \int_E g\,d\mu. \tag{8.4.1}$$

Proof. We prove the claim for nonnegative measurable functions, from which the claim follows immediately for integrable functions.

If f and g are measurable and nonnegative on E, then there are sequences $\{\varphi_n\} \in \Sigma_E(f)$ and $\{\psi_n\} \in \Sigma_E(g)$ such that $\varphi_n \to f$ and $\psi_n \to g$ pointwise in E, by Theorem 8.2.4. Furthermore, we may assume that the convergence is monotone increasing, as indicated in the comment following the proof of that theorem. It follows that $\{\varphi_n + \psi_n\}$ increases pointwise to $f+g$ monotonically in E, and by the monotone convergence theorem,

$$\begin{aligned}\int_E f+g\,d\mu &= \lim_{n\to\infty}\int_E \varphi_n+\psi_n\,d\mu \\ &= \lim_{n\to\infty}\left(\int_E \varphi_n\,d\mu + \int_E \psi_n\,d\mu\right) \\ &= \int_E f\,d\mu + \int_E g\,d\mu,\end{aligned}$$

where the second equality in the first line follows from the fact that (8.4.1) is true for simple functions, by Theorem 8.2.4. □

8.4.2. The Lebesgue integral and the Riemann integral on $[a,b]$

Another important consequence of the monotone convergence theorem is the fact that the Lebesgue integral and Riemann integral agree when both are defined.

Theorem. *If f is Riemann integrable on the interval $I=[a,b]$, then f is Lebesgue integrable there and*

$$\int_a^b f(x)\,dx = \int_I f\,d\mu.$$

We use the familiar notation $\int_a^b f(x)\,dx$ for the Riemann integral, and $\int_I f\,d\mu$ for the Lebesgue integral to distinguish them when necessary. If the distinction is not important, it is common to write $\int_a^b f(x)\,dx$ even when the integrand is Lebesgue integrable on I, but not Riemann integrable there.

Proof. First, if f is Riemann integrable on I, then f is is bounded there and measurable on I by Proposition 8.2.1. So it follows from Lemma 8.3.5 that f is Lebesgue integrable on I, since $\mu(I)=b-a<\infty$.

We next observe that both f^+ and f^- are bounded on I, and f^+ and f^- are both continuous wherever f is, so their sets of points of discontinuity have measure 0. Hence if f is Riemann integrable on I, then f^+ and f^- are also Riemann integrable on I, so it is enough to prove the claim for $f\ge 0$.

By Theorem 7.1.4 and Lemma 7.1.5, there is a sequence of partitions $\{P_n\}$ of I such that for $n\ge 1$, P_{n+1} is a refinement of P_n, $\operatorname{mesh}(P_n)<1/n$ and

$$U(f,P_n)-L(f,P_n)<\frac{1}{n}.$$

Hence, $\int_a^b f(x)\,dx = \lim_{n\to\infty} L(f,P_n)$.

For each partition $P_n=\{t_0,\dots,t_{k_n}\}$ in this sequence, let J_i be the half-open interval $[t_{i-1},t_i)$, for $1\le i<k_n$, let $J_{k_n}=[t_{k_n-1},b]$, and define the simple function φ_n by setting

$$\varphi_n=\sum_{i=1}^{k_n} m_i\cdot\chi_{J_i},$$

where m_i is the infimum of f in the closure, $[t_{i-1}, t_i]$, of J_i. For each n, we have

$$L(f, P_n) = \sum_{i=1}^{k_n} m_i \cdot (t_i - t_{i-1}) = \mathcal{S}_I(\varphi_n) = \int_I \varphi_n \, d\mu,$$

and therefore

$$\int_a^b f \, dx = \lim_{n\to\infty} L(f, P_n) = \lim_{n\to\infty} \int_I \varphi_n \, d\mu. \tag{8.4.2}$$

Since P_{n+1} is a refinement of P_n and $f \geq 0$, it follows that $0 \leq \varphi_n \leq \varphi_{n+1} \leq f$ for all n. In particular, $\{\varphi_n(x)\}$ is bounded, and monotone increasing for each $x \in I$, and so converges to a finite limit. In other words, $\varphi_n \to \tilde{f}$ pointwise, for some function $\tilde{f}$, which is measurable by Corollary 8.2.4. By the monotone convergence theorem and (8.4.2) it follows that

$$\int_a^b f \, dx = \lim_{n\to\infty} \int_I \varphi_n \, d\mu = \int_I \tilde{f} \, d\mu,$$

and to complete the proof, it is therefore enough to show that $f = \tilde{f}$ a.e. in I.

To do this, it is enough to show that $f(x) = \tilde{f}(x)$ at every point of continuity of f in (a, b), since the set of points of discontinuity of f has measure zero in $[a, b]$. If $a < x_0 < b$ is a point of continuity of f, then for every $\varepsilon > 0$ there is a $\delta > 0$ such that $|f(x) - f(x_0)| < \varepsilon$ if $x \in I$ and $|x - x_0| < \delta$. For any $n > 1/\delta$, let $[t_{i-1}, t_i]$ be the subinterval of P_n such that $t_{i-1} \leq x_0 < t_i$. Since the mesh of P_n is less than δ, we have $|f(x_0) - f(x)| < \varepsilon$ for all $x \in [t_{i-1}, t_i]$, so

$$0 \leq f(x_0) - \varphi_n(x_0) = f(x_0) - m_i \leq \varepsilon,$$

for all $n \geq 1/\delta$. Since this is true for any $\varepsilon > 0$, it follows that

$$f(x_0) = \lim_{n\to\infty} \varphi_n(x_0) = \tilde{f}(x_0)$$

at every point of continuity of f in (a, b). □

The converse is not true. For example, if $f = \chi_{\mathbb{Q}}$, the characteristic function of the rationals, then $\int_E \chi_{\mathbb{Q}} \, d\mu = 0$ for any measurable set E, since $\chi_{\mathbb{Q}} = 0$ a.e. On the other hand, f is not Riemann integrable in any interval $[a, b]$ because every point is a point of discontinuity of this function. Unbounded functions may also be Lebesgue integrable.

Example. Let $f(x) = x^{-1/2}$, with $f(0) = \infty$, so f is not bounded in $[0, 1]$. For $n \in \mathbb{N}$, the function $f_n(x) = x^{-1/2} \cdot \chi_{[1/n,1]}$ is Riemann integrable on $[0, 1]$, and so Lebesgue integrable there, and

$$\int_{[0,1]} f_n \, d\mu = \int_0^1 f_n(x) \, dx = \int_{1/n}^1 x^{-1/2} \, dx = 2 - \frac{2}{\sqrt{n}}.$$

On the other hand, $\lim_{n\to\infty} f_n = f$ a.e. in $[0, 1]$, with $0 \leq f_n \leq f$ for all n, and by the monotone convergence theorem

$$\int_{[0,1]} f \, d\mu = \lim_{n\to\infty} \int_{[0,1]} f_n \, d\mu = 2,$$

so f is Lebesgue integrable on $[0, 1]$.

Comment. The limit we used to compute $\int_{[0,1]} f\, d\mu$ in this example is the same we would use to compute the *improper* Riemann integral of f on $[0,1]$.

8.4.3. Fatou's lemma and the dominated convergence theorem

Lemma (Fatou). *Let $\{f_n\}$ be a sequence of nonnegative measurable functions on E, then*

$$\liminf_{n\to\infty} \int_E f_n\, d\mu \geq \int_E \liminf_{n\to\infty} f_n\, d\mu, \tag{8.4.3}$$

where the integrals may be infinite.

Proof. First, by Theorem 8.2.5, $f = \liminf f_n$ is measurable, and $f \geq 0$ on E, since the same is true for each f_n, so the integral on the right of (8.4.3) is defined.

For $n \geq 1$, let $g_n = \inf_{m\geq n} f_m$. Then $g_{n+1} \geq g_n \geq 0$ in E and g_n is measurable there for all n (by Theorem 8.2.5 again). Moreover $g_n \to f$ pointwise in E, so by the monotone convergence theorem and the fact that $\int_E f_n\, d\mu \geq \int_E g_n\, d\mu$ for all n, it follows that

$$\liminf_{n\to\infty} \int_E f_n\, d\mu \geq \lim_{n\to\infty} \int_E g_n\, d\mu = \int_E f\, d\mu.$$

□

Theorem (Fatou-Lebesgue). *Let $\{f_n\}$ be a sequence of measurable functions on E. If there is a function F that is Lebesgue integrable on E, such that $F \geq |f_n|$ in E for all n, then f_n,* $\limsup f_n$ *and* $\liminf f_n$ *are all integrable on E and*

$$\int_E \limsup_{n\to\infty} f_n\, d\mu \geq \limsup_{n\to\infty} \int_E f_n\, d\mu \geq \liminf_{n\to\infty} \int_E f_n\, d\mu \geq \int_E \liminf_{n\to\infty} f_n\, d\mu. \tag{8.4.4}$$

Proof. To begin, $\int_E |f_n|\, d\mu \leq \int_E F\, d\mu < \infty$ for all n, so f_n is integrable on E for all n. Next, the assumption that $F \geq |f_n|$ for all n, implies that $F \geq |\limsup f_n|$ and $F \geq |\liminf f_n|$, so both $\limsup f_n$ and $\liminf f_n$ are integrable on E, by the same argument.

The middle inequality in (8.4.4) is true for any sequence of real numbers. and since $F + f_n \geq 0$ for all n, applying Fatou's lemma gives

$$\int_E F\, d\mu + \liminf_{n\to\infty} \int_E f_n\, d\mu \geq \int_E \liminf_{n\to\infty}(F + f_n)\, d\mu = \int_E F\, d\mu + \int_E \liminf_{n\to\infty} f_n\, d\mu,$$

from which the right-hand inequality follows.

Finally, applying Fatou's lemma to the nonnegative sequence $\{F - f_n\}$, gives the inequality

$$\liminf \int_E F - f_n\, d\mu \geq \int_E \liminf(F - f_n)\, d\mu. \tag{8.4.5}$$

Now, if $\{a_n\} \subset \mathbb{R}$ and $b \geq |a_n|$ for all n, then $\liminf(b - a_n) = b - \limsup a_n$, and it follows from this and (8.4.5) that

$$\int_E F\, d\mu - \limsup \int_E f_n\, d\mu \geq \int_E F\, d\mu - \int_E \limsup f_n\, d\mu,$$

which proves the left-hand inequality of (8.4.4). □

Corollary (The dominated convergence theorem). *Suppose that $\{f_n\}$ is a sequence of measurable functions on E and $f_n \to f$ pointwise a.e. in E. If there is a function F that is integrable on E and $|f_n| \le F$ for all n, then f is integrable on E and*

$$\int_E f\,d\mu = \lim_{n\to\infty} \int_E f_n\,d\mu. \tag{8.4.6}$$

Proof. If $f_n \to f$, then $\liminf f_n = \limsup f_n = f$ a.e. in E, so f is integrable on E by the Fatou-Lebesgue theorem. Furthermore, (8.4.4) becomes

$$\int_E f\,d\mu \ge \limsup_{n\to\infty} \int_E f_n\,d\mu \ge \liminf_{n\to\infty} \int_E f_n\,d\mu \ge \int_E f\,d\mu,$$

from which (8.4.6) follows. □

Exercises

ex8.4.1. Suppose that $\{f_n\}$ is a sequence of nonnegative measurable functions on E, and $f_n(x) \ge f_{n+1}(x)$ a.e. in E for all n. Show that $\{f_n\}$ converges pointwise a.e. to a nonnegative measurable function f and

$$\lim_{n\to\infty} \int_E f_n\,d\mu = \int_E f\,d\mu.$$

ex8.4.2. Let f_n be nonnegative measurable functions on $[0,1]$ and assume that the series $\sum_{n-1}^{\infty} f_n(x)$ converges a.e. to a bounded function F in $[0,1]$. Show that

$$\int_0^1 F(x)dx = \sum_1^{\infty} \int_0^1 f_n(x)dx.$$

ex8.4.3. Suppose that f is integrable on a set E. Show that for any $\varepsilon > 0$ there is a $\delta > 0$ such that if $U \subset E$ has measure $< \delta$, then

$$\int_U |f|\,d\mu < \varepsilon.$$

Hint: Prove this first for positive simple functions.

8.5. The fundamental theorem of calculus for the Lebesgue integral

If f is Riemann integrable on $[a,b]$, then the fundamental theorem of calculus says that the *primitive* function, $F(x) = \int_a^x f(t)\,dt$ is differentiable at every point of continuity of f in $[a,b]$ with $F'(x) = f(x)$ at such points. In particular, F is differentiable a.e. in $[a,b]$, because the Riemann integrable function f is continuous a.e. in $[a,b]$. Conversely, we showed that if F' is Riemann integrable on $[a,b]$, then $F(x) = F(a) + \int_a^x F'(t)\,dt$ for $x \in [a,b]$.

In this section, we extend fundamental theorem to the Lebesgue integral, and in this context, we will express the (Lebesgue) integrals over a closed intervals $[u,v]$ as $\int_u^v f\,d\mu$.

8.5.1. The Cantor function, again

We begin with an observation about something that the fundamental theorem does not claim. Specifically, it is not necessarily true that if f is differentiable a.e. in $[a, b]$ and f' is Lebesgue integrable there, then $f(x) = \int_a^x f'(t)\,dt$. We illustrate this with the Cantor function, $\Phi(x)$.

We studied $\Phi(x)$ in §4.3.3, and showed there that Φ is monotone nondecreasing and (uniformly) continuous on $[0, 1]$, with $\Phi(0) = 0$ and $\Phi(1) = 1$. Furthermore, we showed that there is a collection of disjoint, open intervals $\{I_{n,k}\}$ such that $[0, 1] \setminus \bigcup_{n,k} I_{n,k}$ (this is the Cantor set) has measure zero, and Φ is constant on each of these intervals.

It follows that Φ is differentiable on each $I_{n,k}$, with $\Phi'(x) = 0$ there, and so $\Phi' = 0$ a.e. in $[0, 1]$. More generally, $\Phi' = 0$ a.e. in $[0, x]$ for $0 < x < 1$, so Φ' is Lebesgue integrable in $[0, x]$ for such x. On the other hand, if $x > 0$, then

$$F(x) = \int_0^x \Phi'\,d\mu = 0 \neq \Phi(x).$$

Thus, $\Phi(x)$ is differentiable a.e. in $[0, 1]$ with an integrable derivative, but Φ is *not* equal to the Lebesgue integral of its derivative. The problem is that Φ increases from 0 to 1 on a set of measure zero, and this implies that Φ is not "continuous enough". Specifically, Proposition 7.2.4 states that if $F(x) = \int_a^x f(t)\,dt$ for $x \in [a, b]$, then F is absolutely continuous in $[a, b]$.

Proposition. *The Cantor function is not absolutely continuous in* $[0, 1]$.

Proof. Referring to the definition of the Cantor function in §4.3.3, let $\{J_{n,k} : 1 \leq k \leq 2^n\}$ be the closed intervals left after the nth step in the construction of $\mathscr{C}$. Writing $J_{n,k} = [x_k, y_k]$, we have $y_k - x_k = 3^{-n}$ and $\Phi(y_k) - \Phi(x_k) = 2^{-n}$ for each k, so

$$\sum_{k=1}^{2^n} (y_k - x_k) = (2/3)^n,$$

which can be made arbitrarily small by choosing n large enough, while

$$\sum_{k=1}^{2^n} (\Phi(y_k) - \Phi(x_k)) = 1$$

for all n. □

Proposition 7.2.4 was stated and proved in the case of the Riemann integral of a Riemann integrable function f, but it generalizes to Lebesgue integrals of Lebesgue integrable functions.

Theorem. *If f is Lebesgue integrable on $[a, b]$, and $F(x) = \int_a^x f\,d\mu$ for $a \leq x \leq b$, then F is absolutely continuous in $[a, b]$.*

Proof. Since f is integrable on $[a, b]$, then the same is true for $|f|$. It follows that given $\varepsilon > 0$, there is a $\delta > 0$ such that if $E \subset [a, b]$, and $\mu(E) < \delta$, then $\int_E |f|\,d\mu < \varepsilon$ (see

ex8.4.3). Hence, if $a \le x_1 < y_1 \le x_2 < y_2 \le \cdots \le x_n < y_n \le b$, and $\sum_{j=1}^{n}(y_j - x_j) < \delta$, then,

$$\sum_{j=1}^{n} |F(y_j) - F(x_j)| = \sum_{j=1}^{n} \left| \int_{x_j}^{y_j} f \, d\mu \right| \le \sum_{j=1}^{n} \int_{x_j}^{y_j} |f| \, d\mu = \int_E |f| \, d\mu < \varepsilon,$$

because $E = \bigcup_{j=1}^{n} [x_j, y_j]$ has measure $< \delta$. □

Our goal for the remainder of this section is to establish the converse to this. Namely, that if f is absolutely continuous in $[a, b]$, then f is differentiable a.e. in (a, b), f' is integrable there, and

$$f(x) = f(a) + \int_a^x f' \, d\mu$$

for $x \in (a, b)$.

8.5.2. The differentiability of monotone functions

The first step is showing that monotone functions are differentiable almost everywhere. This is sometime called Lebesgue's differentiation theorem.

Definition. Suppose that the function f is defined in a neighborhood of $x \in \mathbb{R}$. The derivative of f at x, if it exists, is sometimes denoted by $Df(x)$. In this vein, the *upper derivative* of f at x is given by

$$\overline{D}f(x) = \limsup_{t \to x} \frac{f(t) - f(x)}{t - x},$$

and the *lower derivative* of f at x is

$$\underline{D}f(x) = \liminf_{t \to x} \frac{f(t) - f(x)}{t - x}.$$

Clearly, $\underline{D}f(x) \le \overline{D}f(x)$, and f is differentiable at x if and only of $\overline{D}f(x) = \underline{D}f(x)$. It is also clear that if f is nondecreasing around x, then both $\overline{D}f(x)$ and $\underline{D}f(x)$ are non-negative.

Lemma. *If f is monotone nondecreasing on $[a, b]$, and $s > 0$, then*

$$f(b) - f(a) \ge s \cdot \mu^*(\{x \in (a, b) : \overline{D}f(x) > s\}).$$

I.e., the larger the rate of change, the smaller the set on which it occurs.

Proof. For $s > 0$, let $E_s = \{x \in (a, b) : \overline{D}f(x) > s\}$. If $E_s \ne \varnothing$ and $x \in E_s$, then there is a sequence $\{t_n(x)\} \subset (a, b)$ satisfying $0 < |t_n(x) - x| < 1/n$ and

$$\frac{f(t_n(x)) - f(x)}{t_n(x) - x} > s$$

for every $n \in \mathbb{N}$. For each $x \in E_m$ and $n \in \mathbb{N}$, let $I_{x,n} = [x, t_n(x)]$ if $x < t_n(x)$ and $I_{x,n} = [t_n(x), x]$ if $x > t_n(x)$, then the set $\mathscr{V} = \{I_{x,n} : x \in E_s,\ n \in \mathbb{N}\}$ is a Vitali cover of

E_m. Given $\varepsilon > 0$, it follows from Corollary 8.1.9 that there is a finite subset of disjoint intervals $\{[u_1, v_1], \ldots, [u_m, v_m]\} \subset \mathscr{V}$ such that

$$\mu^*(E_s) + \varepsilon > \sum_{k=1}^{m} (v_k - u_k) > \mu^*(E_s) - \varepsilon.$$

For each k, we have $f(v_k) - f(u_k) > m(v_k - u_k)$, and therefore

$$\sum_{k=1}^{N} (f(v_k) - f(u_k)) > s \sum_{k=1}^{N} (v_k - u_k) > s\mu^*(E_s) - s\varepsilon. \tag{8.5.1}$$

Renumbering the intervals if necessary, we may assume that $u_k < v_k < u_{k+1}$ for each k, and setting $b = u_{m+1}$, we have

$$\begin{aligned} f(b) - f(a) &= f(u_1) - f(a) + \sum_{k=1}^{m} [f(v_k) - f(u_k) + f(u_{k+1}) - f(v_k)] \\ &\geq \sum_{k=1}^{m} (f(v_k) - f(u_k)), \end{aligned}$$

because f is nondecreasing. It follows from this and (8.5.1) that for all $\varepsilon > 0$,

$$f(b) - f(a) > s \cdot \mu^*(E_s) - s \cdot \varepsilon,$$

so $f(b) - f(a) \geq s\mu^*(E_s)$, as claimed. □

Theorem (Lebesgue). *If f is monotone on $[a, b]$, then f is differentiable a.e. in (a, b).*

Proof. We may assume that f is nondecreasing, because f and $-f$ are differentiable at the same points.

If f is not differentiable at $x \in (a, b)$, then either $\overline{D}f(x) = \infty$, or $\underline{D}f(x) < \overline{D}f(x)$. It follows from the lemma that $\mu^*(\{x \in (a, b) : \overline{D}f(x) = \infty\}) < (f(b) - f(a))/s$, for all $s > 0$, so $\mu(\{x \in (a, b) : \overline{D}f(x) = \infty\}) = 0$. It remains to show that

$$E = \{x \in (a, b) : \underline{D}f(x) < \overline{D}f(x)\}$$

has measure zero.

If $x \in E$, then there are $r, s \in \mathbb{Q}$, with $0 < r < s$, such that

$$\underline{D}f(x) < r < s < \overline{D}f(x), \tag{8.5.2}$$

so that $E = \bigcup_{r,s \in \mathbb{Q}} E_{r,s}$, where $E_{r,s}$ is the set of points in $[a, b]$ satisfying (8.5.2). Since the union $\bigcup_{r,s \in \mathbb{Q}} E_{r,s}$ is countable, it is enough to show that if $0 < r < s$, then $\mu(E_{r,s}) = 0$. To do this, we begin as in the proof of the preceding lemma, but we start with the condition that $\underline{D}f(x) < r$.

Suppose that $0 < r < s$ and $x \in E_{r,s}$. Then there is a sequence $\{t_n(x)\} \subset [a, b]$ such that $|t_n - x| < 1/n$ and

$$0 \leq \frac{f(t_n) - f(x)}{t_n - x} < r$$

for $n \in \mathbb{N}$. For each $x \in E_{r,s}$ and $n \in \mathbb{N}$, set $I_{x,n} = [x, t_n]$ or $[t_n, x]$, depending on whether $x < t_n$ or $t_n < x$, as before. The set $\mathscr{V} = \{I_{x,n} : x \in E_{r,s}, n \in \mathbb{N}\}$ is a Vitali cover of $E_{r,s}$, and given $\varepsilon > 0$, it follows again from Corollary 8.1.9 that there is a finite subset of disjoint intervals $\{[u_1, v_1], \ldots, [u_m, v_m]\} \subset \mathcal{V}$, such that

$$\mu^*(E_{r,s}) + \varepsilon > \sum_{k=1}^{m} (v_k - u_k) > \mu^*(E_{r,s}) - \varepsilon. \tag{8.5.3}$$

Remembering that $r(v_k - u_k) > (f(v_k) - f(u_k))$ for $1 \leq k \leq m$, and applying the preceding lemma to $E_{r,s} \cap [u_k, v_k]$ for $1 \leq k \leq m$, we have

$$\begin{aligned} r \cdot (\mu^*(E_{r,s}) + \varepsilon) > r \sum_{k=1}^{m} (v_k - u_k) &> \sum_{k=1}^{m} (f(v_k) - f(u_k)) \\ &\geq s \sum_{k=1}^{m} \mu^*(E_{r,s} \cap [u_k, v_k]) > s \cdot \mu^*(E_{r,s} - \varepsilon), \end{aligned}$$

where the first inequality follows from the left-hand side of (8.5.3), and the last inequality follows from the right-hand inequality there, and the additivity of outer measure. Since this is true for all $\varepsilon > 0$, it follows that $r \cdot \mu^*(E_{r,s}) \geq s \cdot \mu^*(E_{r,s})$, which implies that $\mu(E_{r,s}) = 0$ because $r < s$. □

A illustrated by the Cantor function, a monotone function need not be absolutely continuous. On the other hand, the derivative of a monotone function is integrable, and the identity in the fundamental theorem can be replaced by an inequality.

Corollary. *If f is monotone nondecreasing on $[a, b]$, then f' is integrable there and*

$$\int_a^b f' \, d\mu \leq f(b) - f(a).$$

Proof. First, we extend the function f, by setting $f(x) = f(b)$ for $x > b$, and since monotone functions are measurable, it follows that

$$D_n(f)(x) = \frac{f(x + 1/n) - f(x)}{1/n}$$

is measurable in $[a, b]$ for all $n \in \mathbb{N}$, and $D_n(f) \geq 0$ in $[a, b]$ because f is increasing. Next, we observe that

$$\begin{aligned} \int_a^b D_n(f) \, d\mu &= n \int_a^b (f(x + 1/n) - f(x)) \, d\mu \\ &= n \int_{a+1/n}^{b+1/n} f \, d\mu - n \int_a^b f \, d\mu \\ &= n \int_b^{b+1/n} f \, d\mu - n \int_a^{a+1/n} f \, d\mu. \end{aligned}$$

Since $f(x) = f(b)$ for $x > b$ and $f(x) \geq f(a)$ for $x > a$, it follows that $n \int_b^{b+1/n} f \, d\mu = f(b)$ and $n \int_a^{a+1/n} f \, d\mu \geq f(a)$, and hence $\int_a^b D_n(f) \, d\mu \leq f(b) - f(a)$ for all $n \in \mathbb{N}$.

Finally, $f' \geq 0$ and by the preceding theorem, $\liminf_{n\to\infty} D_n(f) = \lim_{n\to\infty} D_n(f) = f'$ a.e. in $[a, b]$, and therefore

$$\int_a^b f'\,d\mu \leq \liminf_{n\to\infty} \int_a^b D_n(f)\,d\mu \leq f(b) - f(a),$$

by Fatou's lemma, which both shows that f' is integrable in $[a, b]$ and establishes the claimed bound. □

8.5.3. The differentiability of absolutely continuous functions

Recall that the total variation of a function f in $[a, b]$ is

$$V_f(a, b) = \sup_P \sum_{j=1}^n |f(t_j) - f(t_{j-1})|,$$

where the supremum is taken over all partitions $P = \{t_0, t_1, \ldots, t_n\}$ of $[a, b]$.[12] If $V_f(a, b)$ is finite, then f is of bounded variation in $[a, b]$. In §7.3.3, we saw that functions of bounded variation are Riemann integrable. More importantly for our present needs, it follows from **ex7.3.2** that f is of bounded variation in $[a, b]$ if and only if $f = g - h$, where g and h are both monotone nondecreasing in $[a, b]$. From this, Theorem 8.5.2 and its corollary we immediately deduce the following proposition.

Proposition. *If f is of bounded variation in $[a, b]$ then f is differentiable a.e in (a, b), and f' is integrable there.*

This proposition is a stepping stone to the following theorem.

Theorem. *If f is absolutely continuous in $[a, b]$, then f is differentiable a.e. in (a, b) and f' is integrable there.*

Proof. To prove the theorem, it is enough to show that absolute continuity implies bounded variation, by the previous proposition.

So suppose that f is absolutely continuous in $[a, b]$, and let $\delta_0 > 0$ be such that $\sum_{j=1}^n |f(y_j) - f(x_j)| < 1$, for any disjoint collection $\{(x_1, y_1), \ldots, (x_n, y_n)\}$ of subintervals of $[a, b]$ such that $\sum_{j=1}^n (y_j - x_j) < \delta_0$. Next, let $m_0 = [(b - a)/\delta_0] + 1$,[13] and for $0 \leq j \leq m_0$, set $u_j = a + \frac{j(b-a)}{m_0}$. Then $P_0 = \{u_0, \ldots, u_{m_0}\}$ is a partition of $[a, b]$, satisfying

$$u_{j+1} - u_j = \frac{b - a}{m_0} < \delta_0,$$

for $0 \leq j \leq m_0 - 1$.

If $P = \{t_0, \ldots, t_n\}$ is any other partition of $[a, b]$, then $P \wedge P_0 = \{x_0, \ldots, x_N\}$ is a refinement of P, so

$$\sum_P |f(t_{j+1}) - f(t_j)| \leq \sum_{P\wedge P_0} |f(x_{i+1}) - f(x_i)| \tag{8.5.4}$$

[12] See §7.3.3.

[13] $[x] = \sup\{m \in \mathbb{Z} : m \leq x\}$ is the integer part of x.

by the triangle inequality. For $0 \le j \le m_0 - 1$, denote the points of $P \wedge P_0$ in $[u_j, u_{j+1}]$ by

$$u_j = x_{j,0} < x_{j,1} < \cdots < x_{j,k_j} = u_{j+1}.$$

Then $\sum_{i=0}^{k_j-1}(x_{j,i+1} - x_{j,i}) = u_{j+1} - u_j < \delta_0$, so

$$\sum_{i=0}^{k_j-1} |f(x_{j,i+1}) - f(x_{j,i})| < 1,$$

and therefore

$$\sum_{P\wedge P_0} |f(x_{i+1}) - f(x_i)| = \sum_{j=1}^{m_0}\sum_{i=0}^{k_j-1} |f(x_{j,i+1}) - f(x_{j,i})| < m_0.$$

It follows from this and (8.5.4) that $\sum_P |f(t_{j+1}) - f(t_j)| < m_0$ for any partition P of $[a, b]$, and hence f is of bounded variation in $[a, b]$. □

8.5.4. The fundamental theorem of calculus, part I

If F is absolutely continuous in $[a, b]$, then F is differentiable almost everywhere in $[a, b]$, and $f = F'$ is integrable there. Conversely, if f is integrable in $[a, b]$, then $F(x) = \int_a^x f\,d\mu$ is absolutely continuous in $[a, b]$. The fundamental theorem of calculus couples these two facts: *F is absolutely continuous in $[a, b]$ if and only if F' is integrable and $F(x) = F(a) + \int_a^x F'\,d\mu$ a.e. in $[a, b]$*. It is common to split this into two statements, beginning with the next theorem.

Theorem. *If f is integrable in $[a, b]$ and $F(x) = \int_a^x f\,d\mu$, then $F' = f$ a.e. in $[a, b]$.*

To prove this theorem, we rely on the following lemma.

Lemma. *If f_1 and f_2 are both integrable in $[a, b]$ and $\int_a^x f_1\,d\mu = \int_a^x f_2\,d\mu$ for all $x \in [a, b]$, then $f_1 = f_2$ a.e. in $[a, b]$.*

Proof. By assumption, $\int_a^x (f_1 - f_2)\,d\mu = 0$ for all $x \in [a, b]$, so it is enough to show that if $\int_a^x f\,d\mu = 0$ for all $x \in [a, b]$, then $f = 0$ a.e. in $[a, b]$. We prove this by contradiction, and assume that $E = \{x \in [a, b] : f(x) > 0\} \cup \{x \in [a, b] : f(x) < 0\} = E^+ \cup E^-$ has positive measure. Without loss of generality, we may assume that $\mu(E^+) > 0$, and hence there is a compact set $K \subset E^+$ such that $\mu(K) > 0$, from which it follows that $\int_K f\,d\mu > 0$.

The set $(a, b) \setminus K$ is open, and is therefore equal to a countable disjoint union of open intervals, $(a, b) \setminus K = \bigcup_n (a_n, b_n)$. For each n, we have

$$\int_{a_n}^{b_n} f\,d\mu = \int_a^{b_n} f\,d\mu - \int_a^{a_n} f\,d\mu = 0,$$

so by Theorem 8.3.3

$$\int_a^b f\,d\mu = \int_K f\,d\mu + \sum_{n=1}^{\infty}\int_{a_n}^{b_n} f\,d\mu = \int_K f\,d\mu > 0,$$

which contradicts the assumption that $\int_a^x f\,d\mu = 0$ for all $x \in [a, b]$. □

Proof (of the theorem). Suppose first that $|f| < M$ in $[a,b]$, and set $f(x) = 0$ for $x > b$. As in the proof of Corollary 8.5.2, let

$$D_n(F)(x) = \frac{F(x+1/n) - F(x)}{1/n},$$

for all $x \in [a,b]$ and $n \in \mathbb{N}$. It follows from the differentiability of F that $\lim_{n\to\infty} D_n(F) = F'$ a.e. in $[a,b]$, and from the bound on $|f|$,

$$|D_n(F)(t)| = n\,|F(t+1/n) - F(t)| = n\left|\int_t^{t+1/n} f\,d\mu\right| < n\int_t^{t+1/n} M\,d\mu = M,$$

for any $t \in [a,b]$. Hence, by the dominated convergence theorem

$$\int_a^x F'\,d\mu = \lim_{n\to\infty}\int_a^x D_n(F)\,d\mu = \lim_{n\to\infty}\left(n\int_x^{x+1/n} F\,d\mu - n\int_a^{a+1/n} F\,d\mu\right). \tag{8.5.5}$$

Now, for any $x \in [a,b]$ and any $\varepsilon > 0$, if n is sufficiently large, then $|F(x) - F(t)| < \varepsilon$ if $|x - t| < 1/n$, since F is continuous in $[a,b]$. Hence,

$$\left|\int_x^{x+1/n} F(t)\,d\mu - \frac{F(x)}{n}\right| < \frac{\varepsilon}{n}$$

for n sufficiently large. From this, (8.5.5) and the fact that $F(a) = 0$, it follows that

$$\int_a^x F'\,d\mu = F(x) - F(a) = F(x) = \int_a^x f\,d\mu$$

for $x \in [a,b]$, so $F' = f$ a.e. in $[a,b]$ by the previous lemma.

In the case that f is unbounded in $[a,b]$, let $F^+ = \int_a^x f^+\,d\mu$ and $F^- = \int_a^x f^+\,d\mu$. If we can show that $F^{+'} = f^+$ and $F^{-'} = f^-$ a.e. in $[a,b]$, then $F' = F^{+'} - F^{-'} = f^+ - f^- = f$ a.e. in $[a,b]$. Hence if f is unbounded, it is enough to prove the claim in the case that $f \geq 0$.

In this case, let $f_n(x) = \min(f(x), n)$ for $x \in [a,b]$ and let $g_n = f - f_n$, whence g_n is nonnegative in $[a,b]$. Since f is integrable, both f_n and g_n are integrable, and we set

$$F_n(x) = \int_a^x f_n\,d\mu \quad \text{and} \quad G_n(x) = \int_a^x g_n\,d\mu.$$

Both F_n and G_n are differentiable by Theorem 8.5.3, $F_n' = f_n$ a.e. in $[a,b]$, because f_n is bounded for each n, and $G_n' \geq 0$ because $g_n \geq 0$, so G_n is nondecreasing. Therefore

$$F' = (F_n + G_n)' = f_n + G_n' \geq f_n$$

a.e. in $[a,b]$, for all $n \in \mathbb{N}$. It follows that $F' \geq \lim_{n\to\infty} f_n = f$ a.e. in $[a,b]$, so

$$\int_a^x F'\,d\mu \geq \int_a^x f\,d\mu$$

for every x in $[a,b]$. On the other hand F is nondecreasing in $[a,b]$ because $f \geq 0$ there, so for every $x \in [a,b]$,

$$\int_a^x F'\,d\mu \leq F(x) - F(a) = \int_a^x f\,d\mu,$$

by Corollary 8.5.2. Thus $\int_a^x F'\,d\mu = \int_a^x f\,d\mu$ for all $x \in [a,b]$, and once again, $F' = f$ a.e. in $[a,b]$ by the previous lemma. □

8.5.5. The fundamental theorem of calculus, part II

Theorem. *If F is absolutely continuous in $[a,b]$, then for all $x \in [a,b]$*

$$F(x) = F(a) + \int_a^x F'\,d\mu.$$

Proof. From the absolute continuity of F, we know that it is differentiable a.e. in $[a,b]$ with F' integrable there, and we set

$$F_1(x) = \int_a^x F'\,d\mu$$

for $x \in [a,b]$. Since F_1 is absolutely continuous in $[a,b]$ and $F_1(a) = 0$, it is enough to show that (the absolutely continuous function) $G = F - F_1$ is constant in $[a,b]$.

By theorem 8.5.4, $F_1' = F'$ a.e. in $[a,b]$, so that $G' = 0$ a.e. in $[a,b]$. Hence, the theorem will follow from the following special case: *if G is absolutely continuous in $[a,b]$ and $G' = 0$ a.e., then $G(x) = G(a)$ for all $x \in [a,b]$.*

Let $x \in (a,b)$, and let $\varepsilon > 0$. Since G is absolutely continuous in $[a,x]$, there is a $\delta_\varepsilon > 0$ such that

$$\sum_{j=1}^{n} |G(v_j) - G(u_j)| < \varepsilon, \tag{8.5.6}$$

for any finite collection of disjoint subintervals $\{[u_1,v_1],\ldots,[u_n,v_n]\}$ of $[a,x]$ satisfying $\sum_{j-1}^{n}(v_j - u_j) < \delta_\varepsilon$. The idea of the proof is to find a collection of subintervals satisfying (8.5.6), such that the variation of G on the remaining pieces of $[a,x]$ is bounded by a multiple of ε due to the fact that $G' = 0$ a.e. in $[a,x]$. We do this "in reverse".

Let $E = \{t \in (a,x) : G'(t) = 0\}$, then for every $t \in E$, and $n \in \mathbb{N}$, there is an interval $I_{t,n} = [s,t] \subset (a,x)$, such that $t - s < 1/n$ and

$$\left|\frac{G(t) - G(s)}{t - s}\right| < \varepsilon. \tag{8.5.7}$$

The set $\mathscr{V} = \{I_{t,n} : t \in E, n \in \mathbb{N}\}$ is a Vitali cover of E, so it follows from Corollary 8.1.9, that there is a finite subset of disjoint intervals $\{[s_k,t_k]\}_{k=1}^m \subset \mathscr{V}$ such that

$$\sum_{k=1}^{m}(t_k - s_k) > \mu(E) - \delta_\varepsilon = x - a - \delta_\varepsilon, \tag{8.5.8}$$

where $\mu(E) = x - a$ because $G' = 0$ a.e. in $[a,b]$. Also, since $[s_k,t_k] \subset \mathscr{V}$, it follows from (8.5.7) that

$$|G(t_k) - G(s_k)| < (t_k - s_k)\varepsilon, \tag{8.5.9}$$

for $1 \le k \le m$.

If we set $t_0 = a$ and $s_{m+1} = x$, then $\{t_0, s_1, t_1, s_2, \dots, s_m, t_m, s_{m+1}\}$ forms a partition of $[a, x]$, and hence

$$\sum_{k=0}^{m}(s_{k+1} - t_k) = (x - a) - \sum_{k=1}^{m}(t_k - s_k) < \delta_\varepsilon, \tag{8.5.10}$$

by (8.5.8). Moreover,

$$G(x) - G(a) = G(s_1) - G(t_0) + G(t_1) - G(s_1) + \cdots + G(t_m) - G(s_m) + G(s_{m+1}) - G(t_m),$$

and therefore

$$\begin{aligned} |G(x) - G(a)| &= \left|\sum_{k=0}^{m}(G(s_{k+1}) - G(t_k)) + \sum_{k=1}^{m}(G(t_k) - G(s_k))\right| \\ &\le \sum_{k=0}^{m}|G(s_{k+1}) - G(t_k)| + \sum_{k=1}^{m}|G(t_k) - G(s_k)|, \end{aligned}$$

by the triangle inequality. The first sum in the second row above is bounded by ε, by (8.5.6) and (8.5.10), and the second sum is bounded by

$$\varepsilon\sum_{k=1}^{m}(t_k - s_k) < \varepsilon(x - a),$$

as follows from (8.5.9) and the fact that $\bigcup_k [s_k, t_k] \subset [a, x]$. Hence, for any $\varepsilon > 0$,

$$|G(x) - G(a)| < \varepsilon(x - a + 1) \le \varepsilon(b - a + 1),$$

so $G(x) = G(a)$ for all $x \in [a, b]$. □

Exercises

ex8.5.1. Let $A \subset \mathbb{R}$ be measurable. For $x \in A$, then density of A at x is given by

$$d_A(x) = \lim_{\delta \to 0^+} \frac{\mu(A \cap (x - \delta, x + \delta))}{2\delta}.$$

Prove the *Lebesgue density theorem*: $d_A(x) = 1$ a.e. in A.

Hint: Suppose that $A \subset [a, b]$, consider the function $F_A(x) = \int_a^x \chi_A \, d\mu$ for $x \in [a, b]$ and apply Theorem 8.5.4. Generalize to any measurable set A.

ex8.5.2. Continuing the previous exercise, show that $d_A(x) = 0$ or 1 a.e. in $\mathbb{R}$.

8.6. The space $\mathscr{L}^1(\mathbb{R})$

In §4.5 we introduced the $C[0, 1]$, the metric space of continuous functions on $[0, 1]$, with the metric $\rho(f, g) = \max_{[0,1]} |f - g|$, and in §5.2 we showed that $C[0, 1]$ is complete and separable with respect to this metric. In this section, we introduce the metric space

$\mathscr{L}^1(\mathbb{R})$ of integrable functions on $\mathbb{R}$. The discussion here only scratches the surface, and the intention is to give the reader another peek at a topic from *functional analysis*.[14]

If $E \subset \mathbb{R}$ is a measurable subset, then $\mathscr{L}^1(E)$ is the space of Lebesgue integrable functions on E, and everything that we discuss in the coming sections generalizes to $\mathscr{L}^1(E)$. As we are focusing on $\mathscr{L}^1(\mathbb{R})$ here, we will shorten $\int_{\mathbb{R}} f\,d\mu$ to $\int f\,d\mu$, and $\mathscr{L}^1(\mathbb{R})$ to $\mathscr{L}^1$.

8.6.1. The $\mathscr{L}^1$ metric and norm

Informally, we think of $\mathscr{L}^1$ as the collection of all functions that are Lebesgue integrable on $\mathbb{R}$. More formally, we need to define $\mathscr{L}^1$ as a space of equivalence classes of integrable functions.

Specifically, for functions f and g on $\mathbb{R}$, we write $f \approx g$ if $f = g$ a.e. in $\mathbb{R}$. This relation is clearly symmetric and reflexive, and if $f \approx g$ and $g \approx h$, then

$$\{x \in \mathbb{R} : f(x) \neq h(x)\} \subseteq \{x \in \mathbb{R} : f(x) \neq g(x)\} \cup \{x \in \mathbb{R} : g(x) \neq h(x)\}$$

has measure zero, so $f \approx h$. Hence $\approx$ is an equivalence relation and we formally define

$$\mathscr{L}^1 = \left\{[f] : \int f\,d\mu < \infty\right\},$$

where $[f]$ is the equivalence class of f modulo the relation $\approx$.

Formality aside, we typically speak of the elements of $\mathscr{L}^1$ as functions (but keep in mind the these functions are representatives of their equivalence classes). In particular if we say that $f \neq g$ in $\mathscr{L}^1$, we mean that $f \not\approx g$.

If f and g are both integrable in $\mathbb{R}$, then $f - g$ and $|f - g|$ are also integrable there. Moreover, if $f_1 \approx f_2$ and $g_1 \approx g_2$, then $|f_1 - g_1| \approx |f_2 - g_2|$, so

$$\int |f_1 - g_1|\,d\mu = \int |f_2 - g_2|\,d\mu. \tag{8.6.1}$$

Definition. For $[f], [g] \in \mathscr{L}^1$, we define

$$\|[f]\|_1 = \int |f|\,d\mu,$$

and $\rho([f],[g]) = \|[f - g]\|_1$, where f and g are any representatives of $[f]$ and $[g]$ respectively. By the identity (8.6.1), $\|[f]\|_1$ and $d([f],[g])$ do not depend on the choice of the representatives, and we typically write $\|f\|_1$ and $\rho(f,g)$ instead of $\|[f]\|_1$ and $\rho([f],[g])$.

Theorem. *$\rho(f, g)$ is a metric on $\mathscr{L}^1$.*

Proof. For any $f, g \in \mathscr{L}^1$, it is clear that $\rho(f,f) = 0$ and $\rho(f,g) = \rho(g,f)$. If $f \neq g$ in $\mathscr{L}^1$, then the set $\{x \in E : |f(x) - g(x)| > 0\}$ has positive measure, and therefore

[14]Very roughly speaking, functional analysis is the study of normed vector spaces and the linear maps (functionals) between them. Both $C[0,1]$ and $\mathscr{L}^1(\mathbb{R})$ are examples of such spaces.

$\rho(f,g) = \int_E |f-g|\,d\mu > 0$. Finally, if $f, g, h \in \mathscr{L}^1$, then

$$\begin{aligned}\rho(f,h) &= \int |f-h|\,d\mu \\ &\le \int |f-g| + |g-h|\,d\mu \\ &\le \int |f-g|\,d\mu + \int |g-h|\,d\mu = \rho(f,g) + \rho(g,h),\end{aligned}$$

so $\rho(\cdot,\cdot)$ satisfies the triangle inequality as well. □

We call $\|f\|_1$ the $\mathscr{L}^1$ norm and $\|f-g\|_1$ the $\mathscr{L}^1$ metric.

8.6.2. The completeness of $\mathscr{L}^1$

We want to show that if $\{f_n\} \subset \mathscr{L}^1$ is a Cauchy sequence, then there is a function $f \in \mathscr{L}^1$ such that $\|f_n - f\|_1 \to 0$. We do this by showing that every Cauchy sequence in $\mathscr{L}^1$ has a subsequence of *bounded variation*, and that a sequence of bounded variation converges to a limit in $\mathscr{L}^1$.

Definition. A sequence $\{x_n\}$ in a metric space (X, ρ) has *bounded variation* if

$$\sum_{n=1}^{\infty} \rho(x_n, x_{n+1}) < \infty.$$

Lemma. *If $\{f_n\} \subset \mathscr{L}^1$ has bounded variation, then $\{f_n\}$ converges pointwise to a function $f \in \mathscr{L}^1$ and $\|f_n - f\|_1 \to 0$.*

Proof. Let $g_n = \sum_{k=1}^{n} |f_{k+1} - f_k|$. Then g_n is measurable and $g_{n+1} \ge g_n \ge 0$ for all n, so $\{g_n\}$ converges monotonically to $g = \sum_{n=1}^{\infty} |f_{n+1} - f_n|$. Therefore, by the monotone convergence theorem,

$$\int g\,d\mu = \lim_{n\to\infty} \int g_n\,d\mu = \lim_{n\to\infty} \sum_{k=1}^{n} \int |f_{k+1} - f_k|\,d\mu = \sum_{n=1}^{\infty} \|f_{n+1} - f_n\|_1 < \infty,$$

so g is integrable in $\mathbb{R}$, and hence $\sum_{n=1}^{\infty} |f_{n+1}(x) - f_n(x)| < \infty$ for almost all x in $\mathbb{R}$, by Proposition 8.3.5. Thus, for almost all $x \in \mathbb{R}$, the sequence $\{f_n(x)\} \subset \mathbb{R}$ has bounded variation (in $\mathbb{R}$) and therefore converges to the (finite) limit

$$f(x) = f_1(x) + \sum_{n=1}^{\infty} (f_{n+1}(x) - f_n(x))$$

a.e. in $\mathbb{R}$ (see **ex2.6.3**). Furthermore $f \in \mathscr{L}^1$ because both f_1 and $\sum_{n=1}^{\infty}(f_{n+1}(x) - f_n(x))$ are integrable.

Given $\varepsilon > 0$, there is an n_ε such that $\sum_{k=n_\varepsilon}^{\infty} \|f_{k+1} - f_k\|_1 < \varepsilon/2$, so if $n > n_\varepsilon$, then

$$\|f_n - f_{n_\varepsilon}\|_1 = \left\| \sum_{k=n_\varepsilon}^{n-1} (f_{k+1} - f_k) \right\|_1 \le \sum_{k=n_\varepsilon}^{n-1} \|f_{k+1} - f_k\|_1 \le \frac{\varepsilon}{2}.$$

Next note that for any $m \in \mathbb{N}$,

$$f(x) = f_1(x) + \sum_{n=1}^{\infty}(f_{n+1}(x) - f_n(x)) = f_m(x) + \sum_{n=m}^{\infty}(f_{n+1}(x) - f_n(x)),$$

because the sum $\sum_{n=1}^{\infty}(f_{n+1}(x) - f_n(x))$ is "telescoping". It therefore follows that

$$\|f_n - f\|_1 = \left\| f_n - f_{n_\varepsilon} - \sum_{k=n_\varepsilon}^{\infty}(f_{k+1} - f_k) \right\|_1$$

$$\leq \|f_n - f_{n_\varepsilon}\|_1 + \sum_{k=n_\varepsilon}^{\infty}\|f_{k+1} - f_k\|_1 < \varepsilon.$$

Hence, $\{f_n\}$ converges to f in $\mathscr{L}^1$. □

Theorem. *The space $\mathscr{L}^1$ is complete with respect to the $\mathscr{L}^1$ metric.*

Proof. If $\{f_n\}$ is a Cauchy sequence in $\mathscr{L}^1$, then for every k, there is an n_k, such that $\|f_n - f_m\|_1 < 2^{-k}$ for all $n, m \geq n_k$. Without loss of generality, we may assume that $n_{k_1} < n_{k_2}$ if $k_1 < k_2$, and hence $\|f_{n_{k+1}} - f_{n_k}\|_1 < 2^{-k}$ for all $k \in \mathbb{N}$. The subsequence $\{f_{n_k}\}_{k=1}^{\infty}$ satisfies the condition of the preceding lemma, so there is a function $f \in \mathscr{L}^1$ such that $\|f_{n_k} - f\|_1 \to 0$.

Given $\varepsilon > 0$, let n_ε be large enough that $\|f_n - f_m\|_1 < \varepsilon/2$ for all $n, m \geq n_\varepsilon$, and let k_ε be large enough that both $n_{k_\varepsilon} \geq n_\varepsilon$ and $\|f_{n_k} - f\|_1 < \varepsilon/2$. If $n > n_{k_\varepsilon}$, then

$$\|f_n - f\|_1 \leq \|f_n - f_{n_{k_\varepsilon}}\|_1 + \|f_{n_{k_\varepsilon}} - f\|_1 < \varepsilon,$$

so $\|f_n - f\|_1 \to 0$. □

8.6.3. Separability of $\mathscr{L}^1$

Recall that a metric space is *separable* if the space contains a countable dense subset. For example, the (countable) collection of polynomials with rational coefficients is dense in $C[0, 1]$. To show that $\mathscr{L}^1$ is separable, we first show that the set of integrable simple functions is dense there, and then we find a countable subset of integrable functions that is dense in the set of integrable simple functions.

Proposition. *If $f \in \mathscr{L}^1$, then there is a sequence $\{\varphi_n\}$ of simple functions such that* $\lim_{n\to\infty} \|f - \varphi_n\|_1 = 0$.

Proof. By Theorem 8.2.4, there are sequences $\{\varphi_n^+\}$ and $\{\varphi_n^-\}$ of simple functions that are converging pointwise to f^+ and f^- in $\mathbb{R}$, respectively, where we may assume that the convergence is monotone nondecreasing in both cases. By the monotone convergence theorem,

$$\lim_{n\to\infty}\int |f^+ - \varphi_n^+|\,d\mu = \lim_{n\to\infty}\int f^+ - \varphi_n^+\,d\mu = \int f^+\,d\mu - \lim_{n\to\infty}\int \varphi_n^+\,d\mu = 0$$

because $f \in \mathscr{L}^1$ implies that $\int f^+\,d\mu < 0$. Hence $\lim_{n\to\infty}\|f^+ - \varphi_n^+\|_1 = 0$, and by the same argument, $\lim_{n\to\infty}\|f^- - \varphi_n^-\|_1 = 0$. Thus,

$$\lim_{n\to\infty}\|f - (\varphi_n^+ - \varphi_n^-)\|_1 = \lim_{n\to\infty}\|(f^+ - \varphi_n^+) - (f^- - \varphi_n^-)\|_1 = 0$$

by the triangle inequality. □

Let $\mathcal{G}$ the set of all (open) sets in $\mathbb{R}$ that are finite disjoint unions of open intervals with rational endpoints. I.e., $G \in \mathcal{G}$ if and only if $G = \bigcup_{j=1}^{n} I_j$, where $I_j = (a_j, b_j)$, with $a_j, b_j \in \mathbb{Q}$ and $I_j \cap I_k = \varnothing$ when $j \neq k$. The collection of all open intervals with rational endpoints is countable, and therefore $\mathcal{G}$ is also countable.[15]

Lemma. *If $A \subseteq \mathbb{R}$ is measurable and $0 < \mu(A) = \alpha < \infty$, then for any $\varepsilon > 0$ there is a $G \in \mathcal{G}$ such that $\mu(A \triangle G) < \varepsilon$.*

Recall that the *symmetric difference* of sets A and B,

$$A \triangle B = (A \setminus B) \cup (B \setminus A),$$

is the set of all elements that belong to exactly one of the two sets.

Proof. For any $\varepsilon > 0$, there is an open set $G_1 \subset \mathbb{R}$ such that $A \subset G_1$ and $\mu(G_1) < \alpha + \frac{\varepsilon}{2}$. By Proposition 3.1.8, $G_1 = \bigcup_{n=1}^{\infty} J_n$, where $\{J_n\}$ is a countable set of pairwise disjoint open intervals, and by the additivity of Lebesgue measure,

$$\sum_{n=1}^{\infty} \ell(J_n) = \sum_{n=1}^{\infty} \mu(J_n) = \mu(G_1) < \infty.$$

Let $N \in \mathbb{N}$ be such that $\sum_{n=N+1}^{\infty} \ell(J_n) < \varepsilon/4$, and for $1 \leq n \leq N$, let I_n be an interval with rational endpoints such that $I_n \subseteq J_n$ and $\ell(I_n) > \ell(J_n) - \varepsilon/4N$. The intervals $\{I_n\}$ are pairwise disjoint because the intervals $\{J_n\}$ are pairwise disjoint, so $G_2 = \bigcup_{n=1}^{N} I_n \in \mathcal{G}$, $G_2 \subset G_1$, and

$$\mu(G_2) = \sum_{n=1}^{N} \ell(I_n) > \sum_{n=1}^{N} \ell(J_n) - \frac{\varepsilon}{4} > \sum_{n=1}^{\infty} \ell(J_n) - \frac{\varepsilon}{2} = \mu(G_1) - \frac{\varepsilon}{2}.$$

It follows that $\mu(G_2 \setminus A) \leq \mu(G_1 \setminus A) < \varepsilon/2$ and $\mu(A \setminus G_2) \leq \mu(G_1 \setminus G_2) < \varepsilon/2$, and hence $\mu(A \triangle G_2) < \varepsilon$. □

Let Q be the set of all functions of the form $\phi = \sum_{j=1}^{n} r_j \chi_{G_j}$, with $r_j \in \mathbb{Q}$ and $G_j \in \mathcal{G}$ for each j. Then Q is countable since $\mathbb{Q}$ and $\mathcal{G}$ are countable, and the separability of $\mathcal{L}^1$ follows from the following theorem.

Theorem. *The set Q is dense in $\mathcal{L}^1$.*

Proof. We first observe that $\int r\chi_G \, d\mu = r\mu(G) < \infty$ for every $r \in \mathbb{Q}$ and $G \in \mathcal{G}$, hence if $\phi = \sum_{j=1}^{n} r_j \chi_{G_j} \in Q$, then $\phi \in \mathcal{L}^1$ as a sum of integrable functions.

It is enough to show that Q is dense in the set of simple functions in $\mathcal{L}^1$, as follows from Proposition 8.6.3. So suppose that $\varphi = \sum_{j=1}^{n} a_j \chi_{E_j}$ is an integrable simple function. Since φ is integrable, it follows that $\mu(E_j) < \infty$ for all j, and we set $\sum_j \mu(E_j) = M$. If $M = 0$, then $\varphi = 0$ a.e. in $\mathbb{R}$, and there is nothing to prove,[16] so we assume that $M > 0$.

[15] The collection of all open intervals with rational endpoints is bijective with a subset of $\mathbb{Q} \times \mathbb{Q}$, and G is bijective with a subset of the set of all finite subsets of $\mathbb{Q} \times \mathbb{Q}$.

[16] Why?

Let $\varepsilon > 0$ be given, and for $1 \leq j \leq n$, let $r_j \in \mathbb{Q}$ be chosen so that $|a_j - r_j| < \varepsilon/2M$, with $r_j \neq 0$. Setting $\varphi_1 = \sum_{j=1}^{n} r_j \chi_{E_j}$, we have $\varphi - \varphi_1 = \sum_{j=1}^{n}(a_j - r_j)\chi_{E_j}$, and therefore

$$\|\varphi - \varphi_1\|_1 = \sum_{j=1}^{n} |a_j - r_j| \mu(E_j) < \frac{\varepsilon}{2M} \cdot \sum_{j=1}^{n} \mu(E_j) = \frac{\varepsilon}{2}. \tag{8.6.2}$$

Now set $R = \max_j |r_j|$ (so $R > 0$), and for $1 \leq j \leq n$, let $G_j \in \mathscr{G}$ be chosen so that $\mu(G_j \triangle E_j) < \varepsilon/2nR$. Such a set exists by the preceding lemma. Let $\phi = \sum_{j=1}^{n} r_j \chi_{G_j}$, then $\phi \in Q$ and

$$\begin{aligned} \int_E |\varphi_1 - \phi| \, d\mu &= \int \left| \sum_{j=1}^{n} r_j(\chi_{E_j} - \chi_{G_j}) \right| d\mu \\ &\leq \int \sum_{j=1}^{n} |r_j| \cdot |\chi_{E_j} - \chi_{G_j}| \, d\mu \leq R \cdot \sum_{j=1}^{n} \mu(G_j \triangle E_j) < \frac{\varepsilon}{2}, \end{aligned}$$

because $|\chi_{E_j} - \chi_{G_j}| = 1$ on $G_j \triangle E_j$ and $|\chi_{E_j} - \chi_{G_j}| = 0$ everywhere else. It follows from this and (8.6.2) that $\|\varphi - \phi\|_1 < \varepsilon$. Since $\varepsilon > 0$ was arbitrary, it follows that Q is dense in the set of simple functions in $\mathscr{L}^1$, and so Q is dense in $\mathscr{L}^1$. □

Exercises

ex8.6.1. Show that if the sequence $\{x_n\}$ has bounded variation in the metric space (X, ρ), then it is a Cauchy sequence there.

ex8.6.2. Show that $C[0,1]$ is dense in $\mathscr{L}^1([0,1])$, and use this to show that the set of polynomials with rational coefficients is also dense in $\mathscr{L}^1([0,1])$, giving another proof of the separability of $\mathscr{L}^1([0,1])$.

Appendix A

Set theory and constructing $\mathbb{R}$

1. Some set theory, in a (very small) nutshell

We assume that the readers of this book have some previous experience with set theory, and present here only a skeleton of the subject, and not a complete skeleton at that. We include just enough to support the earlier chapters of the text. The set-theoretic topic of *cardinality* will get its own small nutshell.

There is more than one version of set theory, depending on the axioms one chooses to include (or exclude). The one we (partially) describe here is founded on the ZFC system of axioms — the Zermelo-Fraenkel axioms, with the axiom of choice. We paraphrase them below, for the sake of completeness, but do not cite them explicitly in what follows.

ZFC 1. *The axiom of extensionality.* Two sets are the same if and only if they contain the same elements.

ZFC 2. *The axiom of foundation.* Every nonempty set A contains an element B such that A and B are disjoint (have no common elements).[1]

ZFC 3. *The axiom schema of separation.* Given a set A and a formula ψ with one variable x,[2] the *subset* $\{x \in A : \psi(x)\} \subset A$ exists (as a set).

ZFC 4. *The axiom of pairing.* If A and B are sets, then there exists a set C with $A, B \in C$.

ZFC 5. *The axiom of union.* Given a set A (of sets), there is a set B that contains every element of every set in A.

ZFC 6. *The axiom schema of replacement.* If A is a set, and f is a function whose domain is A, then $f(A) = \{f(a) : a \in A\}$ (the range of f) is contained as a subset of some set B.

[1] In set theory, everything is a set, including the elements of sets.

[2] The formula in this context describes the properties of the variable x.

ZFC 7. *The axiom of infinity.* There exists an infinite set.

Comment. Technically, this is a consequence of the more formal wording of the axiom that states that there exists a nonempty set A that contains the set $\{x \cup \{x\}\}$, for every element x in A.

ZFC 8. *The axiom of the power set.* For every set A, there exists a set B that contains the subsets of A as elements.

ZFC 9. *The axiom of choice.* For any set A whose elements are all nonempty sets, there exists a function f from A to $\bigcup_A X$ (the union of the elements of A), such that $f(X) \in X$, for all $X \in A$.

Comment. The function f is called a *choice function*, in that it *chooses* one element from each set X in A.

1.1. Basic terminology and facts

A *set* E is a collection of objects, called the *members*, or *elements* of E. The members of a set can be numbers or any other mathematical objects of interest. If x is a member of E, we say that x belongs to E or x is in E, and write $x \in E$.

If E and F are sets and $x \in F$ implies $x \in E$, then F is a *subset* of E, which we denote by $F \subset E$. Note that for any set $E \subset E$. If $F \subset E$ but $F \neq E$, then there is at least one element of E not contained in F. We say in this case that F is a *proper* subset of E.

The sets E and F are equal if and only if they contain the same element. This is often demonstrated by showing that $E \subset F$ and $F \subset E$.[3]

If $F \subset E$, then the *complement* of F in E, is the set $F^c = \{x \in E : x \notin F\}$. We note that the complement of a set F depends on the superset E, of which it is a subset. E.g., the complement of $\mathbb{N}$ in $\mathbb{Q}$ is different from the complement of $\mathbb{N}$ in $\mathbb{R}$.

The (unique) set containing no elements is called the empty set and denoted by $\varnothing$.

The *union* of two sets E and F is $E \cup F = \{x : x \in E \text{ or } x \in F\}$. The *intersection* of E and F is $E \cap F = \{x : x \in E \text{ and } x \in F\}$. We note that E and F are both subsets of their union $E \cup F$, while the intersection $E \cap F$ is a subset of both E and F.

The collection of all subsets of a set E is called the *power set* of E and denoted by $\mathcal{P}(E)$.

If A is a nonempty set and for every $a \in A$, E_a is a set, then the set $\mathcal{F} = \{E_a : a \in A\}$ is called an *indexed family* of sets . The set A is called the *indexing set* in this context.

If $\mathcal{F} = \{E_a : a \in A\}$ is an indexed family of subsets of some set X, then the union of the sets in $\mathcal{F}$ is denoted by $\bigcup_{\mathcal{F}}$ or $\bigcup_{a \in A} E_a$ and defined by

$$\bigcup_{\mathcal{F}} = \bigcup_{a \in A} E_a = \{x \in X : \text{ there is an } a \in A \text{ such that } x \in E_a, \}.$$

Likewise, the intersection of the sets in $\mathcal{F}$ is denoted $\bigcap_{\mathcal{F}}$ or $\bigcap_{a \in A} E_a$ and is defined by

$$\bigcap_{\mathcal{F}} = \bigcap_{a \in A} E_a = \{x \in X : \text{ for all } a \in A, x \in E_a\}.$$

[3] This is the axiom of extensionality.

Proposition. *Let $\{E_a : a \in A\}$ be a family of subsets of a set X. Then*

$$\text{(i)} \quad \left(\bigcup_{a\in A} E_a\right)^c = \bigcap_{a\in A} E_a^c \quad \textit{and} \quad \text{(ii)} \quad \left(\bigcap_{a\in A} E_a\right)^c = \bigcup_{a\in A} E_a^c.$$

Proof. We prove (i) and leave (ii) as an exercise.

If $x \in \left(\bigcup_{a\in A} E_a\right)^c$, then $x \notin \bigcup_{a\in A} E_a$ which implies that $x \notin E_a$ for every $a \in A$. This means that $x \in E_a^c$ for every $a \in A$, and therefore $x \in \bigcap_{a\in A} E_a^c$. □

The *Cartesian product*, $E \times F$ of two nonempty sets, E and F, is the set of *ordered pairs*, $E \times F = \{(x, y) : x \in E \text{ and } y \in F\}$. Here it is understood that $(x_1, y_1) = (x_2, y_2)$ if and only if $x_1 = x_2$ and $y_1 = y_2$.

More generally, given n sets, $E_1, E_2, \ldots, E_n$, the n-fold Cartesian product $E_1 \times E_2 \times \cdots \times E_n$ is the set of ordered *n tuples*,

$$E_1 \times E_2 \times \cdots \times E_n = \{(x_1, x_2, \ldots, x_n) : x_i \in E_i, \text{ for } 1 \le i \le n\}.$$

A (binary) *relation* from a set A to a set B is a subset R of the Cartesian product $A \times B$. If $(x, y) \in R$, we sometimes write $x\overline{R}y$, where $\overline{R}$ is a symbol designating the relation.

Given two relations, f from A to B and g from B to C, we define the *composite relation* (or *composition*) $g \circ f$ from A to C by setting

$$g \circ f = \{(a, c) \in A \times C : (a, b) \in f \text{ and } (b, c) \in g, \text{ for some } b \in B\}.$$

The relations we use most frequently fall into three categories: equivalence relations, orders and functions.

1.2. Equivalence relations and partitions

An equivalence relation on a set A, is a relation R (denoted here by $\equiv$) from A to A, that is *reflexive*, *symmetric* and *transitive*. I.e., for all $a, b, c \in A$,

- $a \equiv a$;
- If $a \equiv b$, then $b \equiv a$;
- If $a \equiv b$ and $b \equiv c$, then $a \equiv c$.

Equivalence relations on a set A are generalizations of equality, and are typically defined by ignoring contextually insignificant differences between different elements.

For example the rational numbers were constructed in Section 1.3 as quotients of integers with positive (i.e., natural) denominators. We observed that if $k, m \in \mathbb{Z}$ and $l, n \in \mathbb{N}$, then the rational numbers m/n and k/l are the same if and only if $ml = kn$. We can rephrase this in terms of an equivalence relation on $\mathbb{Z} \times \mathbb{N}$. Specifically, for $(m, n), (k, l) \in \mathbb{Z} \times \mathbb{N}$, we say that $(m, n) \equiv (k, l)$ if and only if $ml = kn$.

Proposition. *The relation $\equiv$ is an equivalence relation on $\mathbb{Z} \times \mathbb{N}$.*

Proof. We prove that the relation is transitive and leave symmetry and reflexivity as an exercise. Suppose that $(m, n), (k, l)$ and $(i, j) \in \mathbb{Z} \times \mathbb{N}$, with $ml = kn$ and $kj = il$. Then

$$l(mj - in) = lmj - lin = lmj - jkn + jkn - lin = j(ml - kn) + n(kj - il) = 0,$$

so $mj - in = 0$, since $l > 0$. □

If $\equiv$ is an equivalence relation on a set A, then the subsets

$$[a] = \{b \in A : a \equiv b\}$$

are called the *equivalence classes* (or just classes) of $\equiv$ in A. If $a \in A$ then $a \in [a]$ (by reflexivity), which means that A is equal to the union of the classes of $\equiv$. Moreover, if $a, b, c \in A$ and $c \in [a] \cap [b]$, then $a \equiv c \equiv b$ (by transitivity and symmetry), so $[a] = [b]$, i.e., any two classes of $\equiv$ are either disjoint or identical.

A collection of subsets $P \subset \mathcal{P}(A)$, with these two properties

- $A = \bigcup_P X$, and
- For all $X, Y \in P$, either $X = Y$ or $X \cap Y = \varnothing$.

is called a *partition* of A. The subsets in a partition are called the *parts* of the partition. I.e., the equivalence classes of an equivalence relation on A form a partition of A. In fact, this connection goes both ways.

Theorem. *The classes of an equivalence relation on a set A form a partition of A. Moreover, given a partition P of A, the relation $\equiv_P$ defined by $a \equiv_P b$ if and only if a, b belong to the same part of P, is an equivalence relation on A whose equivalence classes are the parts of P.*

Proof. We have already shown that the classes of an equivalence relation on A form a partition of A.

Now suppose that P is a partition and $\equiv_P$ is defined as above. For any $a \in A$, $a \in X$ for some $X \in P$, so $a \equiv_P a$. Similarly, if $a \equiv_P b$, then $a, b \in X$ for some $X \in P$, which means that $b, a \in X$ and $b \equiv_P a$. Finally, if $a \equiv_P b$ and $b \equiv_P c$, then there are sets $X, Y \in P$ such that $a, b \in X$ and $b, c \in Y$. This means that $c \in X \cap Y$, so $X = Y$ and $a \equiv_P c$.

We leave to the reader the task of showing that the classes of $\equiv_P$ are the parts of P. □

Comment. A more formal construction of the rational numbers from the integers starts by defining $\mathbb{Q}$ as the set of equivalence classes of the relation on $\mathbb{Z} \times \mathbb{N}$ discussed in Proposition 1.2. Then one defines addition, and the order on this set of equivalence classes and establishes all the properties of $\mathbb{Q}$ that we did in Chapter 1. The arguments used in the formal construction are essentially the same as the ones we used there.

1.3. Orders

We defined linear orders and well-orders in Chapter 1, in the section on the natural numbers. A *partial order* on a set A is a relation $\leq$ from A to A, that is *reflexive, antisymmetric* and *transitive*. I.e., for all $a, b, c \in A$,

- $a \leq a$;
- If $a \leq b$, and $b \leq a$, then $b = a$;
- If $a \leq b$ and $b \leq c$, then $a \leq c$.

A *strict* partial order $<$ is a relation on a set A is a relation from A to A that is *irreflexive, asymmetric* and transitive. I.e., for all $a, b, c \in A$,

- $a \not< a$;
- If $a < b$, then $b \not< a$.
- If $a < b$ and $b < c$, then $a < c$.

For every strict partial order $<$ on A, we may define a (nonstrict) partial order, by defining $a \leq b$ if and only if $a < b$ or $a = b$. Likewise, if $\leq$ is a partial order on A, then we may define a strict partial order, by defining $a < b$ if and only if $a \leq b$ and $a \neq b$.

A pair of elements $a, b \in A$ are *comparable* by the partial order $\leq$ if $a \leq b$ or $b \leq a$. A partial order $\leq$ on A is a *linear order* if every pair of elements in A are comparable by $\leq$. A linear order on A is a *well-ordering* if every nonempty subset of A has a *least element* with respect to the order. I.e., for every $X \subset A$,with $X \neq \emptyset$, there exists an $x \in X$ such that if $y \in X$ and $y \neq x$, then $x < y$.

The natural order on $\mathbb{N}$ is a well-ordering, while the natural order on $\mathbb{Q}$ is a linear order, but not a well-ordering. If S is a set with at least two elements, then the relation on $\mathscr{P}(S)$ defined by $X \leq Y$ if and only if $X \subseteq Y$, is a partial order, but not a linear order.

Any countable set can be well ordered using a bijection of that set with $\mathbb{N}$. I.e., if A is countable, and $\varphi : A \to \mathbb{N}$ is a bijection, then the relation defined by $a \succeq b$ if and only if $\varphi(a) \geq \varphi(b)$ (in $\mathbb{N}$), is a well-ordering of A. For uncountable sets and assuming the ZF axioms, the axiom of choice is equivalent to the *well-ordering* axiom which states that every nonempty set can be "well-ordered". Thus for example, assuming ZFC, there exists a well-ordering of $\mathbb{R}$, though it should be clear that such a well ordering is *not* the natural order of the real numbers.

1.4. Functions

A function f from the set A to the set B, written $f : A \to B$, is a relation from A to B such that

- for all $x \in A$, there is a $y \in B$ such that $(x, y) \in f$, and
- if $(x, y) \in f$ and $(x, z) \in f$, then $y = z$.

In other words, every $x \in A$ is paired with one and only one $y \in B$. We denote functions in the usual way: if $(x, y) \in f$ then we write $f(x) = y$. The set A is called the *domain* of the function f, and B is called the *codomain*.

Functions are also called *maps* and sometimes *transformations* in the case that the function maps a set to itself.

The *range* of a function $f : A \to B$ is the set of its images in B.[4] I.e.,

$$\text{range}(f) = \{f(x) : x \in A\} \subseteq B.$$

If $f : A \to B$ is a function and $U \subset A$ is nonempty, then the *restriction* of f to U is the function

$$f|_U = \{(u, f(u)) : u \in U\}.$$

A function $f : A \to B$ is *injective* if $f(x_1) = f(x_2)$ implies $x_1 = x_2$ for every $x_1, x_2 \in A$; the function f is *surjective* if range$(f) = B$; and the function f is *bijective* if it is both injective and surjective. Injective functions are also said to be "one-to-one" and surjective functions are also said to be "onto".

If $f : A \to B$ is a function and $U \subset A$, then the *image* of U is the subset $\{f(u) : u \in U\}$ in B.[5] If $V \subset B$, then the *inverse image* of V in A is the *subset* $f^{-1}(V) = \{x \in A : f(x) \in V\}$ of A.

Lemma. *If $f : A \to B$ and $g : B \to C$ are functions, then their composition, $g \circ f$, is a function from A to C.*

Proof. Let $a \in A$. Then there is a unique $b = f(a) \in B$ such that $(a, b) \in f$. Likewise, there is a unique $c = g(b) \in C$ such that $(b, c) \in g$. It follows that

$$(a, c) = (a, g(b)) = (a, g(f(a))) \in g \circ f.$$

Moreover, if $(a, c_1) \in g \circ f$, then $c_1 = g(b) = g(f(a))$, so $c_1 = c$. □

Theorem. *The composition of injections is an injection, the composition of surjections is a surjection and the composition of bijections is a bijection.*

Proof. Suppose that $f : A \to B$ and $g : B \to C$ are both injections. If $a_1, a_2 \in A$ and $(g \circ f)(a_1) = (g \circ f)(a_2)$, then $g(f(a_1)) = g(f(a_2))$, so $f(a_1) = f(a_2)$ because g is injective. Hence $a_1 = a_2$ because f is injective, so $(g \circ f)$ is injective.

If $f : A \to B$ and $g : B \to C$ are both surjections and $c \in C$, then there is a $b \in B$ such that $g(b) = c$ and there is an $a \in A$ such that $f(a) = b$. It follows that $(g \circ f)(a) = g(f(a)) = g(b) = c$, so $(g \circ f)$ is surjective.

Finally, if $f : A \to B$ and $g : B \to C$ are both bijections, then they are both injective and surjective, hence $g \circ f$ is injective and surjective, so $g \circ f$ is bijective. □

If $f : A \to B$ is a bijection, then for every $b \in B$ there exists a unique $a_b \in A$ such that $f(a_b) = b$. In this case, we set $f^{-1}(b) = a_b$ and call $f^{-1} : B \to A$ the *inverse function* of f.

Proposition. *If $f : A \to B$ is a bijection, then $(f^{-1} \circ f)(a) = a$ for all $a \in A$ and $(f \circ f^{-1})(b) = b$ for all $b \in B$.*

Proof. Exercise. □

[4] Some authors use the term *range* to mean *codomain*. We do not.

[5] Which is the range of the restriction $f|_U$.

In other words, the maps $(f^{-1} \circ f)$ and $(f \circ f^{-1})$ are the *identity* maps on A and B, respectively.

Exercises

ex1.1. Prove Propostion 1.1, (ii).

ex1.2. Show that if $f : E \to F$ is a bijection, then $f^{-1} : F \to E$, as defined above, is indeed a function and in fact a bijection.

ex1.3. Show that $A \subset f^{-1}(f(A))$ for any function $f : E \to F$ and any subset A of E. Find an example for which $A \neq f^{-1}(f(A))$

ex1.4. Show that $B = f(f^{-1}(B))$ for any function $f : E \to F$ and any subset $B \subset F$.

ex1.5. Prove Proposition 1.4.

2. Cardinality (in its own small nutshell)

The *cardinality* of a set is its set-theoretic size, in the same sense that the *number* of elements in a *finite* set says how big that set is. We determine the cardinality of a set by comparing it to another set whose cardinality is known, much like the length of a string is determined by comparing it to objects with known lengths (like rulers). In the case of sets, we use bijections, injections and surjections to determine their relative size.

Definition. Two sets X and Y have the same cardinality if there is a bijection $\varphi : X \to Y$. We say in this case that X and Y are *equinumerous* and write $X \sim Y$. We also use the shorthand $|X|$ to denote the cardinality of a set X, so if $X \sim Y$, we may also write $|X| = |Y|$.

If there is an injection $f : X \to Y$, or a surjection $g : Y \to X$, then the cardinality of Y is greater than or equal to that of X and we write $|X| \leq |Y|$.

A set X is *finite* if X is empty or if there is an $n \in \mathbb{N}$ such that $X \sim \{1, 2, \ldots, n\}$, in which case we write $|X| = n$. A set X is *infinite* if it is not finite.

If there is an injection $f : X \to \mathbb{N}$, then the set X is *countable* (so finite sets are countable). If X is countable and infinite, then X is *countably infinite* or *denumerable.*

Finally, a set X is *uncountable* if it is not countable.

Comment. One can also define a set X to be infinite if there is a *proper* subset Y of X such that $X \sim Y$. This version of infinite, called *Dedekind infinite*, is equivalent to "not finite", assuming the ZFC axioms.

Proposition. *If $|X| \leq |Y|$ and $|Y| \leq |Z|$, then $|X| \leq |Z|$. If $X \sim Y$ and $Y \sim Z$, then $X \sim Z$.*

Proof. This follows from the facts that the composition of injections is injective, the composition of surjections is surjective and the composition of bijections is bijective (Theorem 1.4). □

2.1. The Schroeder-Bernstein theorem

It is frequently easier to construct injective functions than bijections, and the theorem of Schroeder and Bernstein is therefore very useful. Cantor stated this theorem first, but without proof, and it is sometimes called the Cantor-Schroeder-Bernstein theorem. The short proof we give here due to Ignace Kolodner,[6] and relies on the following lemma.

Lemma. *Suppose that X is a set and $\Phi : \mathcal{P}(X) \to \mathcal{P}(X)$ has the property that $U \subseteq V$ implies $\Phi(U) \subseteq \Phi(V)$, for all $U, V \in \mathcal{P}(X)$, then there is a set $A \subseteq X$ such that $A = \Phi(A)$.*

Proof. Let $\mathcal{B} = \{B \in X : B \subseteq \Phi(B)\}$ and let $A = \bigcup_{\mathcal{B}} B$. For any $x \in A$, there is a $B \in \mathcal{B}$ such that $x \in B \subseteq \Phi(B)$, and therefore $x \in \Phi(A)$, since $B \subseteq A$, so $\Phi(B) \subseteq \Phi(A)$. It follows that $A \subseteq \Phi(A)$, and therefore $\Phi(A) \subseteq \Phi(\Phi(A))$, which means that $\Phi(A) \in \mathcal{B}$, so $\Phi(A) \subseteq A$. □

Theorem (Schroeder-Bernstein). *If there exist injective functions, $f : X \to Y$ and $g : Y \to X$, then there is a bijection $\varphi : X \to Y$.*

In other words, if $|X| \leq |Y|$ and $|Y| \leq |X|$, then $|X| = |Y|$.

Proof. Let $\Phi : P(X) \to P(Y)$ be defined by

$$\Phi(U) = \left[g\left([f(U)]^{c_y}\right)\right]^{c_x}, \tag{2.1}$$

where the superscripts c_x and c_y indicate the complements in X and Y, respectively. Now, if $U \subseteq V \subseteq X$, then $f(U) \subseteq f(V)$, so that $[f(V)]^{c_y} \subseteq [f(U)]^{c_y}$ and therefore $g\left([f(V)]^{c_y}\right) \subseteq g\left([f(U)]^{c_y}\right)$. Therefore, if $U \subseteq V \subseteq X$, then

$$\Phi(U) = \left[g\left([f(U)]^{c_y}\right)\right]^{c_x} \subseteq \left[g\left([f(V)]^{c_y}\right)\right]^{c_x} = \Phi(V).$$

In other words, Φ satisfies the condition of the lemma above, and it follows that there exists a subset $A \subseteq X$ such that $A = \Phi(A)$.

Let $B = [f(A)]^{c_y} \subseteq Y$, then $f(A) = B^{c_y}$ (by definition of the complement) and $g(B) = A^{c_x}$ because

$$[g(B)]^{c_x} = \left[g\left([f(A)]^{c_y}\right)\right]^{c_x} = \Phi(A) = A.$$

We now use the sets A and B and the injections f and g to define a bijection $\varphi : X \to Y$. Specifically, we set

$$\varphi(x) = \begin{cases} f(x) & : x \in A, \\ g^{-1}(x) & : x \in A^{c_x}. \end{cases}$$

We first note that φ is well-defined (as a function) on all of X, because (i) every $x \in X$ has an image in Y, and (ii) that image is unique because f is injective on X, and g is injective on Y and hence g^{-1} is injective on A^{c_x}. It remains to show that φ is injective and surjective.

To show that φ is injective, suppose that $x_1, x_2 \in X$, with $x_1 \neq x_2$, and consider three cases. If $x_1, x_2 \in A$, then

$$\varphi(x_1) \neq \varphi(x_2),$$

[6] *Amer. Math. Monthly* ***74**, 1967.*

because f is injective by assumption. Likewise, if $x_1, x_2 \in A_1^{c_x}$, then

$$\varphi(x_1) \neq \varphi(x_2),$$

because g is injective, and therefore g^{-1} is injective on $g(B) = A^{c_x}$. Finally, if $x_1 \in A$ and $x_2 \in A^{c_x}$, then $\varphi(x_1) \in B^{c_y}$ and $\varphi(x_2) \in B$, and again $\varphi(x_1) \neq \varphi(x_2)$. For surjectivity, let $z \in Y = B \cup B^{c_y}$. If $z \in B$, and $x = g(z) \in X$, then $x \in A^{c_x}$ and $z = g^{-1}(x) = \varphi(x)$. If $z \in B^{c_y}$, then $z = f(x) = \varphi(x)$, for some $x \in A$, because f is surjective from A to B^{c_y}. □

Corollary. *$\mathbb{Q}$ is countably infinite.*

Proof. By the Schroeder-Bernstein theorem, it is enough to find injections $f : \mathbb{Q} \to \mathbb{N}$ and $g : \mathbb{N} \to \mathbb{Q}$. We begin by noting that the identity map $g(n) = n$ is an injection from $\mathbb{N}$ to $\mathbb{Q}$, because $\mathbb{N} \subset \mathbb{Q}$.

To find an injection from $\mathbb{Q}$ to $\mathbb{N}$, we observe that given $r \in \mathbb{Q}$, there is a unique representation $r = m_r/n_r$, where $n_r \in \mathbb{N}$, $m_r \in \mathbb{Z}$ and m_r and n_r have no common divisors other than 1 (and -1). With this in mind, we define $f : \mathbb{Q} \to \mathbb{N}$ by setting

$$f(r) = f(m_r/n_r) = \begin{cases} 2^{m_r}3^{n_r} & : m_r \geq 0 \\ 2^{|m_r|}5^{n_r} & : m_r < 0. \end{cases}$$

If $r, s \in \mathbb{Q}$ and $r \neq s$, then either $m_r \neq m_s$ or $n_r \neq n_s$, so the map f is injective by the uniqueness of the prime power decomposition of natural numbers (Corollary 1.1.4). □

The proof above relies on the fact that $\mathbb{N}$ contains many infinitely countable subsets. In particular, the set of all powers of one prime number is a countably infinite set. Moreover, the fundamental theorem of arithmetic (Theorem 1.1.4) tells us that sets of prime powers are disjoint for different primes. We make use of this in the proof of the following claim.

Proposition. *If $\{E_a : a \in A\}$ is a countable family of countable sets, then $\bigcup_{a\in A} E_a$ is countable.*

Proof. By assumption, A is countable, so there is either a bijection $g : \mathbb{N} \to A$ (if A is countably infinite) or there is an $N \in \mathbb{N}$ and a bijection $g : \{1, 2, \ldots, N\} \to A$. In either case we may write $A = \{a_n\}_{n\in I}$, where $I = \mathbb{N}$ or $\{1, 2, \ldots, N\}$ and $a_n = g(n)$. Additionally, since E_{a_n} is countable for each $a_n \in A$, it follows that there is an injection $f_n : E_{a_n} \to \mathbb{N}$.

For every $x \in \bigcup_{a_n\in A} E_{a_n}$, let n_x be the smallest natural number such that $x \in E_{a_n}$. Denote the set of prime numbers by $\{p_n\}_{n=1}^{\infty}$ and define $\varphi : \bigcup_{a_n\in A} E_{a_n} \to \mathbb{N}$ by setting

$$\varphi(x) = p_{n_x}^{f_{n_x}(x)}.$$

Let $x, y \in \bigcup_{a_n\in A} E_{a_n}$ and suppose that $x \neq y$. If $n_x \neq n_y$, then $p_{n_x} \neq p_{n_y}$ so $\varphi(x) \neq \varphi(y)$ because they are powers of different prime numbers. If $n_x = n_y = n_0$, then $p_{n_x} = p_{n_y}$ but $f_{n_x}(x) \neq f_{n_y}(y)$, since $f_{n_x} = f_{n_y}$ is an injection and $x \neq y$. In this case, $\varphi(x) \neq \varphi(y)$ because they are different powers of the same prime. It follows that φ is an injection and $\bigcup_{a\in A} E_a$ is therefore countable. □

2.2. The uncountability of $\mathbb{R}$

Georg Cantor showed that the set of real numbers is uncountable. He gave several proofs over the years, and his first was an immediate consequence of the following theorem.

Theorem. *For any map $f : \mathbb{N} \to \mathbb{R}$ and any $a < b$, there is a $z \in (a,b)$ such that $z \notin f(\mathbb{N})$.*

Proof. If $f(\mathbb{N}) \cap (a,b) = \emptyset$, then there is nothing to prove. If not, let $n_{1,1} \in \mathbb{N}$ be the smallest natural number such that $f(n) \in (a,b)$ and write $x_1 = f(n_{1,1})$. If $f(\mathbb{N}) \cap (x_1, b) = \emptyset$, then once again we are done. Otherwise, let $n_{1,2} \in \mathbb{N}$ be the smallest natural number such that $f(n) \in (x_1, b)$ and write $y_1 = f(n_{1,2})$. It follows from these choices that $a < x_1 < y_1 < b$, that $1 \le n_{1,1} < n_{1,2}$ and that $f(n) \notin (x_1, y_1)$ for all $n \le n_{1,2}$.

Next, if $f(n) \notin (x_1, y_1)$ for all $n > n_{1,2}$, then $f(\mathbb{N}) \cap (x_1, y_1) = \emptyset$ and we're done because $(x_1, y_1) \subset (a,b)$. Otherwise, let $n_{2,1} > n_{1,2}$ be the smallest integer n such that $f(n) \in (x_1, y_1)$ and set $x_2 = f(n_{2,1})$. If $f(n) \notin (x_2, y_1)$ for all $n > n_{2,1}$ then $f(\mathbb{N}) \cap (x_1, y_1) = \emptyset$ and as before, we're done. If not, let $n_{2,2}$ be the smallest n such that $f(n) \in (x_2, y_1)$ and write $y_2 = f(n_{2,2})$. We now have $a < x_1 < x_2 < y_2 < y_1 < b$, $1 \le n_{1,1} < n_{1,2} < n_{2,1} < n_{2,2}$ and $f(n) \notin (x_2, y_2)$ for all $n \le n_{2,2}$.

The proof continues in this way. We suppose that $k \ge 1$ and that we have found natural numbers $1 \le n_{1,1} < n_{1,2} < n_{2,1} < \ldots < n_{k,1} < n_{k,2}$ and real numbers $x_j = f(n_{j,1})$ and $y_j = f(n_{j,2})$ for each $1 \le j \le k$, such that

$$a < x_1 < x_2 < \ldots < x_k < y_k < \ldots < y_2 < y_1 < b$$

and $f(n) \notin (x_k, y_k)$ for all $n \le n_{k,2}$.

As before, if $f(n) \notin (x_k, y_k)$ for all $n > n_{k,2}$, then $f(n) \notin (x_k, y_k)$ for all $n \in \mathbb{N}$, and we're done. Otherwise, let $n_{k+1,1}$ be the smallest n such that $f(n) \in (x_k, y_k)$ and set $x_{k+1} = f(n_{k+1,1})$. If $f(n) \notin (x_{k+1}, y_k)$ for all $n > n_{k+1,1}$, then $f(n) \notin (x_{k+1}, y_k)$ for all $n \in \mathbb{N}$, and the proof is over. If not, let $n_{k+1,2} \in \mathbb{N}$ be the smallest n such that $f(n) \in (x_{k+1}, y_k)$ and set $y_{k+1} = f(n_{k+1,2})$. As before, $n_{k,2} < n_{k+1,1} < n_{k+1,2}$ and for all $n \le n_{k+1,2}$, $f(n) \notin (x_{k+1}, y_{k+1})$.

If the process terminates after N steps, then $f(n) \notin (x_N, y_N)$ for all $n \in \mathbb{N}$, (with the possible exception of $f(n_{N+1,1}) = x_{N+1}$, if there is an x_{N+1}). If the process does not terminate, then it produces an unbounded sequence of natural numbers

$$1 \le n_{1,1} < n_{1,2} < n_{2,1} < n_{2,2} < \ldots < n_{k,1} < n_{k,2} < n_{k+1,1} < \ldots,$$

and two (infinite) sequences of real numbers $\{x_m\}$ and $\{y_m\}$ (where $x_m = f(n_{m,1})$ and $y_m = f(n_{m,2})$) satisfying

$$a < x_m < x_{m+1} < y_{m+1} < y_m < b,$$

for all $m \in \mathbb{N}$, and $f(n) \notin (x_m, y_m)$ for all $n \le n_{m,2}$.

The sequence $\{x_n\}$ is monotone increasing and bounded above, because $y_m > x_n$ for all $n, m \in \mathbb{N}$,[7] so it converges to a limit x^*. Likewise $\{y_n\}$ converges to a limit

[7] For any $n, m \in \mathbb{N}$, if $n \le m$, then $x_n \le x_m < y_m$ and if $n > m$, then $x_n < y_n \le y_m$.

y^* because the sequence is monotone decreasing and bounded below by x_m, for each $m \in \mathbb{N}$. It follows that $a < x_m < x^* \leq y^* < y_m < b$ for all $m \in \mathbb{N}$, and we claim that if $x^* \leq z \leq y^*$, then $f(n) \neq z$ for all $n \in \mathbb{N}$.

Indeed, for any $n \in \mathbb{N}$, there is an $m \in \mathbb{N}$ such that $n < n_{m,2}$ and therefore $f(n) \notin (x_m, y_m)$, so $f(n) \neq z$ since $z \in (x_m, y_m)$. □

Corollary. *$\mathbb{R}$ is uncountable.*

Proof. Since there are no surjective maps from $\mathbb{N}$ to $\mathbb{R}$, there can be no bijective maps. □

The same argument can be used to show that intervals are uncountable. On the other hand, we can also show directly that every interval in $\mathbb{R}$ (open, closed or half open) has the same cardinality as $\mathbb{R}$, which is more informative: it tells us what the cardinality of an interval is rather than what it isn't.

Proposition. *For any $a < b$ in $\mathbb{R}$,*

$$|(a, b)| = |[a, b]| = |[a, b)| = |(a, b]| = \mathbb{R}.$$

Proof. We do this in stages. First we show that $|(-1, 1)| = |\mathbb{R}|$, then we show that $|(a, b)| = |(c, d)|$ for any $a < b$ and $c < d$ and finally we show that $|(a, b)| = |[a, b]| = |[a, b)| = |(a, b]|$

Consider the function $f(x) = \frac{x}{|x|+1}$. We first note that $|f(x)| < 1$ for all $x \in \mathbb{R}$, so f is a map from $\mathbb{R}$ to $(-1, 1)$.

We note that $f(x) = 0$ if and only if $x = 0$, $f(x) < 0$ if and only if $x < 0$ and $f(x) > 0$ if and only if $x > 0$, hence if $x < 0 < y$, then $f(x) < 0 < f(y)$. Next, if $0 < x < y$, then $|x| = x$ and $|y| = y$, and therefore

$$f(y) - f(x) = \frac{y}{|y| + 1} - \frac{x}{|x| + 1} = \frac{y - x}{(|x| + 1)(|y| + 1)} > 0.$$

In the same way, if $x < y < 0$, then $f(y) - f(x) > 0$. In other words, if $x \neq y$, then $f(x) \neq f(y)$, so $f : \mathbb{R} \to (-1, 1)$ is an injection. Since the identity map $\iota(x) = x$ is an injection from $(-1, 1)$ to $\mathbb{R}$, it follows that $|\mathbb{R}| = |(-1, 1)|$ by the Schroeder-Bernstein theorem.

Next, if $a < b$ and $c < d$, then the function $g(x) = a + \frac{b-a}{d-c}(x - c)$ is a bijection from (c, d) to (a, b) (see **ex3.1.4**), so $|(a, b)| = |(c, d)|$.

Finally, suppose that $a < b$ and let $c, d \in \mathbb{R}$ be such that $c < a < b < d$. Then (c, d) contains each of the intervals $[a, b]$, $[a, b)$ and $(a, b]$, so the identity map $\iota(x) = x$ is an injection from each of these intervals to (c, d). If $g(x)$ is the map defined in the previous paragraph, then it follows from Theorem 1.4 (the composition of injections is an injection) that $g \circ \iota$ is an injection from each of these intervals to (a, b). Since ι is an injection from (a, b) to each of these intervals, it follows from the Schroeder-Bernstein theorem (again) that all four intervals have the same cardinality. □

Exercises

ex2.1. Find a bijection $\varphi : \mathbb{Z} \to \mathbb{N}$.

Hint: Set $\varphi(0) = 1$ and take advantage of the even and odd natural numbers.

ex2.2. Find a sequence $\{x_n\}_{n=1}^{\infty} \subset \mathbb{R}$ that has every real number as a subsequential limit.

ex2.3. Show that the collection of all open intervals with rational endpoints is countable.

ex2.4. Show that the map $f : \mathbb{R} \to (-1, 1)$ described in Proposition 2.2 is in fact a bijection.

ex2.5. Show that the map $g : (a, b) \to (c, d)$ described in Proposition 2.2 is a bijection.

3. A construction of the real numbers

The set of real numbers is defined as the unique complete ordered field.[8] But a definition does not prove existence, so for the sake of completeness,[9] we describe in the following sections a construction of the real numbers. There are two common ways to do this. Both constructions begin with the rational numbers, and "fill the holes" in $\mathbb{Q}$ to obtain a complete ordered field. The difference between the constructions is how we conceive of the holes and therefore how we fill them.

In Chapter 1, we defined completeness as a property of an ordered field: an ordered field is complete if every nonempty subset that is bounded above has a least upper bound in the field. The rational numbers do not have this property as we illustrated with the set $F = \{r \in \mathbb{Q} : 1 < x \text{ and } x^2 < 2\}$, in Section 1.4.3. With this definition, the missing least upper bounds of bounded sets in $\mathbb{Q}$ are the "holes".

Richard Dedekind filled these holes using *cuts*. A cut is a partition of the rationals into two nonempty subsets, $\alpha, \beta \subset \mathbb{Q}$, such that (i) $a < b$ for all $a \in \alpha$ and $b \in \beta$, and (ii) α does not contain a maximal element. Dedekind defined the real numbers as the set of all such cuts, defined arithmetic and an order on the set of cuts, and showed that with these definitions, the set of cuts is indeed a complete ordered field. The missing least upper bounds in $\mathbb{Q}$ are filled with specific bounded sets (the cuts) that are missing them.

In Chapter 2 we defined completeness as a property of a metric space: (X, ρ) is complete if every Cauchy sequence in X has a limit in X. In the case of the metric space $\mathbb{R}$, the two versions of completeness are equivalent. If (X, ρ) is a metric space that is not complete, then there are Cauchy sequences in X with no limit in X. The missing limits are the holes in this case, and we fill them using the Cauchy sequences that are tending to them.

Both constructions have a common theme. The holes are filled by the "evidence" that they exist in the first place. A detailed description of the construction of $\mathbb{R}$ using

[8]Uniqueness means that all such fields are isomorphic to each other.
[9]Another math pun.

cuts may be found in Walter Rudin's excellent book, *Principles of Mathematical Analysis*.[10] In what follows, we sketch a construction of $\mathbb{R}$ using the second approach. We say "sketch", in that we leave many of the details to the reader to fill in.

We would also like to mention a third way of constructing $\mathbb{R}$, which is to define the real numbers as the set of all "decimals", infinite or finite. I.e.,

$$\mathbb{R} = \left\{ \pm \left(N + \sum_{j=1}^{\infty} \frac{d_j}{10^j} \right) : N \in \mathbb{Z}^+,\ d_j \in \{0, 1, 2, \ldots, 9\} \right\},$$

with the usual equivalence between decimal expansions that end in an infinite sequence of 9s, and the corresponding finite decimal expansion:

$$N + \sum_{j=1}^{n} \frac{d_j}{10^j} + \sum_{j=n+1}^{\infty} \frac{9}{10^j} = N + \sum_{j=1}^{n-1} \frac{d_j}{10^j} + \frac{d_n + 1}{10^n},$$

where $0 \leq d_n \leq 8$. One then extends arithmetic and the natural order on $\mathbb{Q}$ to $\mathbb{R}$ in a straightforward way, and shows that $\mathbb{R}$ is complete. This approach may be the most intuitive, and we leave it as a (long) exercise.

This approach is essentially a special case of the Cauchy sequence approach, since the sum of an infinite series is defined as the limit of its partial sums. The difference is that when constructing the real numbers this way, each real number is represented by a specific Cauchy sequence, whereas in the approach outlined below, we define $\mathbb{R}$ as the set of equivalence classes of Cauchy sequences, with respect to a suitable equivalence relation.

3.1. Equivalence classes of Cauchy sequences

The construction begins with the set of all Cauchy sequences of rational numbers. We denote such sequences by $\{x_n\}$ (or $\{y_n\}$, or $\{u_n\}$, etc., and denote by $\mathscr{R}$ the set of all such sequences.

We want to say that $\{x_n\}, \{y_n\} \in \mathscr{R}$ are equivalent if they converge to the same limit, but we can't define a relation that way, since the limit might not exist in $\mathbb{Q}$. Instead, we define the relation $\sim$ on $\mathscr{R}$ by setting $\{x_n\} \sim \{y_n\}$ if and only if $\lim_n (x_n - y_n) = 0$.

Proposition. *The relation $\sim$ is an equivalence relation on $\mathscr{R}$.*

Proof. We leave the proof that $\sim$ is reflexive and symmetric to the reader, and prove transitivity. Suppose that $\{x_n\}, \{y_n\}, \{z_n\} \in \mathscr{R}$ with $\{x_n\} \sim \{y_n\}$ and $\{y_n\} \sim \{z_n\}$. Then, given $\varepsilon > 0$, there are $N_1, N_2 \in \mathbb{N}$ such that $|x_n - y_n| < \varepsilon/2$ for all $n > N_1$ and $|y_n - z_n| < \varepsilon/2$ for all $n > N_2$. It follows that

$$|x_n - z_n| \leq |x_n - y_n| + |y_n - z_n| < \varepsilon$$

for all $n > \max(N_1, N_2)$, so $\lim(x_n - z_n) = 0$. □

We now define $\mathbb{R}$ to be the set of equivalence classes of the relation $\sim$ in $\mathscr{R}$. The idea is that each real number is represented by a Cauchy sequence that tends to it, but since there are infinitely such sequences for every real number, we lump them all together in an equivalence class.

[10] Aka, "Baby Rudin".

We write $\mathbb{R} = \{[\{x_n\}] : \{x_n\} \in \mathscr{R}\}$, where $[\{x_n\}]$ is the equivalence class of $\{x_n\}$ with respect to $\sim$. If $r \in \mathbb{Q}$, then we denote the equivalence class of the constant sequence $\{r, r, r, \ldots\}$ by $[r]$, and we call $\{[r] : r \in \mathbb{Q}\}$ the set of "rational elements" of $\mathbb{R}$.

Every equivalence class in $\mathbb{R}$ contains uncountably many representatives. In particular, every subsequence of a Cauchy sequence is itself a Cauchy sequence, and $\{x_{n_k}\} \in [\{x_n\}]$ for every subsequence $\{x_{n_k}\}$ of $\{x_n\}$. It will be useful at times to be able to choose a representative Cauchy sequence from each class with additional properties.

Lemma. *Every $[\{x_n\}] \in \mathbb{R}$ contains representatives that are monotone nonincreasing and others that are monotone nondecreasing.*

Proof. If the Cauchy sequence $\{x_n\}$ contains a monotone nonincreasing subsequence and a monotone nondecreasing subsequence, then there is nothing more to be said. If not, then either $\{x_n\}$ is monotone nonincreasing or monotone nondecreasing from some point on, and without loss of generality, we may suppose that $\{x_n\}$ is monotone nondecreasing.

The sequence $\{x_n\}$ is necessarily bounded above (by virtue of being a Cauchy sequence), and we let $y_1 \in \mathbb{Q}$ be an upper bound of this sequence, such that $y_1 - 1$ is not (an upper bound of the sequence).[11] For $n > 1$ we set

$$y_n = \begin{cases} y_{n-1} - 2^{-n} & : y_{n-1} - 2^{-n} \text{ is an upper bound of } \{x_n\} \\ y_{n-1} & : \text{otherwise.} \end{cases}$$

The sequence $\{y_n\}$ is clearly monotone nonincreasing, and we leave it as an exercise to show that $\{y_n\} \sim \{x_n\}$. □

To show that $\mathbb{R}$ has all the desired properties of the real numbers, we need to define addition, multiplication and the order on it, then show that $\mathbb{R}$ is an ordered field with respect to these definitions, and finally prove that $\mathbb{R}$ is a complete ordered field.

3.2. The field structure of $\mathbb{R}$

If $\{x_n\}, \{y_n\} \in \mathscr{R}$, we set $\{x_n\} + \{y_n\} = \{x_n + y_n\}$ and $\{x_n\} \cdot \{y_n\} = \{x_n y_n\}$, and note that $\mathscr{R}$ is closed under these operations. That is to say that sums and products of Cauchy sequences, defined this way, are also Cauchy sequences, as you can verify. Moreover, these operations are commutative and associative and multiplication distributes over addition, as usual.

Next, if $\{x_n\} \sim \{u_n\}$ and $\{y_n\} \sim \{v_n\}$, then $\{x_n\} + \{y_n\} \sim \{u_n\} + \{v_n\}$, and $\{x_n\} \cdot \{y_n\} \sim \{u_n\} \cdot \{v_n\}$, as you can also verify.[12] This means that setting

$$[\{x_n\}] + [\{y_n\}] = [\{x_n + y_n\}] \quad \text{and} \quad [\{x_n\}] \cdot [\{y_n\}] = [\{x_n y_n\}],$$

makes these operations *well-defined* in $\mathbb{R}$, in that sums and products of elements in $\mathbb{R}$ do not depend on the choice of representatives. With this definition, it follows that addition and multiplication in $\mathbb{R}$ are both commutative and associative, and together satisfy the distributive property.

[11] Exercise: show that such a y_1 exists.

[12] We are going to stop saying "as you can verify" now.

We next observe that [0] and [1] are the additive and multiplicative units in $\mathbb{R}$, respectively, and $[\{-x_n\}]$ is the additive opposite of $[\{x_n\}]$. The existence of multiplicative inverses requires (a little bit) more work.

Proposition. *If $[\{x_n\}] \neq [0]$, then $[\{x_n\}]$ has a multiplicative inverse in $\mathbb{R}$.*

Proof. Suppose that $[\{x_n\}] \neq [0]$, and let $\{u_n\} \in [\{x_n\}]$ be a representative nonincreasing sequence. Since $\{u_n\}$ is a nonincreasing Cauchy sequence and $u_n \not\to 0$, it follows that there is an $N \in \mathbb{N}$ and a $\delta > 0$ such that either $u_n > \delta$ for all $n > N$ or $u_n < -\delta$ for all $n > N$. We now set

$$v_n = \begin{cases} 1 & : n \leq N \\ 1/u_n & : n > N. \end{cases}$$

It follows that $\{v_n\} \in \mathscr{R}$ because for $n, m > N$

$$|v_n - v_m| = \left|\frac{u_m - u_n}{u_n u_m}\right| < \frac{|u_m - u_n|}{\delta^2},$$

and $\{u_n\}$ is Cauchy. Hence $[\{v_n\}] \in \mathbb{R}$ and $[\{v_n\}][\{x_n\}] = [\{v_n u_n\}] = [1]$. □

Thus, $\mathbb{R}$ with the operations defined above is a field. Moreover, the subset of rational elements, $\{[r] : r \in \mathbb{Q}\}$, is a subfield of $\mathbb{R}$ that is isomorphic to $\mathbb{Q}$.

3.3. A linear order on $\mathbb{R}$

Intuitively, $[\{x_n\}]$ is greater than $[\{y_n\}]$ if the common limit of the Cauchy sequences in $[\{x_n\}]$ is greater than the common limit of the sequences in $[\{y_n\}]$. Since these limits don't exist yet, we define the order on $\mathbb{R}$ as follows. To minimize confusion, we use the symbol $\succ$ to denote the order on $\mathbb{R}$, and retain $>$ for the natural order on $\mathbb{Q}$.

Definition. For $[\{x_n\}], [\{y_n\}] \in \mathbb{R}$, we say that $[\{x_n\}] \succ [\{y_n\}]$ if and only if there is a $\delta \in \mathbb{Q}$, with $\delta > 0$, such that for every $\{u_n\} \in [\{x_n\}]$ and $\{v_n\} \in [\{y_n\}]$, there is an $N \in \mathbb{N}$ such that $u_n - v_n > \delta$ for all $n > N$.

Theorem. *The relation $\succ$ is a linear order on $\mathbb{R}$.*

Proof. If $[\{x_n\}] \succ [\{y_n\}]$ and $[\{y_n\}] \succ [\{z_n\}]$, then there exist $\delta_1, \delta_2 > 0$ that satisfy the conditions of the definition of $\succ$ for the pairs $[\{x_n\}]$ and $[\{y_n\}]$, and $[\{y_n\}]$ and $[\{z_n\}]$, respectively. It follows that the positive constant $\delta_1 + \delta_2$ satisfies these conditions for the pair $[\{x_n\}]$ and $[\{z_n\}]$. I.e., $[\{x_n\}] \succ [\{z_n\}]$ so the relation is transitive. We observe (without proof) that $\succ$ is also irreflexive and asymmetric, so it is a partial order on $\mathbb{R}$, and it remains to show that $\succ$ is linear: if $[\{x_n\}] \neq [\{y_n\}]$, then either $[\{x_n\}] \succ [\{y_n\}]$ or $[\{y_n\}] \succ [\{x_n\}]$.

If $[\{x_n\}] \neq [\{y_n\}]$, then $\{x_n\} \not\sim \{y_n\}$, so there exists $\delta > 0$ such that for any $N \in \mathbb{N}$, there exists $n > N$ such that $|x_n - y_n| > \delta$. On the other hand, there exists an $M \in \mathbb{N}$ such that $|x_n - x_m| < \delta/4$ and $|y_n - y_m| < \delta/4$ for all $n, m > M$, since both $\{x_n\}$ and $\{y_n\}$ are Cauchy sequences. It follows that there exists an $N_0 \geq M$ such that $|x_n - y_n| > \delta/2$ for all $n > N_0$.

Repeating (a variation of) the argument above shows that there exists an $N_1 \geq N_0$ such that either (i) $x_n - y_n > \delta/4$ for all $n > N_1$ or (ii) $y_n - x_n > \delta/4$ for all $n > N_1$. Without loss of generality, we assume case (i).

Finally, suppose that $\{u_n\} \in [\{x_n\}]$ and $\{v_n\} \in [\{y_n\}]$, so that $\{x_n\} \sim \{u_n\}$ and $\{y_n\} \sim \{v_n\}$. It follows that there exist $N_2, N_3 \in \mathbb{N}$ such that $|x_n - u_n| < \delta/16$ and $|y_m - v_m| < \delta/16$ for all $n > N_2$ and $m > N_3$. Therefore

$$u_n - v_n > (x_n - \delta/16) - (y_n + \delta/16) = (x_n - y_n) - \delta/8 > \delta/8,$$

for all $n > \max(N_1, N_2, N_3)$. Hence the conditions defining $\succ$ are met by the constant $\delta/8$, so $[\{x_n\}] \succ [\{y_n\}]$. □

We note that the proof of the theorem implies the following useful tidbit.

Corollary. $[\{x_n\}] \succ [\{y_n\}]$ *if and only if there is a constant* $\delta > 0$ *and an* $N \in \mathbb{N}$ *such that* $x_n - y_n > \delta$ *for all* $n > N$.

In other words, it is enough to verify the conditions defining $[\{x_n\}] \succ [\{y_n\}]$ for one pair of representative Cauchy sequences from the two equivalence classes.

3.4. ℝ is an ordered field

For ℝ to be an ordered field, the field structure and the linear order we defined need to "play nice" (as described in Section 1.3.3), which they do.

Theorem. *With the field operations and order defined as above,* ℝ *is an ordered field that contains an isomorphic copy of* ℚ *as a sub-ordered field.*

Proof. First, suppose that $[\{x_n\}] \succ [\{y_n\}]$, so that there exist $\delta > 0$ and $N \in \mathbb{N}$ such that $x_n - y_n > \delta$ for all $n > N$. Hence, for any $\{z_n\} \in \mathscr{R}$, we have

$$(x_n + z_n) - (y_n + z_n) > \delta$$

for all $n > N$. Therefore, $[\{x_n\}] + [\{z_n\}] \succ [\{y_n\}] + [\{z_n\}]$ for any $[\{z_n\}] \in \mathbb{R}$, so ℝ satisfies Axiom **ordf 1.** from Section 1.3.3.

Next, suppose that $[\{x_n\}] \succ [0]$ and $[\{y_n\}] \succ [0]$. It follows that there exist $N_1, N_2 \in \mathbb{N}$ and $\delta_1, \delta_2 > 0$ such that $x_n > \delta_1$ for all $n > N_1$ and $y_n > \delta_2$ for all $n > N_2$. Therefore if $n > \max(N_1, N_2)$, then $x_n y_n > \delta_1 \delta_2 > 0$, so

$$[\{x_n\}][\{y_n\}] = [\{x_n y_n\}] \succ [0],$$

and ℝ satisfies Axiom **ordf 2.** as well. Restricting the order and field operations to the set of rational elements in ℝ completes the proof. □

3.5. Order completeness

It remains to show that ℝ is a *complete* ordered field: if $A \subset \mathbb{R}$ is nonempty and bounded above, then A has a supremum in ℝ. To do this, we need to construct a sequence $\{w_n\} \in \mathscr{R}$, such that $[\{w_n\}]$ is an upper bound of A, but $[\{w_n - \delta\}]$ is not, for every $\delta > 0$.

We begin with a lemma, whose proof is based on the same idea as the one used in Lemma 3.1.

Lemma. *If $A \subset \mathbb{R}$ is nonempty and bounded above, then there is an $m_0 \in \mathbb{Z}$ such that $[m_0]$ is an upper bound of A and $[m_0 - 1]$ is not.*

Proof. First, if $[\{u_n\}]$ is an upper bound of A, and $\{d_n\} \in [\{u_n\}]$ is monotone decreasing, then the rational element $[d_1]$ is an upper bound of A. If $d_1 = m'/n'$ (with $(m', n') \in \mathbb{Z} \times \mathbb{N}$), then $[m']$ is also an upper bound of A, and hence the set $U_A = \{m \in \mathbb{Z} : [m] \text{ is an upper bound of } A\}$ is nonempty.

Next, let $[\{x_n\}] \in A$, with $\{y_n\} \in [\{x_n\}]$ monotone increasing. Then U_A is bounded below by y_1, and therefore has a least element, m_0. □

Theorem. *If $A \subset \mathbb{R}$ is nonempty and bounded above, then A has a least upper bound in $\mathbb{R}$.*

Proof. We construct $\sup A \in \mathbb{R}$ analogously to the construction of the sequence $\{y_n\}$ in Lemma 3.1. With m_1 as in the lemma, we set $w_1 = m_1$, and for $n > 1$ we set

$$w_n = \begin{cases} w_{n-1} - 2^{-(n-1)} & : [w_{n-1} - 2^{-(n-1)}] \text{ is an upper bound of } A \\ w_{n-1} & : \text{otherwise.} \end{cases}$$

The sequence $\{w_n\}$ is a monotone nonincreasing by construction,

$$0 \leq w_m - w_n \leq \sum_{j=m}^{n-1} 2^{-j} < 2^{-(m-1)}$$

for $n > m$, so $\{w_n\} \in \mathcal{R}$. Moreover, since $[w_n]$ is an upper bound of A for all n (also by construction), it follows that $[\{w_n\}]$ is an upper bound of A. We remark that $[w_n - 2^{-(n-1)}]$ is *not* an upper bound of A, for all $n \in \mathbb{N}$, as you can show by induction. It remains to show that $[\{w_n\}]$ is the *least* upper bound of A.

Suppose that $[\{v_n\}] \in \mathbb{R}$ and $[\{w_n\}] \succ [\{v_n\}]$. From the definition of $\succ$, there are $k, N \in \mathbb{N}$ such that $w_n - 2^{-k} > v_n$ for all $n > N$, and hence $w_n - 2^{-(n-1)} > v_n$ for all $n > \max(k+1, N)$. From the remark in the last paragraph, it follows that $[v_n]$ is not an upper bound of A for all $n > \max(k+1, N)$, and therefore $[\{v_n\}]$ is likewise not an upper bound of A. □

Exercises

ex3.1. Fill in the missing details in the construction of $\mathbb{R}$, above.

ex3.2. Suppose that (X, ρ) is a metric space. Show that there is a *complete* metric space $(\tilde{X}, \tilde{\rho})$ that "contains" (X, ρ) as a subspace with respect to the same metric,[13] using the following steps (for example).

(i) Let $\mathcal{X}$ be the set of Cauchy sequences in X with respect to the metric ρ. Define the relation $\approx$ on $\mathcal{X}$ by $\{x_n\} \approx \{y_n\}$, if $\rho(x_n, y_n) \to 0$.

(ii) Show that $\approx$ is an equivalence relation on $\mathcal{X}$.

[13] I.e., for $x, y \in X \subset \tilde{X}$, $\tilde{\rho}(x, y) = \rho(x, y)$.

(iii) Define $\tilde{\rho}$ on $\tilde{X} = \mathcal{X}/\approx$, by setting $\tilde{\rho}([\{x_n\}], [\{y_n\}]) = \lim_n \rho(x_n, y_n)$, and show that $\tilde{\rho}$ is a (well-defined) metric on $\tilde{X}$.

(iv) Show that $(\tilde{X}, \tilde{\rho})$ is complete.

(v) Show that $(\tilde{X}, \tilde{\rho})$ contains an isometric copy of (X, ρ).

Index

Selected Published Titles in This Series

65 **Yitzhak Katznelson and Yonatan Katznelson,** An Introduction to Real Analysis, 2024
64 **Johannes A. Buchmann,** Introduction to Quantum Algorithms, 2024
63 **Bettina Richmond and Thomas Richmond,** A Discrete Transition to Advanced Mathematics, Second Edition, 2023
62 **Scott A. Taylor,** Introduction to Mathematics, 2023
61 **Bennett Chow,** Introduction to Proof Through Number Theory, 2023
60 **Lorenzo Sadun,** The Six Pillars of Calculus: Biology Edition, 2023
59 **Oscar Gonzalez,** Topics in Applied Mathematics and Modeling, 2023
58 **Sebastian M. Cioabă and Werner Linde,** A Bridge to Advanced Mathematics, 2023
57 **Meighan I. Dillon,** Linear Algebra, 2023
56 **Lorenzo Sadun,** The Six Pillars of Calculus: Business Edition, 2023
55 **Joseph H. Silverman,** Abstract Algebra, 2022
54 **Rustum Choksi,** Partial Differential Equations, 2022
53 **Louis-Pierre Arguin,** A First Course in Stochastic Calculus, 2022
52 **Michael E. Taylor,** Introduction to Differential Equations, Second Edition, 2022
51 **James R. King,** Geometry Transformed, 2021
50 **James P. Keener,** Biology in Time and Space, 2021
49 **Carl G. Wagner,** A First Course in Enumerative Combinatorics, 2020
48 **Róbert Freud and Edit Gyarmati,** Number Theory, 2020
47 **Michael E. Taylor,** Introduction to Analysis in One Variable, 2020
46 **Michael E. Taylor,** Introduction to Analysis in Several Variables, 2020
45 **Michael E. Taylor,** Linear Algebra, 2020
44 **Alejandro Uribe A. and Daniel A. Visscher,** Explorations in Analysis, Topology, and Dynamics, 2020
43 **Allan Bickle,** Fundamentals of Graph Theory, 2020
42 **Steven H. Weintraub,** Linear Algebra for the Young Mathematician, 2019
41 **William J. Terrell,** A Passage to Modern Analysis, 2019
40 **Heiko Knospe,** A Course in Cryptography, 2019
39 **Andrew D. Hwang,** Sets, Groups, and Mappings, 2019
38 **Mark Bridger,** Real Analysis, 2019
37 **Mike Mesterton-Gibbons,** An Introduction to Game-Theoretic Modelling, Third Edition, 2019
36 **Cesar E. Silva,** Invitation to Real Analysis, 2019
35 **Álvaro Lozano-Robledo,** Number Theory and Geometry, 2019
34 **C. Herbert Clemens,** Two-Dimensional Geometries, 2019
33 **Brad G. Osgood,** Lectures on the Fourier Transform and Its Applications, 2019
32 **John M. Erdman,** A Problems Based Course in Advanced Calculus, 2018
31 **Benjamin Hutz,** An Experimental Introduction to Number Theory, 2018
30 **Steven J. Miller,** Mathematics of Optimization: How to do Things Faster, 2017
29 **Tom L. Lindstrøm,** Spaces, 2017
28 **Randall Pruim,** Foundations and Applications of Statistics: An Introduction Using R, Second Edition, 2018
27 **Shahriar Shahriari,** Algebra in Action, 2017
26 **Tamara J. Lakins,** The Tools of Mathematical Reasoning, 2016
25 **Hossein Hosseini Giv,** Mathematical Analysis and Its Inherent Nature, 2016

For a complete list of titles in this series, visit the
AMS Bookstore at **www.ams.org/bookstore/amstextseries/**.